FASTOVSKY • WEISHAMPEL

DINOSAURS
A Concise Natural History

SECOND EDITION

Updated with the material that instructors want, *Dinosaurs* continues to make science exciting and understandable to non-science majors through its narrative of scientific concepts rather than endless facts. Now with an expanded section on the evolution of the dinosaurs, and new photographs to help students engage with geology, natural history, and evolution. The authors ground the text in the language of modern evolutionary biology, phylogenetic systematics, and teach students to examine the paleontology of dinosaurs exactly as the professionals in the field do using these methods to reconstruct dinosaur relationships. Beautifully illustrated, lively and engaging, this edition continues to encourage students to ask questions and assess data critically, enabling them to think like scientists.

DAVID FASTOVSKY is Professor of Geosciences at the University of Rhode Island. His interest in dinosaurs started as a child when he read about a 1920s paleontologist's adventures in the Gobi Desert. Dinosaurs won out years later when he had the tough decision of choosing between a career in music (he takes his viola on his many field trips) or paleontology, and he has carried out fieldwork in far-flung parts of the world. He's known as a dynamic teacher as well as a respected researcher with a focus on the environments in which dinosaurs roamed.

DAVID B. WEISHAMPEL is Professor in the Center for Functional Anatomy and Evolution at The Johns Hopkins University. His research focuses on dinosaur evolution and how dinosaurs function and he is particularly interested in herbivorous dinosaurs and the dinosaur record of Europe. Among his many publications he is senior editor of *The Dinosauria*, and has contributed to a number of popular publications including acting as consultant to Michael Crichton in the writing of *The Lost World*, the inspiration for Steven Spielberg's film *Jurassic Park*.

JOHN SIBBICK has been creating illustrations of extinct life-forms and their environments for over 20 years, producing numerous books on dinosaurs, as well as pterosaurs, and general books on prehistoric life. His work has appeared in scientific magazines such as the *National Geographic*, and in television documentaries and museums world-wide.

Frontispiece Gideon Mantell (1790–1852), the "father" of modern dinosaur paleontology.

DINOSAURS

A Concise Natural History

DAVID E. FASTOVSKY

DAVID B. WEISHAMPEL

with illustrations by **JOHN SIBBICK**

SECOND EDITION

CAMBRIDGE
UNIVERSITY PRESS

CAMBRIDGE UNIVERSITY PRESS
Cambridge, New York, Melbourne, Madrid, Cape Town,
Singapore, São Paulo, Delhi, Tokyo, Mexico City

Cambridge University Press
The Edinburgh Building, Cambridge CB2 8RU, UK

Published in the United States of America by Cambridge University Press,
New York

www.cambridge.org
Information on this title: www.cambridge.org/9780521282376

First published 2009
Second edition 2012

Printed in the United States by Edwards Brothers Malloy

A catalogue record for this publication is available from the British Library

Library of Congress Cataloging-in-Publication Data

Fastovsky, David E.
    Dinosaurs : a concise natural history / David E. Fastovsky, David B.
Weishampel. -- 2nd ed.
        p. cm.
    Summary: "Updated with the material that instructors want, Dinosaurs
continues to make science exciting and understandable to non-science majors
through its narrative of scientific concepts rather than endless facts. Now
with new material on pterosaurs, an expanded section of the evolution of the
dinosaurs, and new photographs to help students engage with geology, natural
history, and evolution. The authors ground the text in the language of modern
evolutionary biology, phylogenetic systematics, and teach students to examine
the paleontology of dinosaurs exactly as the professionals in the field do using
these methods to reconstruct dinosaur relationships" – Provided by publisher.
    ISBN 978-0-521-28237-6 (pbk.)
    1. Dinosaurs. 2. Dinosaurs–Extinction. 3. Vertebrates–Evolution.
    4. Paleontology–Mesozoic. I. Weishampel, David B., 1952– II. Title.

    QE861.4.F27 2012
    567.9--dc23

2012018472

ISBN 978-1-107-01079-6 Hardback
ISBN 978-0-521-28237-6 Paperback

Additional resources for this publication at www.cambridge.org/dinosaurs

CONTENTS

PREFACE TO SECOND EDITION

Why a second edition? After all, the goals of this edition are the same as those of the first: to give first- or second-year college students insights into the fantastic, long-extinct world of non-bird dinosaurs, while concomitantly using dinosaurs as an attractive vehicle to explore aspects of important themes that have clear relevance for our very-much extant world, themes such as natural history, evolutionary biology, Earth history, and the logic of scientific inference. Our point, then as now, is that knowing who dinosaurs are is really all about knowing who we are: but the latest inheritors of c. 3.8 billion years of uninterrupted life on Earth.

So why write a second edition? Our view is that second editions are needed when the field has changed sufficiently such that the first edition no longer properly treats the subject; second editions ought to be written when, having experience with the first edition, you identify places where the treatment can be refined and improved. These, then, are the twin motivations for this new edition.

In this new edition, we have attempted to retain the light tone of the original; dinosaurs are marvelous creatures and nobody should be crushed by the language that describes them. That said, we remain committed to presenting them in the way that professionals think about them: therefore, a rigorous phylogenetic context is retained.

In the time between the first and this edition, time and resources were devoted to learning users' responses to the first edition. The second edition, therefore, benefits from that effort: errors have been corrected and, in response to requests for a more detailed treatment of the origins of Dinosauria, an entirely new chapter (15) was written. The chapter is particularly timely, since in the years since the first edition, our understanding of transition from proto-dinosaurs to dinosaurs has greatly increased. Likewise, the chapter on dinosaur endothermy (12) has been much updated and completely rewritten. A brief section on plate tectonics has also been included in response to reviewers' suggestions; and the difference between diagnosis and definition in phylogenetic systematics is presented.

The new edition is completely updated. Since the last edition was published, important discoveries have been made about feathers: who has them (*T. rex*, for one!), what colors and patterns they show, and what they tell us about behavior and phylogeny. The power of CT scans and computer-generated reconstructions to elucidate dinosaur morphology is obvious now, and so we have included fully labeled CT scans, emphasizing soft anatomy such as brains and sinuses, to highlight the power of this important work. Likewise, new techniques in geochemistry – particularly stable isotopes – have become important tools for understanding dinosaur food preferences, and the new edition reflects these insights.

All the chapters on individual dinosaur groups have updates and/or added information. Thus along with feathers in theropods, we offer new details on, for example, the power of an ankylosaur club, nesting habits of *Protoceratops*, butting in pachycephalosaurs, breathing in sauropods, computerized muscle mass/weight estimates for dinosaurs generally, dinosaur diseases, and the evolution of large size in tyrannosaurids. Where appropriate, updated citations are also included. As was the case in the first edition, all of these changes are accompanied by illustrations – many from the professional literature and some specially commissioned for this book.

Fundamentally, the book is structured in the same way as the first edition. For this reason, we reprint the Preface to the First Edition in the pages that follow, since the suggestions for both students and faculty are still, we hope, useful. Finally, we remind our readers that, as in the first edition, any errors that appear in this work are entirely Dave's fault.

PREFACE TO FIRST EDITION: WHY A NATURAL HISTORY OF DINOSAURS?

To the student

Dinosaurs: A Concise Natural History has been written to introduce you to dinosaurs, amazing creatures that lived millions of years before there were humans. Along with acquainting you with these magnificent beasts, reading this book will give you insights into natural history, evolution, and the ways that scientists study Earth history.

What were dinosaurs like? Did they travel in herds? What were the horns for? Did the mothers take care of their babies? Was *T. rex* really the most fearsome carnivore of all time? Were they covered with feathers? How fast could brontosaurus run? Why did dinosaurs get so big? Along with getting answers to these and many other questions, you'll also meet legendary and charismatic dinosaur hunters (including the models for Indiana Jones and *Jurassic Park*'s Dr. Alan Grant) whose expeditions have helped to reveal the dinosaurs' stories from fossils and other fragmental clues left behind in the rocks. *Dinosaurs* will help you think like a scientist, while your knowledge of dinosaurs, natural history, and science grows with each chapter you read.

The book is written by authors who are active dinosaur researchers, with between them more than 50 years of experience teaching. It is illustrated by John Sibbick, one of the world's most famous dinosaur illustrators.

DAVID FASTOVSKY is Professor and Chair of Geosciences at the University of Rhode Island. His interest in dinosaurs started as a child when he read about Roy Chapman Andrews in the Gobi Desert (a story that, naturally enough, graces the pages of the book you are holding). Dinosaurs won out years later when he chose paleontology over a career in music. Fastovsky has carried out fieldwork in far-flung parts of the world, including Argentina, Mexico, the western USA and Canada, and Mongolia. He is known as a dynamic teacher as well as a respected researcher with a focus on the extinction of the dinosaurs, as well as the environments in which they roamed. He has made several television documentary appearances, and was a recipient of the Distinguished Service Award by the Geological Society of America in 2006.

DAVID B. WEISHAMPEL is Professor in the Center for Functional Anatomy and Evolution at The Johns Hopkins University. Recipient of three teaching awards, Weishampel teaches human anatomy, evolutionary biology, cladistics, and, of course, a course on dinosaurs. His research focuses on dinosaur evolution and how dinosaurs function, and he is particularly interested in herbivorous dinosaurs and the dinosaur record of eastern Europe and Mongolia. He is the senior editor of the immensely well-received *The Dinosauria*, and has written or co-written six books and very many scholarly articles. Weishampel has contributed to a number of popular publications as well, including acting as consultant to Michael Crichton in the writing of *The Lost World*, the inspiration for Steven Spielberg's film *Jurassic Park*. In 2011, a special symposium on duck-billed dinosaurs was convened by the Royal Ontario and Royal Tyrrell Museums, Canada, to honor Dr Weishampel's contributions in dinosaur paleontology.

JOHN SIBBICK has over 30 years of illustration experience working on subjects ranging from mythology to natural history and is probably best known for his depictions of prehistoric scenes and dinosaurs. In the first stage of any commission he takes the fossil evidence and consults with specialists in their field and works out a number of sketches to build up an overall picture of structure, surface detail, and behavior. From his base in England he has provided images for books, popular magazines such as the *National Geographic*, and television documentaries, as well as museum exhibits and one-man shows of original artwork. For this book he has provided 194 pieces of original art.

To the instructor

Dinosaurs: A Concise Natural History is a new textbook that uses a particularly attractive vehicle – dinosaurs – to introduce students in the early part of their college careers to the logic of scientific inquiry, and to concepts in natural history and evolutionary biology. The perspective and methods introduced through dinosaurs have a relevance that extends far beyond the dinosaurs, engendering in students scientific logic and critical thinking. The text is a fresh, completely rewritten version of our popular *The Evolution and Extinction of the Dinosaurs* (2005), with enhanced accessibility to students and added features to facilitate its utility for teaching.

A unique conceptual approach

Dino factoids – names, dates, places, and features – are available in zillions of books and websites. We depart from a "Who? What? Where?" approach to dinosaurs, instead building a broad understanding of the natural sciences through the power of competing scientific hypotheses.

Unique among dinosaur textbooks, *Dinosaurs* is rooted in phylogenetic systematics. This follows current practice in evolutionary biology, and allows students to understand dinosaurs as professional paleontologists do. The cladograms used in this book have been uniquely drawn in a way that highlights the hierarchical relationships they depict, ensuring that both the methods and conclusions of phylogenetic systematics remain accessible.

Long experience shows that students come to dinosaur courses with many preconceptions about the natural world; *Dinosaurs* asks them to think in new and revolutionary ways. For example, one of the great advances to come out of the past 20 years of dinosaur research is the recognition that *living birds are dinosaurs*. This somewhat startling conclusion leads to a couple of other counter-intuitive conclusions:

1   Birds are reptiles.
2   Dinosaurs didn't go extinct.

In this and in many other ways, our book will challenge students to reconsider their ideas about science and about their world.

Part I introduces the fundamental intellectual tools of the trade. Chapters 1 and 2 treat geology, the geological time scale, fossils, collecting, and what happens after the bones leave the field. The third chapter, a carefully crafted introduction to the logic of phylogenetic systematics, uses familiar and common examples to acquaint students with the method. Chapter 4 takes students from basal Vertebrata to the two great groups of dinosaurs Ornithischia and Saurischia.

Parts II and III cover, respectively, Ornithischia and Saurischia. The chapters within Parts II and III cover the major groups within Dinosauria, treating them in terms of phylogeny and evolution, behavior, and lifestyle. Ornithischia comes before Saurischia to reinforce the fundamental point that, on the cladogram, the ordering of Ornithsichia and Saurischia within a monophyletic Dinosauria makes no difference.

The phylogenetically most complex of dinosaur groups, Theropoda, is treated last in Part III, when students are best prepared to understand it. Three chapters cover the group: one for non-avian theropods, one on the evolution of birds from non-avian theropods, and one on the Mesozoic evolution of birds, since it was during the Mesozoic that birds acquired their modern form.

Part IV covers the aspects of the paleobiology of Dinosauria, from their metabolism, to the great rhythms that drove their evolution, to their extinction. A special chapter is devoted to the history of dinosaur paleontology. Although commonly introduced at the beginning of dinosaur books as a litany of names, dates, and discoveries, our history chapter – a history of *ideas* – is placed toward the end, so the thinking that currently drives the field can be understood in context. Yet we would cheat our readers if we left out accounts of the dinosaur hunters, whose colorful personalities and legendary exploits make up the lore of dinosaur paleontology; so we've included many of their stories as well.

Features

Dinosaurs is designed to help instructors to teach and to help students learn:

- The book is richly illustrated with many new, especially commissioned, art by John Sibbick, one of the world's foremost illustrators of dinosaurs. These images are exciting for the student to learn from and they effectively highlight and reinforce the concepts in the text. Many pages are also graced by research photographs, generously contributed by professional paleontologists.
- The chapters are arranged so that they present the material in order of increasing complexity and sophistication, building the confidence of the student early on, and extending the sophistication of their learning gradually through the book.
- The tone of the text is light, lively, and readable, engaging the student in the science, and dispelling the apprehension many students experience when they pick up a science textbook.
- "Objectives" at the beginning of each chapter help students to grasp chapter goals.
- Boxes scattered throughout the book present a range of ancillary topics, from dinosaur poetry, to extinction cartoons, to how bird lungs work, to colorful accounts of unconventional, outlandish, and extraordinary people, places, and stories.
- A comprehensive series of "Topic Questions," to be used as study guides, are located at the end of each chapter. The questions probe successively deeper levels of understanding, and students who can answer all of the "Topic Questions" will have a good grasp of the material. Variants of these questions can serve as excellent templates for examination questions.
- A Glossary ties definitions of key terms into the page numbers where the term is used.
- There are two indices: an Index of subjects and an Index of genera that includes English translations of all dinosaur names.

- Appendices are included in certain chapters to introduce material that students may need in order to understand chapter concepts, such as the chemistry necessary to understand radioactive decay, and the basic principles of evolution by natural selection.

Online resources to help you deliver your dinosaur course include:

- Electronic files of the figures and images within the book.
- Lecture slides in PowerPoint with text and figures to help you to structure your course.
- Solutions to the questions in the text for instructors.

Amazon.com Marketplace Order Invoice

Transaction number: 112-8373545-9685006

Merchant: the_book_community

Merchant contact e-mail: thebookcommunityus@gmail.com

BUYER INFORMATION	Buyer Name	Mark Pfister
	Buyer contact email	1w4bjpkc3x6v9jf@marketplace.amazon.com
	Order ID	112-8373545-9685006
	Purchase date	2020-07-06
	Product title	Dinosaurs: A Concise Natural History [Paperback] [2012] Fastovsky, David E.; Weishampel, David B.; Sibbick, John
ORDER INFORMATION	ISBN/ASIN	0521282373
	Quantity	1
	Additional product information	New
SHIPPING INFORMATION	Buyer address	Mark Pfister 653 PROVIDENCE LN LINDENHURST, 60046-4942 IL UNITED STATES
	Delivery type	standard
	Product price	$71.13
PAYMENT INFORMATION	Shipping & handling fees	$0.00
	Sales tax	$0.00
	Total cost	$71.13

Thank you for your recent order with The Book Community.

If you are satisfied with the way we handled your order, we would greatly appreciate it if you could take a moment to leave us a positive rating. Receiving excellent feed-back is very important to our business, but also suggestions and comments are much appreciated, as we are constantly striving to improve the quality of our services!

If by any chance something has gone wrong with your order, or if you are not completely satisfied with it, please contact us at thebookcommunityus@gmail.com to let us know about it! We guarantee we will find a prompt and satisfactory solution to any problems you might have, as we want to make sure that your experience with us is a pleasant one.

To **Lesley**, **Naomi**, and **Marieke**,
my family.
To poor **Robert**, because...

To **Sarah** and **Amy**.
Thanks for continuing to remind your dad
that there are things other than dinosaurs!

PART I

REACHING BACK IN TIME

CHAPTER OBJECTIVES

- Understanding fossils and fossilization

- Collecting dinosaur fossils

- Preparing dinosaur specimens

TO CATCH A DINOSAUR

Dinosaur tales

This book is a tale of dinosaurs; who they were, what they did, and how they did it. But more significantly, it is also a tale of natural history. Dinosaurs enrich our concept of the biosphere, the three-dimensional layer of life that encircles the Earth. Our biosphere has a 3.8 *billion*-year history; we and all the organisms around us are merely the most recent products of yet a fourth dimension: its history. To be unaware of the history of life is to be unaware of our organic connections to the rest of life. Dinosaurs have significant lessons to impart in this regard, because, as we learn who dinosaurs really are, we can better understand who *we* really are.

Ours is also a tale of science itself. In an increasingly technical world, an understanding of what science is – and is not – is important. Science is all about imagination and creativity; with a little help from the data. In the following pages, we hope to build a sense of the intellectual richness of science, as well as a feel for what philosopher of science Karl Popper called the "logic of scientific discovery."

The word "dinosaur" in this book   The term "dinosaur" (*deinos* – terrible; *sauros* – lizard) was invented in 1842 by the English naturalist Sir Richard Owen (see Box 14.2) to describe a few fossil bones of large, extinct "reptiles." With modifications (for example, "large" no longer applies to all members of the group), the name has proven resilient. It has become clear in the past 20 years, however, that not all dinosaurs are extinct; in fact, most specialists now agree that *birds are living dinosaurs*. We should use the technically correct term non-avian dinosaurs to specify all dinosaurs except birds, but we'd prefer to use the term "dinosaurs" as shorthand for "non-avian dinosaurs." The distinction between non-avian dinosaurs and all dinosaurs will be most relevant only when we discuss the origin of birds and their early evolution in Chapter 10; there, we will take care to avoid confusing terminology.

Fossils

That we even know there ever were such creatures as dinosaurs is due to dumb luck: some dinosaurs just happened to be preserved as fossils, the buried remains of organic life, in rock. Dinosaurs last romped on Earth 65 million years ago. This means that their soft tissues – muscles, blood vessels, organs, skin, fatty layers, etc. – are, in almost every fossil (but not *every* one: see Box 9.3), long gone. If any vestige remains at all, it is usually hard parts: generally, bones and teeth. Hard parts are not as easily degraded as the soft tissues that constitute most of the body.

Making body fossils

Before burial   Consider what might happen to a dinosaur – or any land-dwelling vertebrate – after it dies (Figure 1.1).

Carcasses are commonly disarticulated (dismembered), often by predators and then by scavengers ranging from mammals and birds to beetles. As the nose knows, most of the heavy lifting in the world of decomposition is done by bacteria that feast on rotting flesh. Some bones might be stripped clean of meat and left to bleach in the sun. Others might get carried off and gnawed. Sometimes the disarticulated remains are trampled by herds of animals, breaking and separating them further. So the sum total of all the earthly remains of the animal will end up lying there: a few disarticulated bleached bones in the grass.

Figure 1.1 Bones. A wildebeest carcass, partly submerged in mud and water and on its way to becoming permanently buried and fossilized. If the bones are not protected from scavengers, air, and sunlight, they decompose rapidly and are gone in 10–15 years. Bones destined to become high-quality fossils must be buried soon after the death of the animal.

If the animal isn't disarticulated right away, it is not uncommon for a carcass to bloat, as feasting bacteria produce gasses that inflate it. After a bit, the carcass will likely deflate (sometimes explosively), and then dry out, leaving bones, tissues, ligaments, tendons, and skin hard and inflexible.

Burial Sooner or later bones are either destroyed or buried. If they aren't digested as some-body's lunch, their destruction can come from weathering, which means that the miner-als in the bones break down and the bones disintegrate. But the game gets interesting for paleontologists when weathering is stopped by rapid burial. At this point, they (the bones, not the paleontologists) become fossils. A body fossil is what is produced when a part of an organism is buried. We distinguish these from trace fossils, which are impressions in the substrate left by an organism. Figure 1.2 shows two of the many paths bones might take toward fossilization.

After burial Bone is made out of calcium-sodium hydroxy apatite, a mineral that weath-ers easily. This means that, after fossilization, many bones no longer have original calcium-sodium hydroxy apatite present. This is especially likely if the bone comes into contact with fluids rich in dissolved minerals, such as commonly occurs after burial. If, however, no fluids are present throughout the history of burial (from the moment that the bone is buried to when it is exhumed by paleontologists, a time interval that could be measured in millions of years), the bone could remain unaltered, which is to say that original bone mineralogy remains. This situation is not that common, and is progressively rarer in the case of older and older fossils.

Ancient, unaltered bones – and even tissue – do exist, and are crucial for our understand-ing of the growth of bone tissue (see Chapter 12) and other soft anatomy (for example, the discovery of genuine red blood cells and connective tissues from *Tyrannosaurus*; see Chapter 9, footnote 3 and Chapter 10).

Most bones are altered to a greater or lesser degree. Since bones are porous, the spaces once occupied by blood vessels, connective tissue, and nerves fill up with minerals. This situation is called permineralization (Figure 1.3).

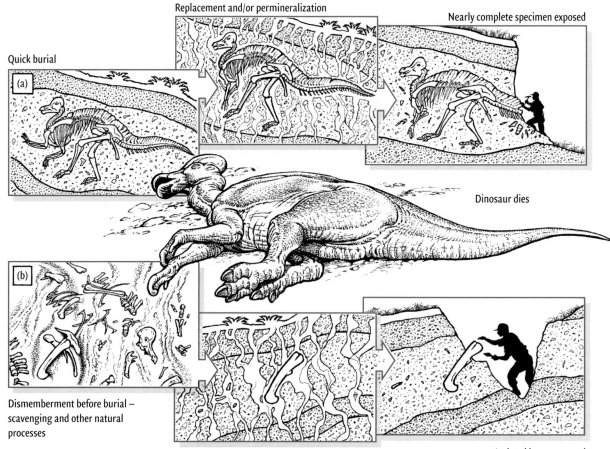

Replacement and/or permineralization

Nearly complete specimen exposed

Quick burial

(a)

Dinosaur dies

(b)

Dismemberment before burial –
scavenging and other natural
processes

Isolated bones buried and mineralized

Isolated bones exposed

Figure 1.2 Two endpoint processes of fossilization. In both cases, the first step is the death of the animal. Some decomposition occurs at the surface. In the upper sequence (a), the animal dies, the carcass undergoes quick burial, followed by bacterial decomposition underground, and permineralization and/or replacement. Finally, perhaps millions of years later, there is exposure. Under these conditions, when the fossil is exhumed, it is largely complete and the bones articulated (connected). This kind of preservation yields bones in the best condition. In the lower sequence (b), the carcass is dismembered on the surface by scavengers and perhaps trampled and distributed over the region by these organisms. The remains may then be carried or washed into a river channel and buried, replaced, and/or permineralized, eventually to be finally exposed perhaps millions of years later. Under these conditions, when the fossil is exhumed, it is disarticulated, fragmented, and the fossil bones may show water wear and/or the gnaw marks of ancient scavengers. Different conditions of fossil preservation tell us something about what happened to the animals after death.

Figure 1.3 Permineralized bone from the Jurassic-aged Morrison Formation, Utah, USA. The fossilized bone is now a solid piece of rock.

Bones can also be replaced, in which case the original bone minerals are replaced with other minerals, retaining the exact original form of the fossil. Most fossil bones undergo a combination of replacement and permineralization. The resultant fossil, therefore, is a magnificent natural forgery: chemically and texturally not bone, but retaining the exact shape and delicate features of the original bone.

Other fossils

Bones are not all that is left of dinosaurs. Occasionally the fossilized feces of dinosaurs and other vertebrates are found. Called coprolites, these sometimes impressive relics can give an intestine's-eye view of dinosaurian diets. Likewise, as we shall see later in this book, fossilized eggs and also skin impressions have been found.

Still, the single most important type of dinosaur fossil, other than the bones themselves, is trace fossils. Dinosaur trace fossils (sometimes also called ichnofossils; (*ichnos* – track or trace)) come as isolated footprints or as complete trackways. Figure 1.4 shows a mold, or impression, of a dinosaur footprint. We also find casts, which are made up of material filling up the mold. Thus a cast of a dinosaur footprint is a three-dimensional object that formed inside the impression (or mold).

In the last 20 years the importance of ichnofossils has been recognized. Ichnofossils have been used to show that dinosaurs walked erect, to reveal the position of the foot, and to reconstruct the speeds at which dinosaurs traveled. Trackways tell remarkable stories, such as that fateful day 150 or so million years ago when a large predatory theropod stalked a herd of smaller animals (Figure 1.5).

Figure 1.4 Theropod dinosaur footprint from the Early Jurassic Moenave Formation, northeastern Arizona, USA. Human foot for scale.

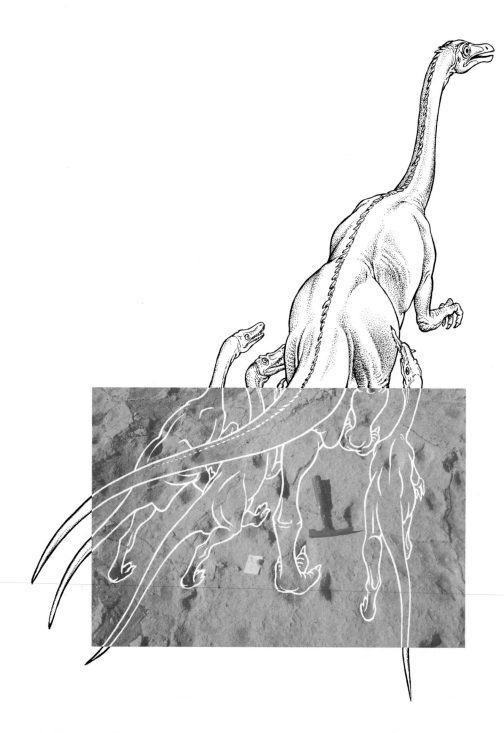

Figure 1.5 Photograph from Shar-tsav, Gobi Desert, Mongolia, showing the tracks of a medium-sized theropod dinosaur among those of a pack of smaller theropods. Our drawing suggests one interpretation, consistent with the evidence: the trackway could record a pack of *Velociraptor* hunting down a single *Gallimimus*.

Finding fossils

So, if the fossils are buried, how is it that we find them? The answer is really in the luck of geology: if fossil-bearing sedimentary rocks happen to be eroded, and a paleontologist happens to be looking for fossils at the moment that one is actively eroding from a rock, the fossil *may* be observed and *may* be collected. Indeed, we may be sure that, throughout their 160 million-year existence on Earth, dinosaurs walked over the exposed fossils of earlier dinosaurs, now lost to eternity (Figure 1.6)!

Figure 1.6 A pair of *Parasaurolophus* walking over some exposed fossilized bones of an earlier dinosaur that are weathering out of the cliff. Fragments of the fossilized bone have fallen at the dinosaurs' feet.

Collecting

The romance of dinosaurs is bound up with collecting: exotic and remote locales, heroic field conditions, and the manly extraction of gargantuan beasts (see Chapter 14). But ultimately dinosaur collecting is a process that draws upon good planning, a strong geological background, and a bit of luck. The steps are:

1   planning;
2   prospecting; that is, hunting for fossils;
3   collecting, which means getting the fossils out of whichever (usually remote) locale they are situated; and
4   preparing and curating them; that is, getting them ready for viewing, and incorporating them into museum collections.

These steps involve different skills and sometimes different specialists.

Planning

Collecting dinosaur fossils is not to be undertaken lightly. Dinosaur bones are – even in the richest sites – quite rare, and the moment they are disturbed the loss of important information becomes a concern. For this reason, most professional paleontologists have advanced degrees – often a Ph.D. in the geological or biological sciences – but before actually leading an expedition themselves, *all* have acquired many years of experience both in the logistical as well as the scientific ends of fieldwork.

Running an expedition The logistical end of an expedition involves keeping one's team fed, watered, healthy, and happy in remote places where, in many cases, these commodities don't come easily. Relentless sun, extreme heat, dust, lack of amenities, subsistence on a limited diet, and isolation from the "real world," all conspire to wear down even the most robust of people. It's all happening in the Great Outdoors, true, but it's nothing like a camping catalog! Add to these, language problems when you are working in other countries and limited access to medical facilities in the event of an accident involving either you or one of your crew, and the potential for disaster increases dramatically.

Many expeditions have to carry everything with them – fuel, water, food, all gear for the maintenance of daily life – as well as all the maps and equipment necessary to successfully carry out the science and safely retrieve heavy, yet delicate, dinosaur bones. This takes some serious planning and experience; you and your crew's lives may depend upon it (Figure 1.7). You have to know what you are doing.

Fossils generally, and dinosaurs in particular, are not renewable resources, which means that collecting a dinosaur is a one-shot deal: it must be done right, because we will never be afforded another chance to do it again. Any information that is lost – any piece of it that is damaged – may be lost or damaged forever. For this reason, there are many regulations associated with collecting vertebrate fossils.

The most basic are the collection permits required for work on public lands. Obtaining the permits requires advanced planning because the agencies in charge of issuing the permits reasonably require detailed accounts of your plans before the process can go forward.

One important part of the permit-obtaining process, especially in the case of dinosaur fossils (which tend to be large and heavy), is the eventual location of the fossils. Who gets them?

Figure 1.7 Supplies for one of the American Museum of Natural History's 1920s expeditions to the Gobi Desert. In the intervening 90 years, nobody has found a way to get around hauling the basic necessities into the field.

Does that person or place have the proper resources – or even the space – to store, preserve, and make them accessible to scientists and the general public? How is all this to be accomplished? Most of the truly great collections and many of the most important dinosaur fossils are housed in major museums, such as the American Museum of Natural History (New York), the Yale Peabody Museum (New Haven, CT), Tyrrell Museum (Alberta), the Smithsonian (Washington, DC), the Natural History Museum (London), and the *Musée National d'Histoire Naturelle* (Paris). These institutions have the resources required for the care of important specimens and the data associated with them.

Paleontology commonly involves a lot of foreign travel, which requires a whole new level of administrative preparation. All of the problems described above are compounded by language barriers, by the necessity to obtain visas along with permits, by the logistics of preparing a field expedition in a foreign country, and by the necessity of arranging for the eventual disposition of the fossils. What country, after all, would gladly see its fossil resources dug up and exported elsewhere? It's a delicate balance, sometimes requiring the skills of a diplomat.

Science All of that care expended upon all those logistics is meaningless unless our planning extends to the science as well. Paleontologists don't just go to weird places and grab bones. If they did, they'd lose, forever, essential information bearing upon four major problems:

1   What kind of environment was it in which the dinosaur was preserved (because it might have lived somewhere else)?
2   Where did it live?
3   When it did it live?
4   How did it die?

Oryctodromeus, a small herbivorous dinosaur first described in 2007, is a perfect example of the importance of geological context (Figure 1.8).

Here was an animal found fossilized in its own burrow. Had the important geological context not been properly interpreted, the burrow would not have been recognized and this animal's unusual behavior (for a dinosaur, at least), would have gone unappreciated.

So before even collecting the fossil, the locality – the area in which the fossil or fossils occur – must be mapped geologically, in a way that records the most information possible about the setting in which the fossil was found. This kind of information requires specialized geological study of the **paleoenvironments**,

Figure 1.8 Fossil burrow of the dinosaur *Oryctodromeus*. Careful study of the sedimentary context of this dinosaur revealed the burrow.

that is, the ancient environments represented by the rocks in which the fossils are found, as well as the geological context above and below the fossil. Usually this is accomplished by geological mapping and by detailed study of the sedimentary geology of the locality. Interpreting the ancient environment in which the bones came to rest commonly involves teaming up with sedimentologists – geoscientists with specialized knowledge of sedimentary rocks and the ancient environments that they preserve. This kind of teamwork allows paleontologists to develop the most complete picture of the fossils and the conditions in which they lived and died.[1]

A question that is commonly asked is "How do you know where the dinosaur fossils are?" The simplest answer is "We don't." There is no secret, magic formula for finding dinosaurs, unless it is long, hard hours of pre-expedition library time and careful assessment of potentially productive – that is, fossil-bearing – regions. On the other hand, a well-educated guess rooted in knowing something about the kinds of environment in which dinosaurs lived can greatly increase the odds of finding fossils.

Some basic criteria help the success of the search. These are:

1   Right rocks: the rocks must be sedimentary.
2   Right time: the rocks must be of the right age.
3   Living on the land: the rocks must be terrestrial.

Right rocks Sedimentary rocks have the best potential to preserve dinosaur fossils. Indeed, sedimentary rocks form in, and represent, sedimentary environments, many of them places where dinosaurs lived and died. Dinosaurs are known from other types of rocks, but their fossils are most likely found in sedimentary rocks.

Right time If the rocks you search were *not* deposited sometime between the Late Triassic and the Late Cretaceous, you won't find dinosaurs. Older and younger rocks may yield amazing fossil creatures, but not dinosaurs.

Living on land Dinosaurs were terrestrial, that is, non-marine, beasts through and through, which means that their bones will generally be found in rocks that preserve the remnants of ancient river systems, deserts, and deltas. However, dinosaur remains are also known from lake deposits and from near-shore marine deposits.

Many of the richest fossil localities in the world are in areas with considerable rock exposure, such as badlands. Fossil localities are common in deserts: plant cover on the rocks is low, and the dry air slows down the rates of weathering so that, once a fossil is exposed, it isn't chemically destroyed or washed away. Paleontologists, therefore, don't often find themselves in the jungle looking for fossils; the weathering rates are too high and the rocks are covered by vegetation. The chances of finding fossils are best in deserts, or at least fairly dry regions. Still, not all dinosaur material has been found in deserts. As long as the three criteria above are met, there is a possibility of finding dinosaur fossils, and that's usually reason enough for going in and taking a look.

---

[1] The study of all that happens to an organism after its death is called "taphonomy," and is a specialized field combining sedimentology and paleontology. Understanding the taphonomy of a fossil is the best way to know whether the animal actually lived in the environment in which its fossils were found, or whether its carcass was just deposited there after death.

Because fossils are a non-renewable resource, collecting them should be treated with the utmost circumspection. Poor planning, indifference to their significance, lack of training, and ignorance when retrieving them from the ground, will at best lose important data, and at worst place your and your team's lives in jeopardy.

Prospecting

Once we've planned properly, and we've got ourselves and our equipment to outcrops that match the criteria above, and once we have developed a concept of their geological context, what then? Simply put, we drop our eyes and start searching for bone weathering out of the rock.[2] If we're lucky and/or have a good eye, we'll spot something. Some collectors fare better than others, and a "feel" born of experience is surely involved in finding bone, as well as an experienced eye and dumb luck (Figure 1.9).

Figure 1.9 Paleontologists prospecting along an outcrop, eastern Montana, USA.

Collecting

Collecting is the arena in paleontology in which finesse meets brute force. Delicacy is required in preparing the fossils for transport; raw power is required for lifting blocks of bone and matrix (the rock which surrounds the bone) – commonly weighing many hundreds of pounds – out of the ground and back to civilization. Because of their size and delicacy, most dinosaur fossils are encased in a rigid jacket, or protective covering. Figure 1.10 shows how this is done.

Transport out of the field can be difficult, depending upon the size and weight of the jackets. A small jacket (soccer-ball size) can be carried out easily enough; but large jackets can require braces, hoists, winches, cranes, flatbed trucks, front-end loaders, helicopters, and even freight cars.

---

[2] Notice that the term dinosaur "dig" is a misnomer. Nobody digs into sediment to find bones; bones are found because they are spotted weathering out of sedimentary rocks.

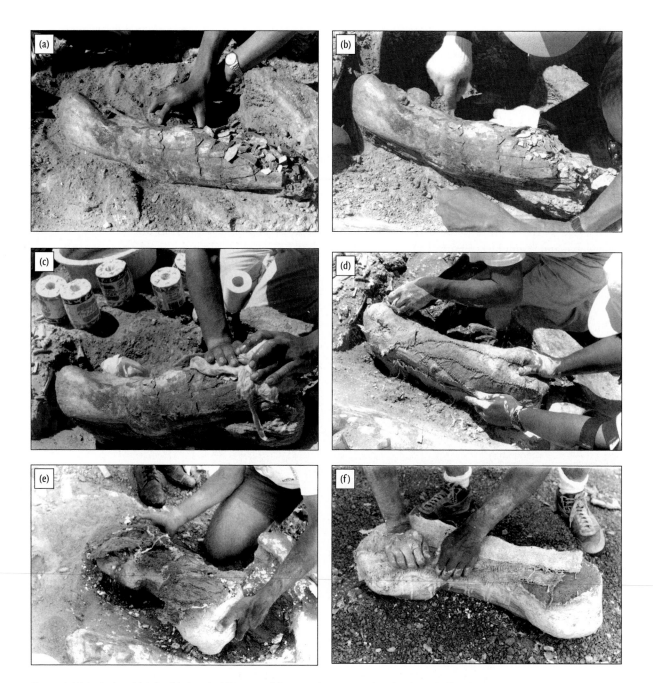

Figure 1.10 Jacketing. (a) A fossil is found sticking out of the ground; now it needs to be cleaned off so that its extent can be assessed. Exposing bone can be done with a variety of tools, from small shovels, to dental picks, to fine brushes. As the bone is exposed, it is "glued"; that is, impregnated with a fluid hardener that soaks into the fossil and then hardens. (b) The pedestal. When the surface of the bone is exposed, the rock around it is then scraped away. For small fossils, this can be quite painless; however, for large fossils, this can mean taking off the face of a small hill. Anyway, this process continues until the bone (or bones) is sitting on a pedestal, a pillar of matrix underneath the fossil. (c) Toilet paper cushion. Padding is placed around the fossil to cushion it. The most cost-effective cushions are made from wet toilet paper patted onto the fossil. It takes a lot of toilet paper: for example, a 1 m thigh bone (femur) could take upward of one roll. On the other hand, this is not a step where we should cut corners, because returning from the field with a shattered specimen, or one in which the plaster jacket is stuck firmly to the fossil bone, is not so good. (d) Plaster jacket. Jackets are made of strips of burlap cloth soaked in plaster, and then applied to the toilet-paper-covered specimen. A bowl of plaster is made up, and then pre-cut, rolled strips of burlap are soaked in it and then unrolled onto the specimen and the pedestal. (e) Turning the specimen. After the plaster jacket is hardened, the bottom of the pedestal is undercut, and the specimen is turned; that is, separated at the base of the pedestal from the surrounding rock and turned over. This is a delicate step in which the quality of the jacket is tested. (f) The top jacket. More plaster and burlap are then applied to the open (former) bottom of the jacket, now its top. At this point, the fossil is fully encased in the plaster-and-burlap jacket, and is ready for transport from the field.

Figure 1.11 Scenes from a prep lab, in this case, that of the Museum of Northern Arizona in Flagstaff, USA. (a) Foreground: a large plaster jacket containing the theropod dinosaur *Coelophysis*. Background: sand tables for stabilizing specimen fragments (to be glued), open jackets, storage shelves, and grinding equipment. (b) The jacket shown in (a). The *Coelophysis* specimen is visible in the foreground. The arms and hands are to the left; the pelvis, legs, and tail are visible to the right. (c) A dinosaur skull (*Pentaceratops*) laid out for study. In the background are the large bays in which specimens are stored.

Back at the ranch

Once the fossil dinosaur bone is out of the field and back where it can be studied, the jacket must be cut open, and the fossil prepared, or freed from the matrix. This runs from simple brushing, to scraping with dental needles, to sophisticated treatments such as acid removal of the rock matrix. These techniques are generally carried out in a preparation laboratory (or prep lab; Figure 1.11).

We often expect that the final result will be a free-standing display in a museum. Mounts of real fossil bone are attractive, but also time-consuming and costly, and the metal frames that support the bones can be destructive to the fossil. Moreover, mounted specimens commonly undergo damage over time; slight shifts in the mounts because of the extraordinary weights of the fossil bones, or vibrations in the buildings in which the bones are housed, or museum patrons lifting presumably "insignificant" bits all diminish the quality of mounted specimens. In addition, when the specimens are assembled and mounted, they can be hard to examine for study.

Many museums, therefore, have begun to cast the bones in fiberglass and other resins, and display the casts. Such displays are virtually indistinguishable from the originals. With their light weight, and the possibility of internal frames, they can be spectacular and dynamic (Figure 1.12).

Leaving the bones disarticulated, properly curated, and available for study maximizes returns on the very substantial investments that are involved in collecting dinosaur remains. Paleontology is carried out in large part by public support, and mounted casts give the public the best value for its money.

Figure 1.12 A spectacular mount of the sauropod *Barosaurus* and the theropod *Allosaurus*. This mount is made of fiberglass and epoxy resin, cast from the bones of the original specimens. A dynamic pose like this would not have been possible using the original fossil bones.

Summary

Fossils, the buried remains of organic life, are divided into two types: body and trace fossils. The body fossils include bones, shells, and other organic remains; trace fossils consist of tracks, trackways, and other impressions in the form of molds and casts.

Fossilization is a process that occurs after the organism dies. It consists of burial, and commonly involves a variety of types of replacement, in which the original organic and mineral material of the once-living organism is naturally replaced by other minerals while buried.

Obtaining fossils, particularly dinosaur fossils, requires rigorous training and preparation, along with perhaps a bit of educated guessing. Four steps are involved: planning, prospecting, collecting, and laboratory preparation and curation. The planning ranges from figuring out where to look, to getting the legal permission to carry out the study, to outfitting an expedition properly to safely meet its goals. The prospecting requires a well-trained eye, always enhanced by experience. Collecting is a process designed to bring delicate fossils safely back to where they can be prepared, which involves cleaning, reconstruction, and protection. Curation makes it possible for fossils to be safely stored in the long term, and to be accessible to researchers and to an interested public.

Selected readings

Behrensmeyer, A. K. and **Hill, A. P.** (eds.). 1980. *Fossils in the Making.* University of Chicago Press, Chicago, IL, 338pp.

Cvancara, A. M. 1990. *Sleuthing Fossils: The Art of Investigating Past Life.* John Wiley and Sons, New York, 203pp.

Gillette, D. D. and **Lockley, M. G.** (eds.). 1989. *Dinosaur Tracks and Traces.* Cambridge University Press, New York, 454pp.

Horner, J. R. and **Gorman, J.** 1988. *Digging Dinosaurs.* Workman Publishing, New York, 210pp.

Kielan-Jaworowska, S. 1969. *Hunting for Dinosaurs.* Maple Press Co. Inc., York, PA, 177pp.

Martin, R. E. 1999. *Taphonomy: A Process Approach.* Cambridge University Press, Cambridge, 524pp.

Sternberg, C. H. 1985 (originally published 1917). *Hunting Dinosaurs in the Badlands of the Red Deer River, Alberta, Canada.* NeWest Press, Edmonton, Alberta, 235pp.

Topic questions

1 Define: fossil, dinosaur, replacement, preparation, coprolite, biosphere, body fossil, molds, casts, and ichnofossils.

2 What kind of training does it take to be a paleontologist?

3 What steps are involved in the collection of dinosaur bones?

4 What criteria maximize the likelihood of finding dinosaur fossils?

5 What kinds of paleoenvironments are most likely to preserve dinosaur bones? Why?

6 Why is it that, generally, the older the fossil, the less likely it is that original bone material will be preserved?

7 Why is understanding the geological context so important when collecting dinosaurs?

8 Why do dinosaur fossils tend to be preserved better in deserts than elsewhere?

9 What kinds of conditions might be required to find a fully articulated dinosaur fossil?

10 What kinds of geological activity are important to carry out before a fossil is extracted from the ground? Why?

CHAPTER 2
DINOSAUR DAYS

CHAPTER OBJECTIVES

- Introducing geological time and stratigraphy

- Learning about continental drift during the time of the dinosaurs

- Learning about ancient climates during the time of the dinosaurs

When did dinosaurs live and how do we know?

Fossils, including dinosaur remains, are found in layers of rock, commonly called strata. The field of stratigraphy is the geological specialty that tells us how old or young particular strata are; thus, stratigraphy is a means of learning the age of dinosaur fossils. Stratigraphy is divided into geochronology or Earth time stratigraphy (*geo* – Earth; *chronos* – time), lithostratigraphy or rock stratigraphy (*lithos* – rock), and biostratigraphy or stratigraphy as indicated by the presence of fossils (*bios* – organisms).

Geochronology

Geologists generally signify time in two ways: in numbers of years before present, and by reference to blocks of time with special names. For example, we say that the Earth was formed 4.6 billion *years before present*, meaning that it was formed 4.6 billion *years ago* and is thus 4.6 billion years *old*. Unfortunately, determining the precise age in years of a particular rock or fossil is not always easy, or even possible. For this reason, geologists have divided time into intervals of varying lengths, and rocks and fossils can be referred to these intervals, depending upon how exactly the age of the rock or fossil can be estimated. For example, you might not know that a fossil was 92.3 million years old, but you might be able to determine that it was within the interval of time known as the Late Cretaceous, meaning that its age is somewhere between 99.6 and 65.5 million years old, dates about which you have more information.

We start our discussion intuitively, by discussing the age in years, or the numerical age. Later we will address the division of time into blocks of varying lengths.

Geochronology: the ages of the ages Geoscientists are happiest when they can learn the numerical age of a rock or fossil; that is, its age in years before present. Ages in years before present are reckoned from the decay of unstable isotopes found in certain minerals. The unstable isotopes spontaneously decay from an energy state that is not stable (that is, that "wants" to change) to one that is more stable (that is, that will not change, but rather remain in its present form). The decay of an unstable isotope to a stable one occurs over short or long amounts of time, depending upon the isotope. The basic decay reaction runs as follows:

$$\text{unstable "parent" isotope} \rightarrow \text{stable "daughter" isotope} + \text{nuclear products}$$
$$+ \text{ energy (generally, heat)}$$

The element carbon provides a good example. In the decay of the unstable isotope of carbon ^{14}C, a neutron splits into a proton and an electron, in the following reaction:

$$^{14}C \rightarrow {}^{14}N + \text{energy}$$

Note that the atomic number in the decay reaction changes; it increases from 6 to 7. Now, with 7 protons and 7 electrons, the stable daughter has an atomic number of 7, which means that the element in question has become nitrogen (see Background 2.1 for a quick review of the chemistry underlying these concepts).

The *rate* of the decay reaction is the key to obtaining an numerical age. If we know:

1   the original amount of parent isotope at the moment that the rock was formed or the animal died (before becoming a fossil);
2   how much of the parent isotope is left; and
3   the rate of the decay of that isotope,

we can estimate the amount of time that has elapsed. For example, suppose we know that 100% of an unstable isotope was present when a rock was new, but now only 50% remains.

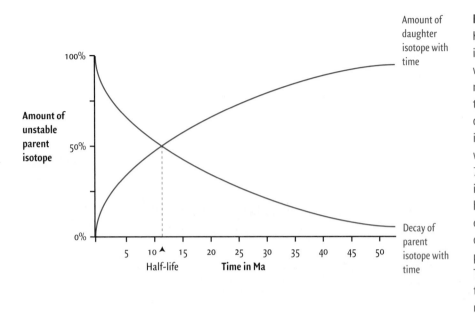

Figure 2.1. An isotopic decay curve. Knowing the amount of unstable isotope that was originally present, as well as the amount of unstable isotope now present and the rate of decay of the unstable isotope, it is possible to determine the age of a rock with that isotope in it. Suppose we found a rock with a ratio of 25% unstable parent : 75% stable daughter of a particular isotope. That would mean *two* half-lives had elapsed (half-life no. 1 = 50% of 100% parent (50% parent : 50% daughter); half-life no. 2 = 50% of 50% parent (25% parent : 75% daughter)). The amount of time represented by two half-lives can be read on the axis marked "Time"; in this case, about 25 million years. The rock would thus be about 25 million years old.

If we know the rate at which the element decayed, we can estimate the amount of time that has elapsed since the rock was formed; that is, the *age of the rock*. This is shown in Figure 2.1.

Choosing the right isotope Since each unstable isotope has its own constant rate of decay,[1] it is convenient to summarize that rate by a single number. That number is called the half-life, which is the amount of time that it takes for 50% of the atoms of an unstable isotope to decay (leaving half as much parent as was originally present). The half-life, then, is an indicator of decay rate, and provides guidance about which isotope is appropriate for which amount of time. For example, to date human remains, not likely more than several thousand years old, the rubidium/strontium isotopic system (^{87}Rb/^{87}Sr), with a half-life of 48.8 billion years, would hardly be the ideal isotopic system. This would be a bit like timing a 100 m dash with a sundial. Likewise, dating dinosaur bones (ages that would be in hundreds of millions of years) using ^{14}C, which has a half-life of 5,730 years, would be like giving your own age in milliseconds. The ages that involve dinosaurs are 10s to 100s of millions years old, abbreviated Ma.[2]

Unstable isotopes are powerful dating tools, but they cannot be used directly on dinosaur bone. There has to be a source of unstable isotopes, which occur commonly in certain minerals, some of which form as lava cools and the minerals crystallize. The decay process begins when the unstable isotope is first formed (that is, when it crystallizes in the lava), so, age of the rock can be obtained from the moment the crystal formed as the lava cooled.

No dinosaur ever lived *within* hot lava – at least not for very long – so how can we get the age of the dinosaur bone when all we have is a date from some lava? This – the relationship between one body of rock (in this case, containing dinosaur bone) to another (here, the cooled lava) – is the province of lithostratigraphy.

---

[1] In fact the rate that individual molecules decay fluctuates in the short term but is statistically constant over long periods of time.

[2] We use the expression Ma, from *mille annos* – a million years. Thus, 65 Ma is 65 million years ago.

Lithostratigraphy

Sedimentation and sedimentary rocks Sediments – sand, silt, mud, dust, and other less-familiar materials – are deposited on scales of meters to 100s of kilometers, and are the direct result of sedimentation such as flowing water, wind, or explosion from a volcano, to name a few more or less common processes. Virtually every geographical location we can think of – a river, a desert, a lake, an estuary, a mountain, the bottom of the ocean, the *pampas* – has sedimentary processes peculiar to it that will produce distinctive sediments and, with time and burial, distinctive sedimentary rocks. These geological processes – most commonly, wind and water – deposit sediments in strata. The strata pile up on one another, stacking into a thick sequence of sedimentary rocks (Figure 2.2).

Relative dating It is a fact that younger sediments are deposited upon older sediments (exemplified in Figure 2.2), and yet this apparently self-evident insight is the fundamental basis of all correlations of sedimentary strata in time. Ascertaining the relative ages of two strata is termed relative dating and, while not providing the age in years before present, provides the age of one stratum *relative* to another stratum. Here, then, is part of the solution to dating dinosaur bone. Suppose that a stratum containing a dinosaur bone is sandwiched between two layers of volcanic ash. Ideally, a numerical age date could be obtained from each of the ash layers. We would know that the bone was younger than the lower layer but older than the upper layer. Depending upon how much time separates the two layers, the bone between them can be dated with greater or lesser accuracy (Figure 2.3).

But how can one tell that two geographically separated deposits were deposited at the same time if numerical ages are unknown? In this, fortunately, stratigraphers are aided by one last, extremely important tool: biostratigraphy.

Figure 2.2. Superposition of strata in Petrified Forest National Park, Arizona, USA. Thick stacks of red mudstones were deposited by rivers 210 million years ago, to produce the succession of layers visible in this photograph.

Biostratigraphy

Biostratigraphy is a method of relative dating that utilizes the presence of fossil organisms. It is based upon the idea that a particular time interval can be characterized by a distinctive assemblage, or group, of organisms. For example, if one knows that dinosaurs lived from 228 to 65 Ma, then any rock containing a dinosaur fragment must fall within that age range. Although biostratigraphy cannot provide ages in years before present, the fact that many

Figure 2.3. Bone between two dated horizons. As we know the ages of the two horizons, the age of the bone can be interpolated between them (see the text).

species of organisms have existed on Earth for 1–2 million-year intervals enables them to be used as powerful dating tools. For example, *Tyrannosaurus rex* lived for only about 2 million years, from 67 to 65 Ma. Therefore if we found a *T. rex* fossil (a good find, indeed), we would know that, no matter where that dinosaur was found, it could be correlated with *T. rex*-bearing sediments in North America that have been well dated at 67–65 Ma.

Eras and Periods and Epochs, Oh My!

Geological time is divided into a hierarchy, much as our time is divided into years, months, weeks, days, hours, minutes, and seconds. We'll begin with large blocks of time called **Eras**. The Eras are, from oldest to youngest, the Paleozoic (*paleo* – ancient; *zoo* – animal), the Mesozoic (*meso* – middle), and the Cenozoic (*cenos* – new). Within each Era are smaller subdivisions (still consisting of 10s of millions of years each) called **Periods**, and within these are yet smaller sub-divisions of time called **Epochs** (consisting of several millions of years each). Figure 2.4 shows the part of the geological time scale during most of which dinosaurs roamed the Earth.

Continents and climates

Although the idea is less than 100 years old, most people are aware that throughout its history, somehow the continents have moved around the Earth's surface; the Earth has been anything other than static. The process is commonly called "**continental drift**" but is in fact part of a much more significant concept: **plate tectonics** (see Appendix 2.2). Plate tectonics, when it was first conceived in the late 1960s, revolutionized our understanding of the Earth, because it ultimately provided explanations for phenomena as different as the rise of mountains, oceanic circulation, and even the distribution of organisms.

Here, because we are interested in dinosaurs, we take the luxury of bypassing a mere *3.7 billion* years of continental evolution and zip right up to the Late Triassic, when all the continents coalesced into a single landmass, now known as **Pangaea** (Figure 2.5).

Late Triassic

Pangaea was the dominant land feature of the Late Triassic Earth. Like any large landmass, Pangaea had great mountain ranges; however, it was at least theoretically possible to walk on land from any place to any other. Unlike today, where the faunas and floras on different continents differ, Late Triassic faunas and floras all around the world were somewhat similar.

Early Jurassic

The initial break-up of Pangaea took place in the Early Jurassic. The effect was like unzipping the great supercontinent from south to north. Sediments in the eastern seaboard and Gulf Coast regions of North America, and in Venezuela and western Africa, record the opening and widening of a seaway.

Also at this time, some of the earliest **epicontinental** (or "epeiric") seas of the Mesozoic Era first made their appearances. Epicontinental seas are shallow marine waters that cover parts of continents. In the past, epicontinental seas were considerably more widespread than they are today, because then **eustatic** (or global) sea levels were higher than they are now.

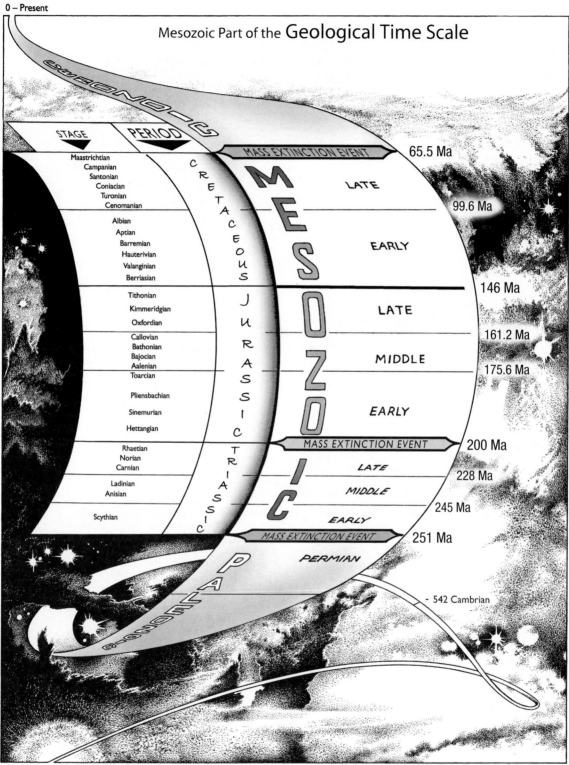

0 – Present

Mesozoic Part of the Geological Time Scale

CENOZOIC

STAGE	PERIOD

MASS EXTINCTION EVENT — 65.5 Ma

MESOZOIC

Maastrichtian	
Campanian	
Santonian	
Coniacian	**C** LATE
Turonian	**R**
Cenomanian	**E** — 99.6 Ma
Albian	**T**
Aptian	**A**
Barremian	**C** EARLY
Hauterivian	**E**
Valanginian	**O**
Berriasian	**U** **S** — 146 Ma
Tithonian	
Kimmeridgian	**J** LATE
Oxfordian	**U** — 161.2 Ma
Callovian	**R**
Bathonian	**A** MIDDLE
Bajocian	**S**
Aalenian	**S** — 175.6 Ma
Toarcian	**I**
Pliensbachian	**C** EARLY
Sinemurian	
Hettangian	

MASS EXTINCTION EVENT — 200 Ma

Rhaetian	**T**
Norian	**R** LATE
Carnian	**I** — 228 Ma
Ladinian	**A** MIDDLE
Anisian	**S** — 245 Ma
Scythian	**S** EARLY — 251 Ma

MASS EXTINCTION EVENT

PERMIAN

PALAEOZOIC

- 542 Cambrian

Origin of the Earth 4.6 Billion years ago

Figure 2.4. The Mesozoic part of the geological time scale. The Mesozoic constitutes only a rather tiny fraction of the expanse of Earth time. If you compacted Earth time into a single year, from January 1 (the formation of the Earth) to December 31 (the past 100,000 years, which, by this way of measuring Earth history, would occur in less than a day!), then dinosaurs were on Earth from about December 11 to December 25. (Dates from Gradstein, F., Ogg, J. and Smith, A. 2004. *A Geological Time Scale 2004.* Cambridge University Press, Cambridge, 589 pp.)

EARLY TRIASSIC 237 MA

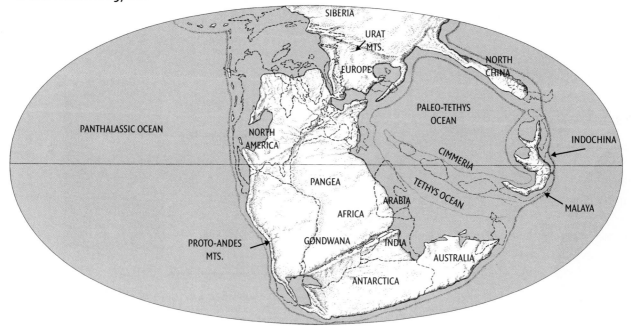

Figure 2.5. The positions of the present-day continents during the Early Triassic (237 Ma). Earth was dominated by the unified landmass Pangaea.

LATE JURASSIC 152 MA

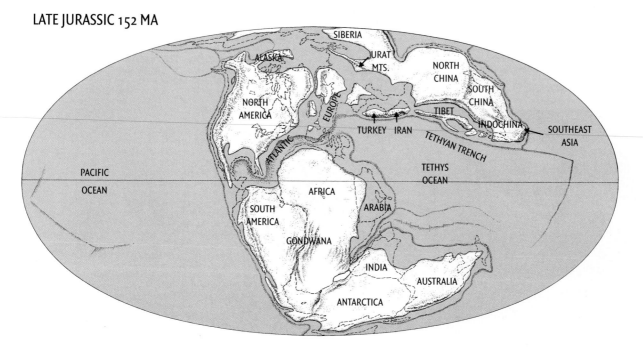

Figure 2.6. The positions of the continents during the Late Jurassic (152 Ma). Pangaea has begun its dismemberment while the southern continent of Gondwana remains together.

Middle and Late Jurassic

In the Late Jurassic (Figure 2.6) and Early Cretaceous, continental separation was well under-way. A broad seaway, the *Tethyan Seaway* (after the Greek goddess *Tethys*, Goddess of the Sea), ran between two supercontinents, one in the north known as Laurasia and one in the south called Gondwana.

Early Cretaceous

The first half of the Cretaceous was a time of active mountain-building, sea-floor spreading, high eustatic sea levels, and broad epeiric seas. The Tethys Ocean, a sea that eventually became the Atlantic Ocean, remained a dominant geographical feature, as Europe continued to separate from North America.

In Gondwana, a stable continental marriage dating back to the Early Paleozoic Era finally underwent rifting involving two of its largest constituents – Africa and South America – as well as two smaller constituents, India and Madagascar. While India and Madagascar were in the first bloom of unconfinement, Australia and Antarctica remained together (a union that would not end until about 50 Ma), and a land connection remained, as it *almost* does today, between South America and Antarctica.

Late Cretaceous

The global positions of continents during the Late Cretaceous would be almost familiar to us (Figure 2.7). North America became nearly isolated, connected only by a newly emergent land connection across the modern Bering Straits to the eastern Asiatic continent. Although best known from the last Ice Age (100,000 years ago, this land bridge has come and gone several

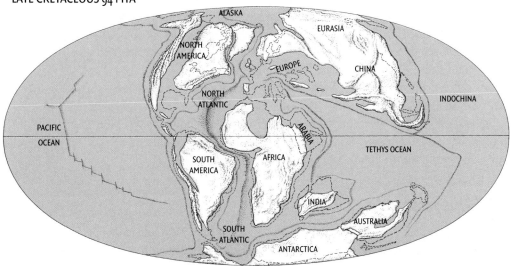

Figure 2.7. The positions of the continents during the Late Cretaceous (94 Ma). The positions of the continents did not differ significantly from their present-day distribution. Note the land bridge between Asia and North America, as well as the European archipelago. By this point in time, both of the supercontinents, Gondwana and Laurasia, had disintegrated.

times since the Cretaceous. Africa and South America were fully separated, the former retaining its satellite, Madagascar, and the latter retaining a land bridge to the Antarctica/Australia continent. India was by now well on its way towards colliding with southern Asia.

Climates during the time of the dinosaurs?

Earth has recorded traces that allow us to infer at least *aspects* of past climates, and indeed the flavor of the Mesozoic would be lost without some general sense of Mesozoic paleoclimates (ancient climates). Distributions of the landmasses as well as the amount and distribution of the oceans on the globe drastically modify temperatures, humidity, and precipitation patterns. In the following, we explore this, comparing the extreme case of the continents coalesced into a single landmass (see Figure 2.5) with the equally extreme (but more familiar) case of the continents widely distributed around the globe (see Figure 2.7).

Heat retention in continents and oceans

Continents (land) and oceans (bodies of water) respond very differently to heat from the sun: you can jump into a swimming pool and be comfortable long after the air and land have gotten chilly! This is due to differences in respective heat capacity[3]: how easily the temperature of a given material can be changed. Water has a higher heat capacity than continental crust, which means that it takes more energy to change its temperature than it takes to change the temperature of continents. The result is that the water takes more time to heat up, and more time to cool down. This is also why temperatures at the coast are almost always more mild than those found inland.

Consider how these properties of continents and oceans might modify climates at the dawn of the Mesozoic, when the continents were united into the single landmass Pangaea (see Figure 2.5). Here, "continental effects" – more rapid warming and cooling of continents than oceans – would have been more intense than today. Pangaea must have experienced wide temperature extremes. It would have heated up quickly and got hotter, and then cooled off more rapidly and got colder faster than modern continents, whose continental effects are mitigated by the broad, temperature-stabilizing expanses of oceans between them.

The post Late Triassic break-up of Pangaea weakened the strong continental effects. With the rise in eustatic sea level and supercontinental dismemberment, the effects of the large epeiric seas were superimposed upon the diminishing continental effects. These large bodies of water would have stabilized global temperatures, decreasing the magnitude and rapidity of the temperature fluctuations experienced on the continents during times of lower sea level.

Climates through the Mesozoic

The Late Triassic and Early Jurassic were times of heat and aridity. They also were times of marked seasonality; that is, well-defined seasons, strongly affected by the Pangaea continental mass. By the latter two-thirds of the Jurassic, however, as well as most of the Cretaceous, Earth is thought to have been *without* polar ice or glaciers on the northern parts of the the continents. This is quite beyond our own experience; now, glaciers occur at high latitudes at both poles,

---

[3] Defined as the amount of heat required to change the temperature of a mole of material 1 °C.

and the poles themselves are covered in ice. The conclusion that there were no ice or glaciers above the Arctic and Antarctic Circles (latitudes 66° N and 66° S) is based largely upon the presence of warm climate indicators such as the fossils of warmth-loving plants and certain fish at high latitudes, and upon the absence of any evidence of continental glaciation from this time.

The absence of polar ice had an important consequence for climates: water that would have been bound up in ice and glaciers was instead in ocean basins. This in turn meant higher eustatic sea levels, which lead to extensive epeiric seas. The increased abundance of water on the continents as well as in the ocean basins had a stabilizing effect on temperatures (because it decreased continental effects), and decreased the amount of seasonality experienced on the continents.

Continental climates are enormously variable, however, and in North America Upper Jurassic terrestrial deposits, features preserved in the rocks, such as oxidized sediments and calcium carbonate deposits, suggest that the Late Jurassic there was was marked by seasonally arid conditions. So much for dinosaurs in steamy, swampy jungles!

Paleoclimates in the Cretaceous are somewhat better understood than those of the preceding periods. During the first half of the Cretaceous at least, global temperatures remained warm and equable. The poles continued to be ice-free, and the first half of the Cretaceous saw far less seasonality than we see today. This means that, although equatorial temperatures were approximately equivalent to those we experience today, the temperatures at the poles were somewhat warmer. Temperatures at the Cretaceous poles have been estimated at 0–15 °C, which means that the temperature difference between the poles and the equator was only between 17 and 26 °C, considerably less than the ±41 °C of the modern Earth.

The first half of the Cretaceous was synergistic: **tectonic** activity, such as mountain building and sea-floor spreading caused an increase in atmospheric CO_2 and a decrease in the volume of the ocean basins, which in turn increased the area of epeiric seas. The seas thus stabilized climates already warmed by enhanced absorption of heat in the atmosphere.

Sound familiar? The Early to mid-Cretaceous experienced the notorious "greenhouse" conditions that are currently of such concern today. Because several times in its past history, including in the Early Cretaceous, the Earth has "experimented" with greenhouse conditions, Earth history has a lot to offer to the dialog about global warming.

The last 30 million years of the Cretaceous produced a mild deterioration in the equable conditions of the mid Cretaceous. A pronounced withdrawal of the seas took place, and evidence exists of more pronounced seasonality.

Summary

Determining the ages of dinosaurs is accomplished by a mixture of biostratigraphy, lithostratigraphy, and geochronology. These allow paleontologists to date rocks and fossils in relation to each other, as well as to obtain estimates of their ages in years before present. Using these techniques, geoscientists have constructed and refined a geological time scale for the entire history of the Earth. The time scale is hierarchically divided into successively more refined time intervals: Eras, Periods, and Epochs.

The earliest dinosaurs appeared during the Late Triassic, a time in which the Earth's continents were united into a single supercontinent called Pangaea. Since then, the continents have separated, moving to their present positions.

The presence of a single supercontinent had major implications for climates, which were strongly seasonal. These mitigated throughout the Jurassic, although the Late Jurassic appears to have had strong seasonality, at least in North America. By mid-Cretaceous time, high sea levels, melted polar ice, and high levels of atmospheric CO_2, appear to have acted synergistically to produce global warming and greenhouse conditions.

Selected readings

Barron, E. J. 1983. A warm, equable Cretaceous: the nature of the problem. *Earth Science Reviews*, **19**, 305–338.

Berry, W. B. N. 1987. *Growth of a Prehistoric Time Scale Based on Organic Evolution*. Blackwell Scientific Publications, Boston, MA, 202pp.

Crowley, T. J. and **North, G. R.** 1992. *Paleoclimatology*. Oxford Monographs in Geology and Geophysics, no. 18. Oxford University Press, New York, 339pp.

Gradstein, F., Ogg, J. and **Smith, A.** (eds.). 2004. *A Geological Time Scale 2004*. Cambridge University Press, Cambridge, 549pp.

Lewis, C. 2000. *The Dating Game: One Man's Search for the Age of the Earth*. Cambridge University Press, Cambridge, 253pp.

Robinson, P. L. 1973. Palaeoclimatology and continental drift. In Tarling, D. H. and Runcorn, S. K. (eds.), *Implications of Continental Drift to the Earth Sciences*, vol. I. NATO Advanced Study Institute, Academic Press, New York, pp. 449–474.

Ross, M. I. 1992. Paleogeographic Information System/Mac Version 1.3: Paleomap Project Progress Report no. 9, University of Texas at Arlington, 32pp.

Topic questions

1   Define: eustatic, biostratigraphy, stratigraphy, geochronology, half-life.

2   What is the difference between "numerical" dates and relative ages?

3   If a dinosaur bone is found between two layers dated at 90 million and 70 million years, respectively, what is the age of the dinosaur bone?

4   Referring to question no. 3, how precisely can that bone be dated?

5   What is the numerical value for the half-life shown in Figure 2.1?

6   If there were dinosaur-bearing rocks in North America and Africa, and the African dinosaur remains could be dated biostratigraphically, how could you correlate the North American deposits with the African ones?

7   Compare the Late Triassic Earth with the Late Cretaceous Earth.

Background 2.1 – Chemistry quick 'n dirty

Earth is made up of elements, such as are seen on a Periodic Table. Many of these, such as hydrogen, oxygen, nitrogen, carbon, and iron, are familiar, while others, such as berkelium, iridium, and thorium, are probably not. All elements are made up of atoms, an atom being the smallest particle of any element that still retains the properties of that element. Atoms, in turn, are made up of protons, neutrons, and yet smaller electrons, which are collectively termed subatomic ("smaller-than-atomic") particles. Protons and neutrons reside in the central core, or nucleus, of the atom. The electrons are located in a cloud surrounding the nucleus. Some electrons are more tightly bound around the nucleus and others are less tightly bound. Those that are less tightly bound are, as one might expect, more easily removed than those that are more tightly bound (Figure A2.1).

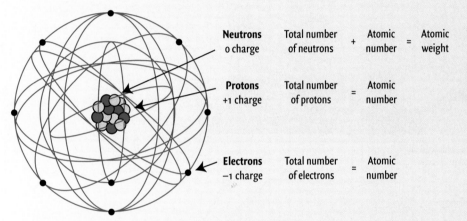

Figure A2.1. Diagram of a carbon atom. In the nucleus are the protons and neutrons. In a cloud around the nucleus are the electrons, whose position relative to the nucleus is governed by their energy state.

Keeping that in mind, let us further consider the subatomic particles. Protons and electrons are electrically charged; electrons have a charge of –1 and protons have a charge of +1. Neutrons, as their name implies, are electrically neutral and have no charge. To keep a charge balance in the atom, the number of protons (positively charged) must equal the number of electrons (negatively charged). This number – which is the same as the number of protons – is called the atomic number of the element, and is displayed to the lower left of the elemental symbol. For example, the element carbon is identified by the letter C, and it has 6 protons. Its atomic number is thus 6, and it may be written $_6C$.

Along with having an electrical charge, some subatomic particles also have mass. Rather than force us to work with the extremely small mass of a proton (one of them weighs about 6.02×10^{23} grams!), it is assigned a mass of 1. Neutrons have a mass of 1 as well. Because relative to protons and neutrons the masses of electrons are negligible, the mass number of an element is composed of the total number of neutrons *plus* the total number of protons. In the case of the element carbon, for example, the mass number equals the total number of neutrons (6) plus the total number of protons (6); that is, 12. This is usually written ^{12}C and is called carbon-12. Note that ^{12}C has 6 protons and therefore must also have 6 electrons, so its atomic number remains 6. Because the atomic number is always the same for a particular element, it is commonly not included when the isotope is discussed. Thus $^{12}_6C$ is usually abbreviated to ^{12}C.

Variations in elements that have the same atomic number but different mass numbers are, as we've seen, called isotopes. For example, a well-known isotope of ^{12}C (carbon-12) is ^{14}C (carbon-14). Since $_{14}C$ is an isotope of carbon, it has the same atomic number as ^{12}C (based upon 6 electrons and 6 protons). The change in *mass* number results from additional neutrons. ^{14}C has 8 neutrons, which, with the 6 protons, increase its atomic mass to 14. Because it is carbon, of course, its atomic number remains 6.

PLATE TECTONICS!

SUBDUCTION LEADS TO OROGENY...

EARTH IS DYNAMIC AND NOTHING ON IT IS PERMANENT EXCEPT THE CHANGE, ITSELF.

...SO THE STRUCTURE OF THE EARTH IS LIKE AN ONION, GOING FROM THE IRON-RICH CORE, THROUGH THE THICK LOWER MANTLE (MESOSPHERE), THE THIN, FLUID UPPER MANTLE, TO THE RIGID CRUST AND LITHOSPHERE RIDING ATOP THE UPPER MANTLE (ASTHENOSPHERE).

UPPER MANTLE (LITHOSPHERE)
UPPER MANTLE (ASTHENOSPHERE)
LOWER MANTLE (MESOSPHERE)
OUTER CORE (LIQUID IRON)
INNER CORE (SOLID IRON)

CRUST
LITHOSPHERE

THE EARTH HASN'T BEEN STATIC LO THESE 4.6 BILLION YEARS OF EXISTENCE. THE PLATES HAVE BEEN SLIDING AROUND, AND THE CONTINENTS, THOSE BITS OF THICKENED CRUST LAMINATED TO THE LITHOSPHERIC PLATES, HAVE BEEN MOVING AROUND WITH THEM.

THE LITHOSPHERE ISN'T A SINGLE RIGID SPHERE ~ INSTEAD IT IS DIVIDED INTO 10 OR SO PLATES, THAT FLOAT AND MOVE AROUND ON THE PLASTIC ASTHENOSPHERE LIKE SCUM ON A POT OF BOILING HOT CHOCOLATE.

THE PLATES OF LITHOSPHERE HAVE A TOP LAYER CALLED THE "CRUST"; THE CONTINENTS HAVE THICKENED CRUST WHILE OCEANIC LITHOSPHERE HAS THINNER CRUST.

CRUST

THE MOVEMENT OF THE PLATES IS REFERRED TO AS THE THEORY OF PLATE TECTONICS AND IT IS AS IMPORTANT FOR UNDERSTANDING THE WAY EARTH WORKS AS THE THEORY OF EVOLUTION HAS BEEN FOR UNDERSTANDING LIFE!

WE'LL RETURN TO THE BOILING HOT CHOCOLATE ANALOGY TO UNDERSTAND THE MECHANICS OF PLATE TECTONICS.

BOILING ON A STOVE, HOT CHOCOLATE DEVELOPS A SCUM, WHICH CONTINUOUSLY MOVES AS HEAT CIRCULATES THROUGH THE POT OF LIQUID. NEW HOT SCUM IS FORMED, AND AS THIS OCCURS, OLDER, COOLER SCUM PUCKERS AND WRINKLES TO ACCOMMODATE THE NEW MATERIAL.

SO IT IS WITH THE LITHOSPHERE: NEW ADDITIONS OF LITHOSPHERE OCCUR ALONG ELONGATE GEOGRAPHIC FEATURES – SPREADING CENTERS – PLACES WHERE HOT MOLTEN MATERIAL COMES UP FROM THE ASTHENOSPHERE TO COOL AND MAKE MORE LITHOSPHERE. AS NEW HOT ASTHENOSPHERE IS ADDED, OLDER, COOLER LITHOSPHERE IS PUSHED AWAY.

OBVIOUSLY THE LITHOSPHERE ALREADY NEXT TO THE SPREADING CENTER HAS TO GO SOMEWHERE, AND WHEN IT DOES, IT COLLIDES WITH OTHER LITHOSPHERE, LIKE TWO CARS CRASHING HEAD-ON. THE CARS CAN:

SLIDE OVER THE OTHER – THIS CAN OCCUR WHEN OCEANIC LITHOSPHERE MEETS OCEANIC LITHOSPHERE ~ BUT BECAUSE OCEANIC LITHOSPHERE IS DENSER THAN CONTINENTAL LITHOSPHERE, OCEANIC LITHOSPHERE RARELY OVERRIDES CONTINENTAL LITHOSPHERE WHEN THERE IS A OCEANIC-CONTINENTAL LITHOSPHERE COLLISION.

SLIDE UNDER THE OTHER – ON THE OTHER HAND, THIS IS THE CLASSIC RESULT OF AN OCEANIC-CONTINENTAL LITHOSPHERIC COLLISION: THE OCEANIC LITHOSPHERE SLIDES UNDER THE CONTINENT. WE SAY THAT THE OCEANIC LITHOSPHERE HAS BEEN **SUBDUCTED**.

CRUMPLE AGAINST EACH OTHER – IN THIS CASE, CONTINENTAL LITHOSPHERE CRUMPLING AGAINST CONTINENTAL LITHOSPHERE – IS HOW BIG MOUNTAINS ARE MADE. THE CONTINENTAL LITHOSPHERE OF INDIA COLLIDED WITH THE CONTINENTAL LITHOSPHERE OF ASIA TO PRODUCE THE HIMALAYAS, THE WORLD'S HIGHEST MOUNTAIN RANGE.

SO HERE'S THE COMPLETE GEOLOGICAL PACKAGE: HOT MANTLE MOVES UP, FORMING AN OCEANIC SPREADING CENTER; FAR AWAY, AT THE EDGE OF A CONTINENT, THIS CAUSES THE OLDER OCEANIC CRUST THAT FORMED EARLIER TO BE SUBDUCTED UNDER THE CONTINENT. BECAUSE PIECES OF THE SUBDUCTING SLAB ARE HOT, THEY TEND TO MELT AND RISE THROUGH THE OVERRIDING CONTINTENTAL LITHOSPHERE. WHEN THEY FINALLY REACH THE SURFACE, ALL THAT MANTLE-DERIVED HEAT ENERGY, AND GAS WANTS TO ESCAPE, AND SO VOLCANIC MOUNTAIN RANGES ARE BUILT; FOR EXAMPLE, THE ANDES IN SOUTH AMERICA AND IN NORTHWESTERN UNITED STATES, THE CASCADES.

PLATE TECTONICS REALLY EXPLAINS HOW THE WORLD *WORKS* ~ FROM WHY THE CONTINENTS HAVE MOVED THROUGH TIME, TO WHY VOLCANOES ARE WHERE THEY ARE, TO THE ORIGIN OF MOUNTAIN RANGES (TERMED "OROGENY"), TO THE REASONS FOR AND LOCATIONS OF EARTHQUAKES (MOVEMENTS IN THE CRUST).

THIS IS WHY GEOLOGISTS LIKE TO SAY, "SUBDUCTION LEADS TO OROGENY!"

CHAPTER 3

WHO'S RELATED TO WHOM – AND **HOW DO WE KNOW?**

Who are you? Who? Who? PETE TOWNSEND, The Who

CHAPTER OBJECTIVES

The goal of this chapter is to get comfortable with the following subjects, because we'll revisit them again and again throughout this book:

- Phylogeny

- Evolution

- Phylogenetic systematics

- Cladograms

- Logic of science and hypothesis testing

Who are you?

Identity is *the* fundamental question in all animals – including humans. Knowing who we are provides the essential context for us, and, for this reason, a favorite dramatic technique (think soap operas) is to have a key character lose some aspect of his or her identity. If you think about it, who we are governs every aspect of our behavior.

"Who are you?" is really the same question as "To whom are you related?" because *relationship*, as in human families, is the key to identity. So to understand who dinosaurs are, we need to know their relationships – to each other and to other animals.

Those relationships – who we are and where we come from – are the special province of phylogeny: the history of the descent of organisms. And here is where evolution comes in: evolution is the cause of the fundamental genealogical connection among organisms.

Evolution

Evolution refers to *descent with modification*: the concept that organisms have changed and modified their morphology (*morph* – shape; *ology* – the study of) through each succeeding generation (see Appendix 3.1 for a brief refresher on the theory of evolution). That makes each new generation the most recent bearer of the unbroken genetic thread that connects life. Each new generation is forward-looking in that its members potentially contain new features that might be useful for whatever the organism encounters in its life; but it is also connected to its ancestors by features that it has inherited. That connection is relationship.

Homology

If there are genetic relationships among organisms, then there must be genetic relationships among their parts. For example, the five "fingers" in the human "hand" and the five "toes" in the front "foot" of a lizard didn't just occur independently. They're all digits of the forelimb, a particular feature that happens to have been retained in these two lineages (humans and lizards). Ideally, the digits on the forelimbs of lizards and humans can be traced back in time to digits in the forelimb of the common ancestor of humans and lizards. We call two anatomical structures homologous when they can, at least in theory, be traced back to a single original structure in a common ancestor (Figure 3.1). Thus we infer that the digits in the forelimbs of all mammals are homologous with those of, for example, dinosaurs. The wings of a fly, however, are not homologous with those of a bird, since they cannot be traced to a single structure on a common ancestor. Because the wings of a fly and the wings of a bird do the same thing (enable flight), they are said to be analogous (Figure 3.2).

Chopping down the "tree of life"

This brings us to the "tree of life." We've all seen "trees of life"; they begin with a pulsating, primordial slime-blob that eventually evolves into everything else as you trace the branches outward (Figure 3.3 is an example). Such trees show who came from whom, and when that occurred. They are common in textbooks and museum displays, and deeply influence our ideas about evolution. But how do we know who gave rise to whom? After all, no human witnessed the zillions of speciation events that constitute the "tree of life."

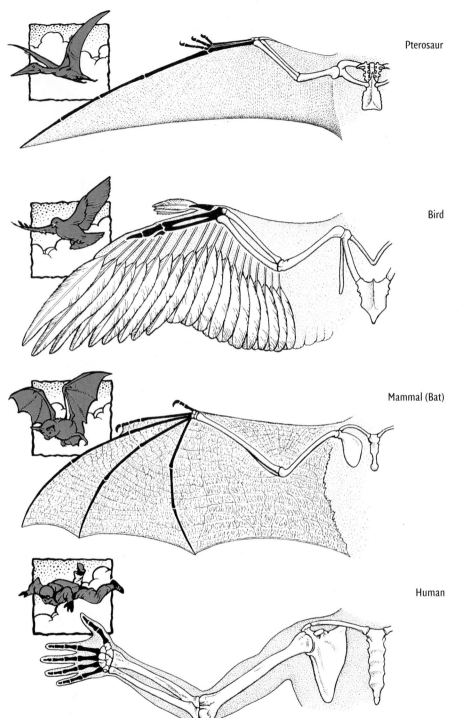

Pterosaur

Bird

Mammal (Bat)

Human

Figure 3.1. Homologs. Homologs are anatomical structures that can, at least theoretically, be traced back to a single structure in a common ancestor. The front limbs of humans, bats, birds, and pterosaurs are all homologous, and retain the same basic structure and bone relationships even though the appearance of these forelimbs may be outwardly different. Homology forms the basis for hypotheses of evolutionary relationships.

The problem is worse than that. We learned in Chapter 1 that fossil preservation is a rare event indeed. Is it likely that some fossil that we find is the actual ancestor of some other? The chances of this occurring are vanishingly small. Thus the oldest fossil in the human family is very unlikely to be the great great great granddaddy of all subsequent humanity.

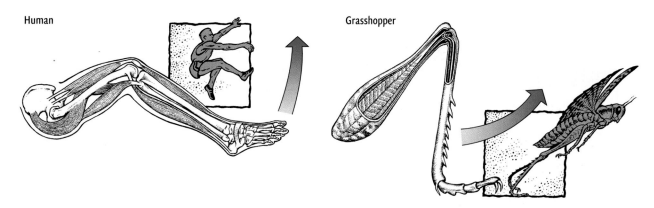

Human   Grasshopper

Figure 3.2. Analog. Analogs may perform similar functions, and may even look outwardly similar, but internally they can be very different. Here a human leg is contrasted with that of a grasshopper. Although both have legs, the two structures are different. For starters, human muscles are on the outside of the skeleton, whereas grasshopppers' muscles are on the inside of their skeleton!

Still, that fossil *is* likely to have many features that the real great great great granddaddy possessed. And therein lies the key to recognizing who is related to whom: while we'll never find the *actual* ancestor, we have a really good chance of finding out more or less how that ancestor looked. Because trees of life specify actual ancestors, and because, as we have seen, we're not likely to actually have the real ancestor in hand, we'll avoid trees of life, and instead use a revolutionary method to understand who is related to whom or, thinking of it another way, the course of evolution. That method is called phylogenetic systematics.

Phylogenetic systematics – the reconstruction of phylogeny

Phylogenetic systematics was developed in the mid 1900s specifically for reconstructing the course of evolution. As we shall see later in this chapter, it is the only *scientific* means of determining relationship, and for this reason it is used by virtually all evolutionary biologists and paleontologists. And although it hasn't yet trickled down into the popular literature, the only way to really understand dinosaurs is through the lens of phylogenetic systematics.

To reconstruct phylogeny, we need a way to recognize how closely two creatures are related. Superficially this is very simple: *things that are more closely related tend to share specific features.* We know this intuitively because we can see that organisms that we believe are closely related (for example, a dog and a coyote) share many similarities. Breeders have taken advantage of this for thousands of years. They've depended upon the fact that offspring look, and sometimes act, very much like their parents to obtain plants and animals with the qualities for which they select. Phylogenetic systematics is the technique by which relationships between organisms can be inferred using unique features of organisms. It depends first and foremost upon the hierarchical distribution of features in the natural world.

Hierarchy

All features in the natural world are organized in a hierarchy, which can be understood to be a successive ranking of subsets within sets. A familiar hierarchy, for example, is ranks within the

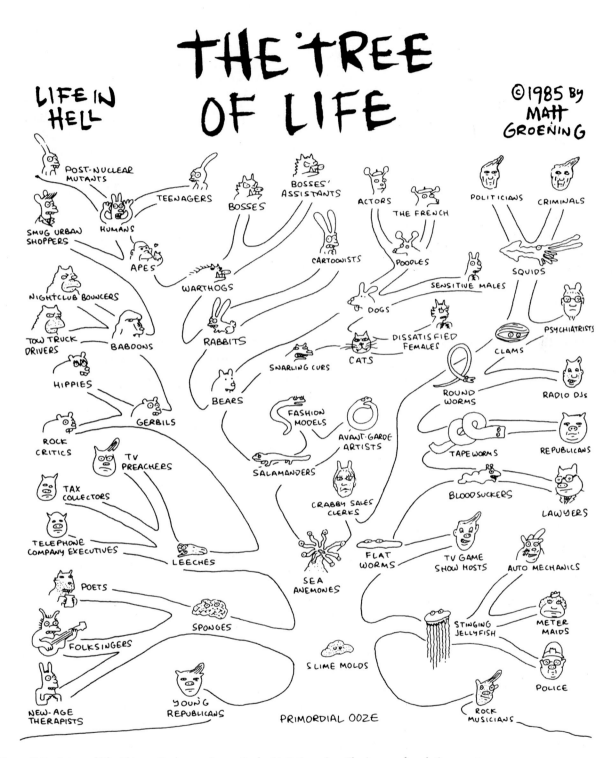

Figure 3.3. A tree of life. This particular one is a satire by Matt Groening. The image of evolution as a tree, however, is completely familiar. From the *Big Book of Hell* © Matt Groening. All Rights Reserved. Reprinted by permission of Pantheon Books, a division of Random House, Inc., NY.

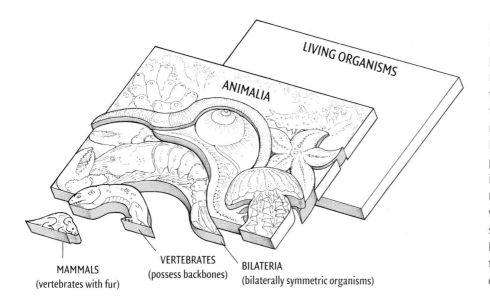

MAMMALS
(vertebrates with fur)

VERTEBRATES
(possess backbones)

BILATERIA
(bilaterally symmetric organisms)

Figure 3.4. The natural hierarchy illustrated as a wooden jigsaw puzzle. The different organisms represent the larger groups to which they belong. For example, the mouse, representing Mammalia, and the lizard, representing Reptilia, together fit within the puzzle to represent Vertebrata, itself a subset of bilaterally symmetrical organisms (Bilateria), which would include invertebrates such as a lobster or a mosquito. Bilateria and other groups constitute the group of organisms we call Animalia (animals).

military. To choose a biological example, all creatures possessing fur[1] are a subset of all animals possessing a backbone, which are in turn a subset of all living organisms (Figure 3.4); these features are distributed hierarchically. *In fact, all features of living organisms are hierarchically distributed in nature*, from the possession of DNA – which is almost ubiquitous – to highly restricted features such as the possession of a brain capable of producing a written record of culture.

Always, however, unmodified or slightly modified vestiges of the original plan remain, and these provide the keys to the fundamental hierarchical relationships that reveal who's related to whom.

Characters

Identifying the features themselves is a prerequisite to establishing life's hierarchy, so we need to look more closely at what we mean by "features." Observable features of anatomy are termed characters. Unique bones or unusual morphologies would all be characters. On the other hand, "observable features" would not include what something does – or how it does it. So, for example, "bites hard" is not a character, but perhaps "big jaw muscles" would be.

Characters acquire their meaning not as a single feature on a particular organism but when their *distribution* among a selected group of organisms is considered. For example, modern birds are generally linked on the basis of having feathers. All living feathered animals are birds and all birds have feathers. Thus, not only is a penguin a bird but so are eagles, ostriches, and kiwis: they all have feathers. And if someone told us that something is a bird, we could confidently predict that it too has feathers.

Because characters are distributed hierarchically, their *position* in the hierarchy is obviously dependent upon the groups they are being used to identify. Consider again the simple

---

[1] A number of organisms in the world are fur-covered besides mammals; for example, bees and some spiders (like tarantulas) have a fur-like covering. But the fur on mammals is unique: it is made of unique proteins, grows in a unique way, and thus is different from these other furry or hairy coatings. When we refer to "fur" in this chapter, it is the mammalian type of fur to which we refer.

example of mammalian fur. Because mammals uniquely have this type of fur, it follows that, if you wanted to tell a mammal from a non-mammal (that is, any other organism), you need only observe that the mammal is the one that has that type of fur. On the other hand, the character "possession of fur" is not useful for distinguishing a bear from a dog; both have mammalian fur. To distinguish a bear from a dog, you'd need some character other than fur.

These distinctions are extremely important in establishing the hierarchy, and for this reason, characters function in two ways: as diagnostic characters and as non-diagnostic characters. The word "diagnostic" here has the same meaning as in medicine. Just as a doctor diagnoses a malady by distinctive and unique properties, so a group of organisms is diagnosed by distinctive and unique characters.

The same character may be diagnostic in one group, but non-diagnostic in a smaller subset of that group (because it is now being applied at a different position in the hierarchy). We saw that fur allowed us to tell a mammal from a non-mammal, but it can't distinguish one mammal from another: it wouldn't tell a bear from a dog.

Cladograms

Cladograms (*clados* – branch; *gramma* – letter) are simply branching diagrams that show hierarchies of diagnostic characters. But, as we'll see, they're not just visual aids, they're the keys to understanding who's related to whom.

To understand how a cladogram works, we begin with two familiar animals; say, a cat and a dog. A cladogram of a cat and a dog is shown in Figure 3.5.

So we're looking for diagnostic characters for these animals. Here, we choose:

1   possession of fur;
2   possession of a backbone; and
3   possession of carnivorous teeth of a unique design.

The cladogram links two separate objects – the cat and the dog – based upon the characters that they *share.* The features are listed on the cladogram adjacent to the node, which is a split point (bifurcation) in the diagram (see Figure 3.5).

The issue becomes more complicated (and more interesting) when a third animal is added to the group (Figure 3.6), in this case a monkey. Now, for the first time, because none of the three animals

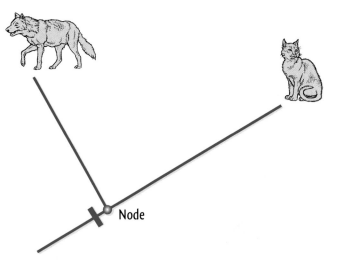

Figure 3.5. A cladogram. The cat and dog are linked by the characters listed at the hatch mark (or bar), just below the node. The node itself defines the things to be united; commonly a name is attached to the node that designates the group. Here, such a name might be "mammalian carnivores."

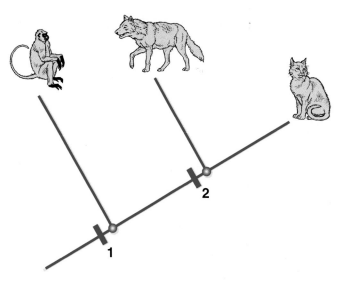

Figure 3.6. One possible distribution of three mammals. Members of the group designated by node **1** are united by the possession of fur and a backbone; that group could be called Mammalia. Within the group Mammalia is a subset united by possession of carnivorous teeth (for example, "mammalian carnivores"). That subset is designated at node **2**.

is identical, two of the three will have more in common with each other than either does with the third. It is in this step that the hierarchy is established. The group that contains all three animals is diagnosed by certain features shared by all three (fur and posession of a backbone). Notice that a subset containing two animals (the cat and the dog) has also been established, linked together by a character (uniquely designed carnivorous teeth) that diagnoses them as being exclusive of the third animal (the monkey).

How the characters, and even the animals, are arranged on the cladogram, is controlled by the *choice* of characters. Let's try something different:

- shortened snout
- large eyes

Based upon these characters, the cladogram in Figure 3.7 contradicts the cladogram in Figure 3.6.

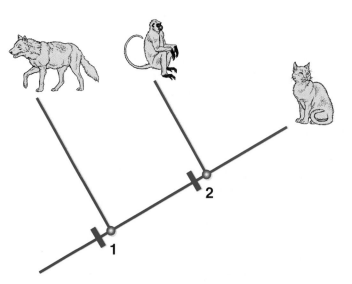

Figure 3.7. An alternative distribution of three mammals. The characters selected at node *2* suggest that the cat and monkey have more in common with each other than either of them does with the dog (see the text).

How do we choose? *The cladogram that is most likely correct is the one that doesn't change when new characters are added.* All characters that apply to these mammals support the cladogram in Figure 3.6. And we can infer from the cladogram in Figure 3.6 that a dog has much more in common with a cat than it does with a monkey.

Cladograms as tools in understanding the evolution of organisms

So how does this apply to evolution? Using character hierarchies portrayed on cladograms, we establish **clades** or **monophyletic groups**[2]: *groups that have evolutionary significance because the members of each group are more closely related – by genealogy – to each other than they are to any other creature.* If a group is monophyletic, it also implies that all members of that group share a more recent common ancestor with each other than with any other organism. The cladogram in Figure 3.6 suggests that the cat and the dog are more closely related to each other than either is to the monkey.

Now that we're speaking in terms of organic evolution, the specific characters that we said characterize groups can now be treated as homologous among the groups that they link. Mammalian fur once again (!) provides a convenient example. We conclude that mammals are monophyletic based upon the fact that mammals all share a unique type of fur (among many other characters). If all mammals are fur-bearing (and mammals are monophyletic), the implication is that the fur found in bears and that found in horses can in fact be traced back to fur that must have been present in the most recent common ancestor of bears and horses.

Because now we're working with the evolution of organisms, the word "specific" (or "diagnostic") is generally replaced by the terms **derived**, or **advanced**, and "general" ("non-diagnostic") characters are termed **primitive** or **ancestral**. "Primitive" certainly does

---

[2] To add to the nomenclature, these are sometimes termed "natural groups."

not mean worse or inferior, just as "advanced" certainly does *not* mean better or superior; these refer only to how much the character has been changed by evolution. Primitive specifies the condition of a particular feature in the ancestor; advanced specifies an evolved condition of that character in its descendant.

Mapping the course of evolution with the cladogram

Because cladograms are hierarchical, they are an excellent way to map the hierarchical distributions of characters in nature. *Derived characters are evidence of monophyletic groups because, as newly evolved features, they are potentially transferable from the first organism that acquired them to all its descendants*: in short, they characterize the bifurcations at each node on the cladogram. Primitive characters – those with a much more ancient history – provide no such evidence of monophyly.

To illustrate this, we resort for the last time (!) to mammals and their fur. Mammalian fur, we said, is among the shared, derived characters that unite mammals as a monophyletic group. On a cladogram, therefore, we look for characters that mark a node in the diagram. All organisms characterized by shared, derived characters are linked by the cladogram into monophyletic groups. Reflecting the hierarchy

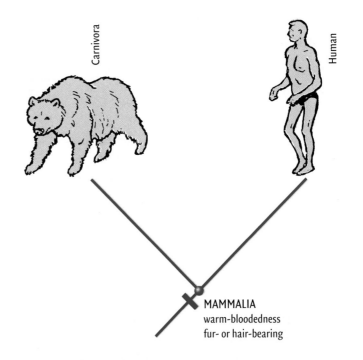

Figure 3.8. A cladogram showing humans within the larger group Mammalia. Mammalia is diagnosed by warm-bloodedness and possession of fur (or hair); many other characters unite the group as well. Carnivora, a group of mammals that includes bears and dogs (among others) is shown to complete the cladogram. Carnivores all uniquely share a special tooth (the carnassial) and humans all uniquely share, among many other features of the skull and skeleton, a large cranium. Note that all mammals (including humans and carnivores) are warm-blooded and have fur (or hair), but only humans have the gracile skeletal features, and only members of Carnivora have the carnassial tooth.

of character distributions in nature, the cladogram documents monophyletic groups within larger monophyletic groups. In Figure 3.8, a small part of the hierarchy is shown: humans (a monophyletic group, possessing shared, derived characters) are nested within mammals (another monophyletic group possessing other shared, derived characters). Notice that the character of warm-bloodedness is primitive for *Homo sapiens*, but derived for mammals. As we have seen, features can be derived or primitive, all depending upon what part of the hierarchy one is investigating.

The cladogram need not depict every organism within a monophyletic group. If we are talking about humans and carnivores, we can put them on a cladogram and show the derived characters that diagnose them, but we might (or might not) include other mammals (for example, a gorilla). So we said with regard to Figures 3.6 and 3.7, if the hierarchical relationships that we have established are valid, the addition of other organisms into the cladogram should not alter the basic hierarchical arrangements established by the cladogram. Figure 3.9 shows the addition of one other group into the cladogram from Figure 3.8. The basic relationships established in Figure 3.8 still hold, even with the new organism added. The cladogram is likely correct.

How to read evolution in the cladogram

We identified monophyletic groups using derived characters, and that the hierarchies of characters designate hierarchies of groups. So, looking at Figure 3.9, the distribution of shared, derived characters suggests that humans and gorillas are more closely related to each other than either is to a bear. It also suggests that all three are more closely related to each other than they are to something that does not posess the derived character of bearing fur or hair.

And how does that apply to evolution? The evolution of the derived character of fur is associated with the evolution of the group Mammalia. As we currently understand their relationships, what we call mammals were invented when that character – among others – first evolved. And, the cladogram tells you that sometime thereafter – the cladogram does not specify when – a character that unites both humans and gorillas evolved, a character that we now recognize diagnoses a new group within Mammalia.[3]

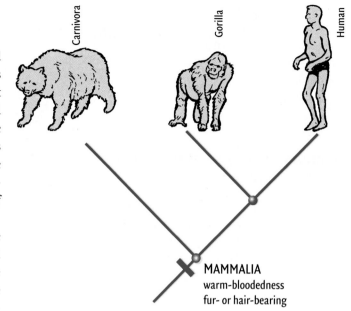

Figure 3.9. Addition of the genus *Gorilla*. The addition of gorillas to the cladogram does not alter the basic relationships outlined on the cladogram shown in Figure 3.8.

Here is a fundamental difference between a cladogram and the more familiar tree of life that we discussed above. The cladogram does not incorporate time, nor does it tell you who the ancestors were. Instead, it can indicate the sequence of the evolutionary events and, more importantly, specify the characteristics that the ancestor possessed. So, in the case of the cladogram in Figure 3.9, we aren't told who was the ancestor of bears, gorillas, and humans, but we infer that the earliest mammal was fur-bearing (among the other characters that diagnose Mammalia). In Box 3.1 cladograms are used to reconstruct the evolution of wristwatches, demonstrating their power to reveal the underlying relationships of even inanimate objects.

Used as a tool to reconstruct evolution, then, a cladogram is actually a hypothesis of relationship; that is, a hypothesis about how closely (or distantly) organisms are related, and about what the sequence of the appearance of different diagnostic characters must have been.

Parsimony

As we have seen, it is possible to construct several possible cladograms which would represent different evolutionary sequences. Which to choose? We choose using the principle of parsimony. Parsimony, a sophisticated philosophical concept first articulated by the fourteenth-century English theologian William of Ockham, states that *the explanation with the least necessary steps is probably the best one*. Why resort to complexity when simplicity is equally informative? Why suppose more steps took place when fewer can provide the same information?

Figure 3.10 shows two cladograms that are possible with birds, a human, and a bat and the characters of wings, fur, feathers, and mammary glands. In (b), the bird has to lose

---

[3] That new group is called "Hominoidea," as it happens, and is diagnosed by lots of characters, among which are a series of specializations in the arms and trunk associated with walking bipedally and swinging through trees.

Box 3.1 Wristwatches: when is a watch a watch?

We've used cladistic techniques to infer the history of the biota. Here we'll try something different: we'll use cladistic techniques to infer the evolutionary history of watches. Analog and digital timepieces are comonly called "watches." Implicit in the term "watches" is some kind of evolutionary relationship: that these instruments have a common heritage beyond merely post-dating a sundial. But is this really so?

Consider three types of watch: a wind-up watch, a digital watch, and a watch with a quartz movement. Six cladograms are possible for these instruments (Figure 3B.1.1), but it can be seen that, by the definition of a cladogram, a and b

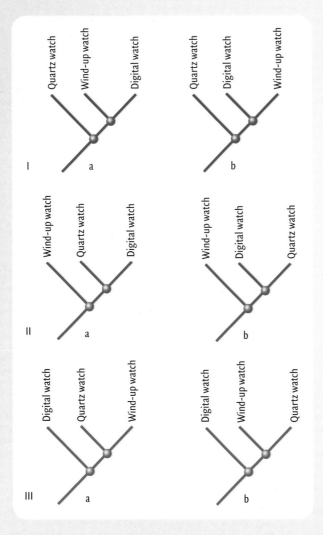

Figure B3.1.1. Six possible arrangements of three timepieces on cladograms. Note that each pair is redundant: the order in which the objects on each "V" is presented is irrelevant.

for each type are identical. This is because the groups at a node share the characters listed at that node, regardless of order. For this reason, we really have only three cladograms to consider (Figure 3B.1.2).

One might at first wish to place the digital watch in the smallest subset, in the most derived position (as in types I and II), since it is the most modern, technologically advanced, and sophisticated of the three. Remember, however, how the cladogram is established: on the basis of *shared*, *derived* characters. Cladogram types I and II say that the digital watch shares the most characters in common with either a wind-up watch (type I) or a quartz watch (type II). A look at the characters themselves suggests that this is not correct: wind-up and quartz watches are both analog watches (have a dial with moving, mechanical hands) and their internal mechanisms consist of complex gears and cogs to drive the hands at an appropriate speed. The digital watch, on the other hand, consists of microcircuitry and a microchip, with essentially no moving parts. It is apparently something very different and, from its characters, bears little relationship to the other "watches."

What *is* the digital watch? In an evolutionary sense, it is really a computer masquerading (or functioning) as a timepiece. The computer has been put in a case, and a watchband has been added, but fundamentally this "watch" is really a computer. In our hypothesis of relationship, the watchbands and cases of watches have evolved independently two times (once in computers and once in watches), rather than the guts of the instrument, itself, having evolved twice. That the watchbands and cases evolved independently two times is a more parsimonious hypothesis than arguing that the distinctive and complex internal mechanisms (themselves consisting of many hundreds of characters) of the watches evolved independently twice.

What, then, is a watch? If the term "watch" includes digital watches as well as the other two more conventional varieties, then it should also include computers, since a digital watch has the shared, derived characters of computers. The cladogram suggests that the term "watch" does not describe an evolutionarily meaningful (monophyletic) group, in the sense that a cladogram that includes digital watches, wind-up watches, and quartz watches must also include computers, as well as a variety of more conventional mechanical timing devices (such as stop watches). Rather, the term "watch" may be thought of as some other kind of category: it describes a particular function (time-keeping) in conjunction with size (relatively small).

3.1 cont.

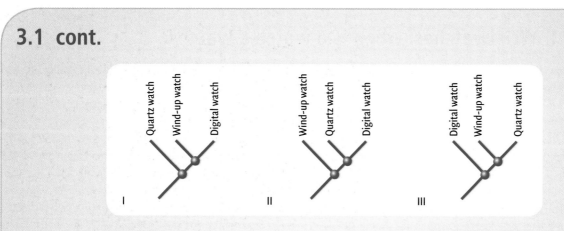

Figure B3.1.2. Because each pair of cladograms in Figure B3.1.1 is redunant, there are really only three cladograms under consideration.

In this example, we are fortunate in that, should we so choose, we can test the cladogram-based conclusions by studying the historical record and find out about the evolution of wrist watches, digital watches, and quartz watches. Obviously this is not possible to do with the record of the biota, because there is no written or historical record with which to compare our results. The characters of each new fossil find, however, can be added to existing cladograms and the hypothesis of relationship that shows the least complexity will be favored according to the principle of parsimony. In our discussions of the biota, we attempt to establish categories that are evolutionarily significant (monophyletic groups), and avoid groups that have less in common with each other than with anything else.

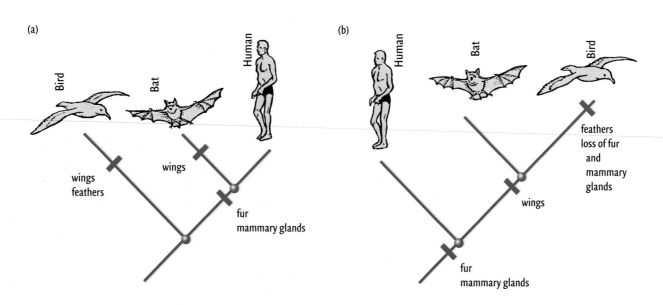
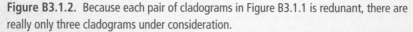

Figure 3.10. Two possible arrangements for the relationships of birds, bats, and humans. (a) The left-hand cladogram requires wings to have evolved two times; (b) the right-hand cladogram requires birds to have lost fur and mammary glands. These as well as many other characters suggest that (a) is the more parsimonious of the two cladograms.

ancestral mammary glands and it has to replace fur with feathers. In (a), wings must be invented by evolution twice. Cladogram (a) is the simpler of the two because it requires fewer evolutionary events or steps. It is uncomplicated by the addition of more characters. In contrast to cladogram (a), the addition of virtually any other characters that are shared by humans and bats to cladogram (b) (for example, the arrangment, shape, and number of bones, particularly those in the skull and forelimbs, the structure of the teeth, the biochemstry of each organism) requires that each of these shared characters evolved independently: once in bats and once in humans. That considerably complicates the number of evolutionary steps, leaving us with the conclusion that cladogram (b), the hypothesis that birds and bats are more closely related to each other than either is to a human, is a less parsimonious alternative.

Indeed, as a hypothesis about the *evolution* of these vertebrates, it is much more likely that bats and humans share a more recent common ancestor than that either does with a bird (which, obviously, is why bats and humans are classified together here as mammals). In this case, the use of shared, derived characters has led us to the most parsimonious conclusion with regard to the evolution of these three creatures.

Science = testing hypotheses

Science is the business of constructing testable hypotheses and then testing them. An example of a simple scientific hypothesis is: "The sun will rise tomorrow." This hypothesis makes specific predictions. Most importantly, the hypothesis that the sun will rise tomorrow is testable; that is, it makes a prediction that can be assessed. The test is relatively straightforward: we wait until tomorrow morning and either the sun rises or it doesn't.

So what's *not* science? Important questions that you can't get at using science might be "Is there a God?", "Does she love me?", and "Why don't I like hairy men?" In *Music Man*, Marian "the librarian" Paroo asks "What makes Beethoven great?" She'll never learn through science. Some types of question are far better suited to science than others.

The "proof" is in … the test!

In our example of a scentific hypothesis (above), if the sun does not rise, the statement has been falsified, or demonstrated to be not correct, and the hypothesis can be rejected. On the other hand, if the sun rises, the statement has *failed falsification*, and the hypothesis cannot be rejected. For a variety of relatively sophisticated philosophical reasons, scientists do not usually claim that they have *proven* the statement to be true; rather, the statement has simply been tested and not falsified. It turns out that, in a philosophical sense, it is very difficult to "prove" anything. For this reason, scientists rarely use words like "prove" or "true" when discussing their work.

But we can test hypotheses, and o*ne of the basic tenets of science is that it consists of hypotheses that have predictions which can be tested.* We will see many examples of hypotheses in the coming chapters; to be valid, all must involve testable predictions. Without testability, it may be very interesting; it may be very exciting, but it is not science!

Cladograms are science

Cladograms are hypotheses concerning phylogenetic relationships. They make predictions about the distributions of characters in organisms. Any organism – living or extinct – can test an existing cladogram-based phylogenetic hypothesis. With living organisms, not only do we use their anatomy on the cladogram; we can also use their genetic material. Parsimony is then used to determine which cladogram most likely approximates to the course of evolution. As will become evident, cladograms are among the most powerful tools available for learning about what occurred in ages long past.

Science in the popular media

"Science" is everywhere in the popular media, from Animal Planet to National Geographic, to the Weather Channel. That's good; our lives are increasingly intersected by science, and the more savvy we are scientifically, the better the decisions that we can make. But the problem is that not all of the "science" that one sees in the popular media is very scientific!

Among scientists, work is recognized and accepted only after it is peer-reviewed; that is, when it has been vetted and carefully scrutinized by other scientists qualified to evaulate the work. Very commonly, the work goes through multiple cycles of review and revision by the authors in response to the critiques of professionals in the field. At that time – and only at that time – it can be published. This is the time-honored – and time-proven – way of producing the very highest quality scientific research.

In the popular media, the goals are somewhat different. In movies, on television, and on the radio, the goals are generally entertainment as well as immediacy: publicizing the newest discoveries as quickly as possible. Both of these have very little to do with high-quality science. In the case of immediacy, the rush to publicize can mean that half-baked theories and interpretations are popularized long before they have seen anything like peer review. And when peer review finally comes, the results can be (and have been) devastating.

The goal of entertainment can be equally destructive, in terms of the science. For example, television and radio reports commonly present "the other side" of a particular scientific issue, even when the vast preponderance of scientists take a particular position and the "other side" has very little legitimacy. People can always be found who will voice heterodox opinions, even when their opinions are plainly way outside of the mainstream of peer-reviewed science. It's always *entertaining* to present controversy.

In paleontology, the quest for entertainment brings a variety of excesses, beyond mere controversy. Size sells. Teeth and claws sell. Destructo asteroids sell; extinction sells. But ultimately, we hope in this book to show that the real riches – and entertainment – lie in a balanced, undistorted treatment of dinosaurs: a more magnificent, fascinating group of creatures could hardly have been invented, and the ongoing scientific process of revealing who they were and what they did is compelling drama.

The web, an increasingly popular and important source of information, is driven by different imperatives, but suffers from a similar fundamental weakess: the lack of peer review. Accessability and ease is its great strength, but without peer review, the likelihood that the quality of information available on the web will be compromised is extremely high. The web is a powerful tool, to be sure; but *caveat emptor*: Not all sources of information are equivalently authoritative. The web is a useful place to begin to learn about something; however, digging deeper almost always brings greater rewards.

Summary

Relationship is essential to understanding the identities of organisms. To reconstruct relationships, branching diagrams called cladograms are used. Organisms are grouped on these using the presence of shared, derived (or diagnostic) characters; the groups of organisms that result are all inferred to be more closely related to each other than to anything else; that is, they are monophyletic. The evolution of new types of organisms is represented on the cladogram with the novel features, which make the descendants unlike their ancestors, being the shared, derived characters.

Cladograms differ from "trees of life" in several fundamental aspects. Cladograms are based upon shared derived characters; they do not show time; they do not show ancestors (although they specify what the ancestral condition of an organism must have been like); and, most importantly, they are testable.

Testability is a key part of science; any scientific inference must be able to be tested. Scientists thus "prove" nothing; scientific hypotheses merely fail falsification via careful testing. Something that is not testable, no matter how significant to us, is not science. In the case of cladograms, falsification consists of the prediction on the cladogram that a character will be present that is not (or vice versa). Because cladograms are falsifiable, they are merely hypotheses of the relationships of organisms.

With several hypotheses of relationship (cladograms) to choose among, the cladogram requiring the least number of steps (that is, the most parsimonious) is the preferred one.

Appendix 3.1 What is "evolution"?

Charles Darwin, through his book *On the Origin of Species* (1859), is generally regarded as the father of modern evolutionary theory. Yet, the most common understanding of the word "evolution," that organisms on Earth have changed through time, was *not* the point of Darwin's work. That organisms have changed through time had been well established by savvy natural historians (or natural philosophers, as they were sometimes called) for well over 200 years before Darwin. Darwin's contribution was to propose the *means* by which such changes occurred. His hypothesis was constructed in the following way:

1   Domesticated animals and plants show a wide range of variation.
2   A similarly wide range of variation exists among wild animals and plants as well.
3   All living creatures are engaged in a "struggle" to survive and ultimately reproduce, and that struggle is most severe among those individuals that are most closely related.
4   The struggle to survive in combination with the variation that exists among individuals leads to the survival and, most importantly, successful reproduction of some variants as opposed to some others, a process that Darwin called natural selection.
5   The reproductive success of some variants as opposed to others ensures that the characteristics of the successfully reproducing variants make it into the next generation.
6   This process, repeated over hundreds or even thousands of generations, is evolution by natural selection, sometimes called "Darwinian evolution."

The variants that survived to produce viable offspring are said to be more fit than those that did not. And successive generations of "fit" offspring would, in a manner analogous to breeding, eventually produce a descendant very different from its ancestor (for example, a new species). So, if fitness in some hypothetical lineage of organisms meant longer legs, then that lineage might show an evolutionary trend toward increasing leg length until the long-legged descendant was sufficiently different from its ancestor to be considered a different species. So Darwin's contribution to ideas about evolution, therefore, was actually the hypothesis of *evolution by natural selection.*

Because the science of genetics was not known to Darwin (having been invented, as it were, during his lifetime), Darwin had no mechanism to explain what exactly was meant by "closely related," although he knew that in some physical way parents, for example, are closely related to their children. The explicit meaning of relationship came with the understanding of chromosomes, genes, alleles, and, some 70 or so years later, DNA.

Nonetheless, all of those ideas were, from the 1920s onward, integrated into Darwin's original hypothesis (except, of course the molecular basis of inheritance, which came somewhat later) in an intellectual movement called the "Evolutionary Synthesis." The Evolutionary Synthesis applied the then-burgeoning fields of population ecology, genetics, paleontology, and statistics to Darwinian evolution, to understanding the precise mechanisms by which particular combinations of genes (genotypes) are selected and passed on to succeeding generations. This led to the hypothesis of an unbroken genetic chain of successive changes in the physical appearance (phenotype) and behavior of organisms, to the origin of new species, as life perpetuated itself on Earth.

In the intervening years since the Evolutionary Synthesis, the theory of evolution by natural selection has been refined and now incorporates very significant advances in our understanding of genetics, embryology, and molecular biology. The origin of new features is recognized as gradual in some cases, as postulated by Darwin, but also as abrupt in other cases. Sometimes changes in genotype produce new species, but sometimes they may produce phenotypic changes that are either smaller- or larger-scale than the development of merely new species. Natural selection then acts upon phenotypes in the current generation favoring, as Darwin hypothesized, the reproductive viability of some phenotypes and not others.

The fitness – or lack of it – of an organism is determined by an immense number of variables, including, but not limited to, the environment in which it evolves, the other organisms (plant and animal) with which it must interact, and how viable are its progeny. Obviously these variables are utterly unpredictable, and so evolution by natural selection, as currently understood by evolutionary biologists, does not involve the possibility of predicting future evolutionary events.

As we cannot use it to predict the future, should we consider the hypothesis of evolution by natural selection to be unscientific? Recall that science requires a hypothesis with explicit predictions that can be tested. Although the hypothesis of evolution does not predict the future, it certainly makes testable predictions. For a few examples:

- It predicts that we will find organisms that contain mixes of characters of older organisms and younger organisms occurring intermediate in time between them. (We do.)
- It predicts that life's diversity takes the form of many variations on a basic design, with modifications upon modifications that take us to the present. (It does.)
- It predicts that the biochemical building blocks of life will be present – if slightly modified – in all organisms. (They are.)
- Of particular relevance in this book, it predicted that a creature that mixed bird and "reptile" features would exist. (It does; see Chapter 10.)

Soviet-born geneticist Theodosius Dobzhansky, one of the pioneers of the Evolutionary Synthesis, once wrote in *The American Biology Teacher*, "Nothing in biology makes sense except in the light of evolution." We concur.

Yet for all that there is an irony. The word "evolution" refers to an unfolding to a predetermined and inevitable end, such as the *evolution* of a tragedy. Since the evolution of life – organic evolution – does *not* unfold along predetermined or inevitable pathways, it is not surprising that the word "evolve" was avoided by Darwin until the very last page of *On the Origin of Species* (1859).

Selected readings

Cracraft, J. and **Eldredge, N.** (eds.). 1981. *Phylogenetic Analysis and Paleontology*. Columbia University Press, New York, 233 pp.

Darwin, C. 2008 [1859]. *On the Origin of Species: A Facsimile of the First Edition*. Harvard University Press, Cambridge, MA, 540 pp.

Eldredge, N. and **Cracraft, J.** 1980. *Phylogenetic Patterns and the Evolutionary Process: Method and Theory in Comparative Biology*. Columbia University Press, New York, 349 pp.

Nelson, G. and **Platnick, N.** 1981. *Systematics and Biogeography, Cladistics and Vicariance*. Columbia University Press, New York, 567 pp.

Futuyma, D. J. 2005. *Evolution*. Sinauer Associates, Sunderland, MA, 603pp.

Stanley, S. M. 1979. *Macroevolution*. W. C. Freeman and Company, San Francisco, 332 pp.

Wiley, E. O., Siegel-Causey, D., Brooks, D. and **Funk, V. A.** 1991. *The Compleat Cladist: A Primer of Phylogenetic Procedures*. University of Kansas Museum of Natural History Special Publication no. 19, Lawrence, KS, 158 pp.

Topic questions

1 Define: phylogeny, morphology, homologous, analogous, a tree of life, hierarchy, characters, diagnostic, cladogram, node, derived, advanced, monophyletic groups, primitive, ancestral, parsimony, test, falsify, science.

2 Why is it that the direct ancestor of any organism isn't easy to identify?

3 If you have several possible cladograms, how do you determine which is the preferred one?

4 Construct a cladogram of something that particularly interests you (examples: guitars, music, sports, shoes, etc.). Be sure to show the diagnostic characters in their correct hierarchical locations.

5 If you don't know the ancestor, how can you hope to understand how something evolved?

6 On p. 44 we said that we could "reconstruct" the course of evolution with a cladogram. How do cladograms allow us to do this?

7 Construct a scientific hypothesis.

8 Construct a non-scientific hypothesis.

9 How is a cladogram a "hypothesis of relationship?"

10 Contrast a "tree of life" with a cladogram.

11 What is meant by "peer review?" Why is it important?

CHAPTER 4

WHO ARE THE DINOSAURS?

CHAPTER OBJECTIVES

- Learn basic relationships among tetrapods – particularly amniotes

- Understand something about the course of tetrapod evolution

- Learn who dinosaurs are (and are not); and

- Become familiar with the characters that diagnose Dinosauria

Finding the history of life

In the preceding chapter, we explored the methods that scientists use to learn the identity and origin of all organisms. Now we will apply those techniques – diagnostic characters hierarchically distributed on cladograms – to properly position dinosaurs within the biota. The history of life will unfold as we systematically encounter each bifurcation in the cladistic road, reconstructing the path of evolution until we reach Dinosauria. We will be looking at increasingly small subgroups, each characterized by a suite of diagnostic characters. The appearance of each of those suites of diagnostic characters represents new features, forged by evolutionary processes. We'll go with Glinda the Good Witch's suggestion that "it's always best to start at the beginning."

In the beginning

Modern life is generally understood to be monophyletic. It's united by the possession of RNA, DNA, cell membranes with distinctive chemical structure, a variety of amino acids (proteins), the metabolic pathways (that is, chemical reaction steps) for their processing, and the ability to replicate itself (not simply grow).[1] Notice we said "modern" life – for who knows how many forms of molecular life arose, proliferated, and died out very early in Earth's history – before the thing that we now call "life" finally prevailed?

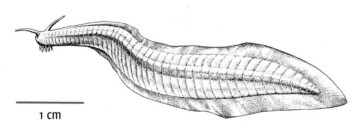

While it is possible to construct a cladogram to develop the full history of modern life, we'd have to summarize about 3.8 *billion* years of organic evolution. Instead, let's cut to the chase, to a mere 510 Ma, where we first meet *Pikaia gracilens*, a

1 cm

Figure 4.1. *Pikaia gracilens*, a presumed chordate from the Middle Cambrian of Canada.

5 cm, flattened, miniature anchovy fillet of a creature that represents an early data point in the ancestry of vertebrates (Figure 4.1).[2]

Chordata

Pikaia reveals characters that are diagnostic of the clade to which we (and the dinosaurs) belong: Chordata ("nerve cord-bearing"). Although *Pikaia* provides an inkling about our distant relatives, what we know about the early evolution of vertebrates and their forebears among Chordata comes principally from living organisms, with some input from a few fossils. The living chordates consist of living urochordates (*uro* – tail; popularly called "sea squirts"),

---

1 This statement is not strictly true, because viruses don't have intrinsic membranes, amino acids, and metabolic pathways (they hijack the molecules and mechanisms of the cells they invade), nor do they reproduce themselves; retroviruses don't even have DNA. The origin of viruses remains shrouded in mystery; but one hypothesis is that they lost these early in their evolutionary history.

2 But not the first. That honor is shared by two primitive chordates from Chengjiang, China, which are believed to date to about 520 Ma.

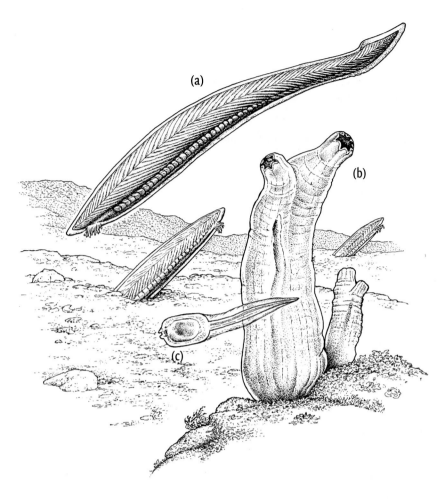

Figure 4.2. Two primitive chordates. (a) Cephalochordate (the lancet *Amphioxus*); (b) urochordate (the sea squirt *Ciona*); and (c) larval sea squirt. Cephalochordates (a) and urochordates (b) share with the vertebrates a host of derived features, including segmentation of the muscles of the body wall, separation of upper and lower nerve and blood vessel branches, and many newly evolved hormone and enzyme systems. The juvenile sea squirt (c) is free-swimming and has a notochord running down its tail. When it metamorphoses into the stationary adult, it parks on its nose and rearranges its internal and external structures.

cephalochordates (sometimes called "lancets") and, most important for our story, vertebrates (Figure 4.2). All of these groups are united within Chordata on the basis of the following diagnostic characters:

1 Features of the throat (pharyngeal gill slits).
2 The presence of a notochord, a stiffening rod that gives the nerve cord support.
3 The presence of a dorsal, hollow nerve cord.
4 Upper and lower muscle masses, repeated *en echelon*, like a series of "V's".

This distinctive suite of characters – pharyngeal gills, notochord, and nerve cord – appears to have evolved only once, thus uniting these animals as a monophyletic group. We see Chordata and its diagnostic characters at the base of the cladogram in Figure 4.3. We – and the dinosaurs – appear to have chordate relatives as far back as the Cambrian.

Vertebrata

Most interesting for us, within Chordata is also found the familiar Vertebrata (*vertere* – to turn referring to joints in the spine that bend). The diagnostic characters of vertebrates, with one exception, include a calcified internal skeleton (that is, bone) divided into discrete pieces called elements,[3] and a variety of other characters (Figure 4.3).

---

[3] In anatomy, the word "element" confusingly has a different meaning from its meaning in chemistry (see Chapter 2 and Glossary).

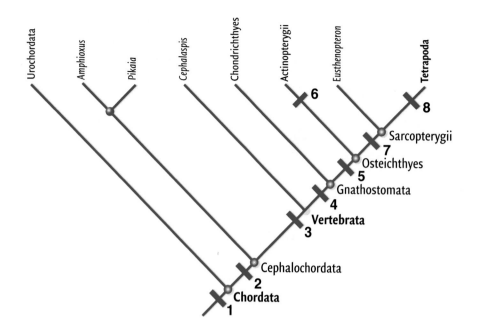

Figure 4.3. Cladogram of the Chordata. Because this is a book about dinosaurs (and not all chordates), we have provided diagnoses for only some of the groups on the cladogram. Bars denote the shared, derived characters of the groups. The characters are the following: at **1**, pharyngeal gill slits, a notochord, and a nerve cord running above the notochord along its length; at **2**, segmentation of the muscles of the body wall, separation of upper and lower nerve and blood vessel branches, and new hormone and enzyme systems; at **3**, bone organized into elements, neural crest cells, the differentiation of the cranial nerves, the development of eyes, the presence of kidneys, new hormonal systems, and mouthparts; at **4**, true jaws; at **5**, bone in the endochondral skeleton; at **6**, ray fins; at **7**, distinctive arrangement of bones in fleshy pectoral and pelvic fins (see Figure 4.4); at **8**, skeletal features relating to mobility on land – in particular, four limbs. Consistent with a cladistic approach, only monophyletic groups are presented on the cladogram. Some of the groups may not be familiar: for example, *Cephalaspis* and *Eusthenopteron* are not discussed in the text. *Cephalaspis* was a primitive, jawless, bottom-dwelling, swimming vertebrate, and *Eusthenopteron* was a predaceous lobe-finned fish, bearing many characters present in the earliest tetrapods. *Cephalaspis* and *Eusthenopteron* are included here to complete the cladogram as monophyletic representatives of jawless vertebrates and lobe-finned fishes, respectively.

Among Vertebrata, we're naturally interested in the subset Gnathostomata (*gnathos* – jaw; *stome* – mouth), vertebrates with true jaws (among other diagnostic features). So, who are these gnathostomes? Well, we certainly are, as are most of the vertebrates that might come readily to mind. But, we'll focus directly on the subset of gnathostomes that we call the bony fishes.

Here, now, is a major split in the cladogram, representing a major evolutionary branching point. On the one hand is the lineage of bony fish (Osteichthyes; *osteo* – bone; *ichthys* – fish) leading to familiar forms such as goldfish, tuna, and salmon. But on the other branch is Sarcopterygii (*sarco* – flesh; see Figure 4.3), a not-so-familiar group that is diagnosed by, among other things, the presence of distinctive *lobed* fins, fleshy places where the fin attaches to the body.

Here too is demonstrated the power of homology: those lobes contain bones that are recognizable as – and thus homologous with – bones in the limbs of Tetrapoda (*tetra* – four;

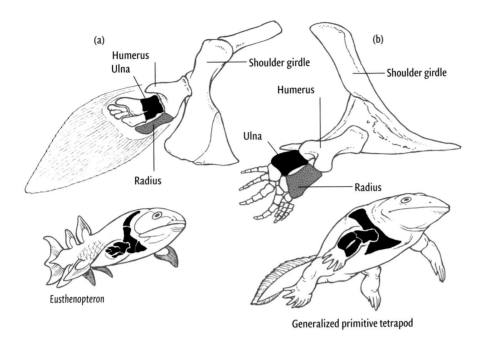

Figure 4.4. Some homologous features between lobe-finned fishes and tetrapods. (a) The shoulder girdle of *Eusthenopteron*, an extinct lobe-fin; (b) the shoulder girdle in early tetrapods. Because aspects of the forelimb in different early tetrapods are incomplete, the forelimb shown here is a composite prepared from two early tetrapods (*Acanthostega* and *Ichthyostega*). Key homologous bones are labeled in both drawings.

pod – foot): vertebrates with four limbs, including dinosaurs and ourselves (Figure 4.4). What's more, bones of the pelvis, vertebral column, and even bones in the skulls of lobe-finned fishes can also be recognized within tetrapods. All these diagnostic characters strongly indicate that it is here, among the lobe-fins, that the ancestry of Dinosauria – as well as our own ancestry – is to be found (Box 4.1).

Tetrapoda

Tetrapoda is diagnosed by the appearance of limbs with the distinctive arrangement of bones shown in Figure 4.4. So now let's take a closer look at Tetrapoda and, because we're interested in dinosaurs, we'll try to understand the part that's generally best preserved: the skeleton.

The tetrapod skeleton made easy

Figure 4.5 shows a typical tetrapod skeleton – in this case a prosauropod dinosaur – blown apart. Not surprisingly (because Tetrapoda is monophyletic), they are all built in the same way: a vertebral column is sandwiched by paired forelimbs and paired hindlimbs. The limbs are attached to the vertebral column by groups of bones called girdles. At the front end is the head, composed of a skull and mandible, or lower jaw. At the back end is the tail. It's that simple!

Vertebral column The vertebral column is composed of distinct, repeated structures (the vertebrae), which consist of a lower spool (the centrum), above which, in a groove, lies the spinal cord (Figure 4.5, inset on p. 59). Planted on the centrum and straddling the spinal cord is the neural arch. Various processes, that is, parts of bone that are commonly ridge-, knob-, or blade-shaped, may stick out from each neural arch. These can be for muscle and/or ligament attachment, or they can be sites against which the ends of ribs can abut. The repetition of vertebral structures, a relic of the segmented condition that is primitive for chordates, allows flexibility along the length of the animal.

Figure 4.5. Exploded view of a tetrapod skeleton exemplified by the saurischian dinosaur *Plateosaurus*.

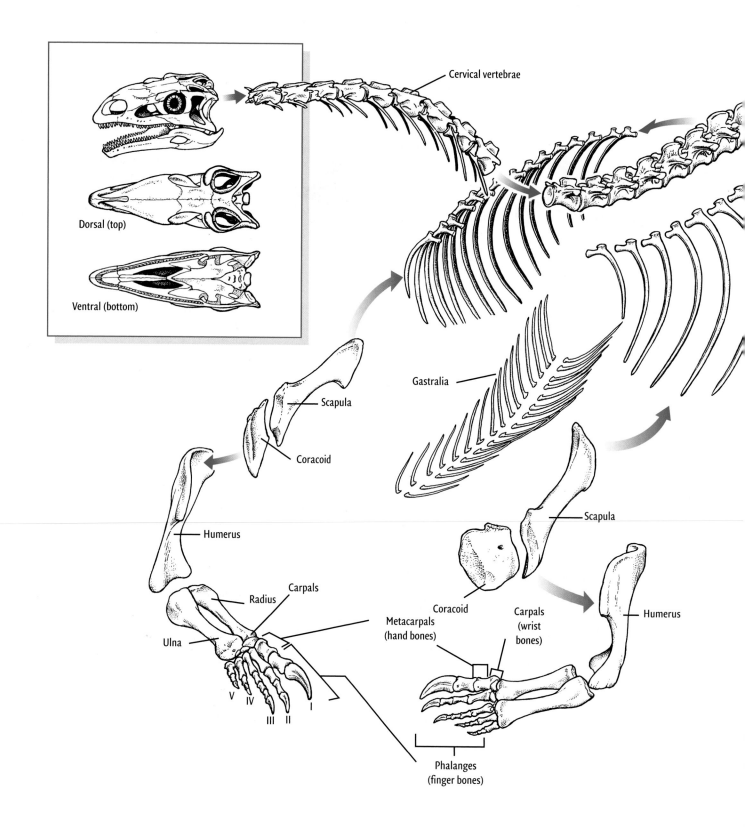

Dorsal (top)

Ventral (bottom)

Cervical vertebrae

Gastralia

Scapula

Coracoid

Scapula

Humerus

Radius

Carpals

Ulna

Coracoid

Metacarpals (hand bones)

Carpals (wrist bones)

Humerus

V   IV

III   II

I

Phalanges (finger bones)

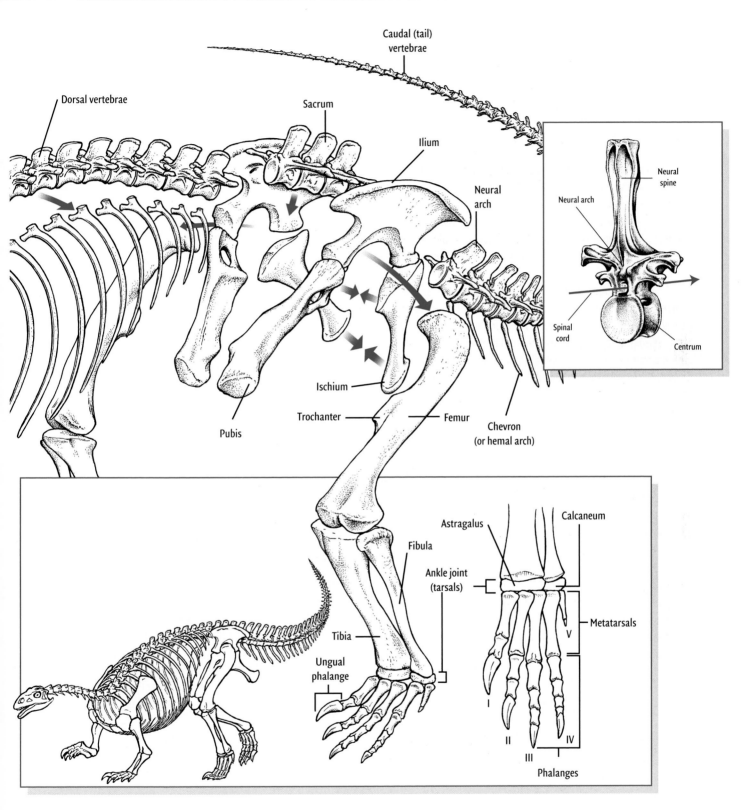

Caudal (tail) vertebrae

Dorsal vertebrae

Sacrum

Ilium

Neural arch

Neural spine

Neural arch

Spinal cord

Centrum

Ischium

Trochanter

Femur

Chevron (or hemal arch)

Pubis

Calcaneum

Astragalus

Fibula

Ankle joint (tarsals)

Metatarsals

V

Tibia

Ungual phalange

I

II

III

IV

Phalanges

Box 4.1 Fish and chips

As 1978 turned to 1979, a provocative and entertaining letter and reply were published in the scientific journal *Nature*, discussing the relationships of three gnathostomes: the salmon, the cow, and the lungfish.[1] English paleontologist L. B. Halstead argued that, obviously, the two fish must be more closely related to each other than either is to a cow. After all, he argued, they're both *fish*! A coalition of European cladists disagreed, pointing out that, in an evolutionary sense, a lungfish is more closely related to a cow than to a salmon. In their view, if the lungfish and the salmon are both to be called "fish," then the cow must also be a fish. Can a cow be a fish?

The vast majority of vertebrates are what we call "fishes." They all make a living in either salt or fresh water and, consequently, have many features in common that relate to the business of getting around, feeding, and reproducing in a fluid environment more viscous than air. But, as it turns out, even if "fishes" describes creatures with gills and scales that swim, "fishes" is not an evolutionarily meaningful term because there are no shared, derived characters that unite all fishes that cannot also be applied to all non-fish gnathostomes. The characters that pertain to fishes are either characters present in all gnathostomes (that is, primitive in gnathostomes) or characters that evolved independently.

The cladogram in Figure B4.1.1 is universally regarded as correct for the salmon, the cow, and the lungfish. In light of what we have discussed, this cladogram might look more familiar using groups to which these creatures belong: salmon are ray-finned fish, that is, fish with long rays made of a distinct protein supporting their pectoral and pelvic fins; cows are tetrapods; and lungfishes are lobe-finned fishes. Clearly, lobe-finned fishes share more derived characters in common with tetrapods than they do with ray-finned fishes. Thus there are two clades on the cladogram:

1 lobe-finned fishes + tetrapods; and
2 lobe-finned fishes + tetrapods + ray-finned fishes.

1 Halstead, L. B. 1978. The cladistic revolution – can it make the grade? *Nature*, **276**, 759–760.
Gardiner, B. G., Janvier, P., Patterson, C., Forey, P. L., Greenwood, P. H., Mills, R. S. and Jeffries, R. P. S. 1979. The salmon, the cow, and the lungfish: a reply. *Nature*, **277**, 175–176.

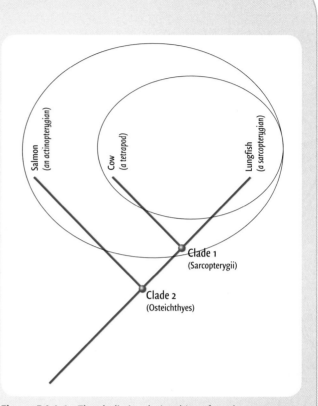

Figure B4.1.1. The cladistic relationships of a salmon, a cow, and a lungfish. The lungfish and the cow are more closely related to each other than either is to the salmon.

Clade 1 is familiar as Sarcopterygia. Clade 2 occurs at the level of all fish (and the descendants of fish) and looks like part of the cladogram presented in Figure 4.3 for gnathostome relationships. If only the organisms in question are considered, the only two monophyletic groups on the cladogram must be (1) lungfish + cow; and (2) lungfish + cow + salmon (that is, representatives of the sarcopterygians and Osteichthyes, respectively).

Which are the "fishes?" Clearly the lungfish and the salmon. But the lungfish and the salmon do not in themselves form a monophyletic group unless the cow is also included. The cladogram is telling us that the term "fishes" has *phylogenetic* significance only at the level of Osteichthyes (or even below). But we can and do use the term "fishes" informally. Fish and chips will never be a burger and fries.

Girdles   Sandwiching the backbone are the pelvic and pectoral girdles (Figure 4.5). These are each sheets of bone (or bones) against which the limbs attach for the support of the body. The pelvic girdle is the attachment site of the hindlimbs; the pectoral girdle is the attachment site of the forelimbs.

Each side of the pelvic girdle is made up of three bones: (1) a flat sheet of bone, called the ilium, that is fused to the sacrum, which is a block of vertebrae between the iliac blades; (2) a piece that points forward and down, called the pubis; and (3) a piece that points backward and down, called the ischium. Primitively, the three bones come together in a depressed area of the pelvis called the acetabulum: the hip socket.

By contrast, the pectoral girdle consists of a flat sheet of bone, the scapula (shoulder blade), on each side of the body, attached to the outside of the ribs by ligaments and muscles.

Chest   Some chest elements deserve mention. The breastbone (sternum) is generally a flat or nearly flat sheet of bone that is locked into its position on the chest by the tips of the thoracic (or chest) ribs. The rib cage is supported at its front edge by the clavicles, themselves connected to the coracoids, a pair of shield-like bones that contact the scapula.

Legs and arms   Limbs in tetrapods show the arrangement pioneered in their sarcopterygian ancestors (see Figure 4.4). All limbs, whether fore or hind, have a single upper bone connecting to a pair of lower bones. In a forelimb, the upper arm bone is the humerus, and the paired lower bones (forearms) are the radius and ulna. The joint in between is the elbow. In a hindlimb, the upper bone (thigh bone) is the femur, the joint is the knee, and the paired lower bones (shins) are the tibia and fibula.

Beyond the paired lower bones of the limbs are the wrist and ankle bones, termed carpals and tarsals, respectively. The bones in the palm of the hand are called metacarpals, the corresponding bones in the foot are called metatarsals and collectively they are termed metapodials. Finally, the small bones that allow flexibility in the digits of both the hands (fingers) and the feet (toes) are called phalanges (singular phalanx). At the tip of each digit, beyond the last joint, are the ungual phalanges.

Tetrapods primitively had as many as eight digits on each limb. Early in the evolutionary history of tetrapods, this number rapidly reduced to, and stabilized at, five digits on each limb, although many groups of tetrapods subsequently reduced that number even further (Figure 4.5).

Head   At the front end of the vertebral column of chordates are the bones of the head, composed, as we have seen, of the skull and mandible (Figure 4.6). Primitively, the skull has a distinctive arrangement: the braincase, a bone-covered box containing the brain, is located centrally and toward the back of the skull. At the back of the braincase is the occipital condyle, the knob of bone that connects the braincase (and hence the skull) to the vertebral column. A rear-facing opening in the braincase, the foramen magnum, allows the spinal cord to attach to the brain. Located on each side of the braincase are openings for the stapes, the bone that transmits sound from the tympanic membrane (eardrum) to the brain. Finally, covering the braincase and forming much of the upper rear part of the skull is a curved sheet of interlocking bones, the skull roof (inset to Figure 4.6).

The skull has two familiar pairs of openings. Located midway along each side of the skull is a large, round opening – the eye socket, or orbit. At the anterior tip of the skull is another pair

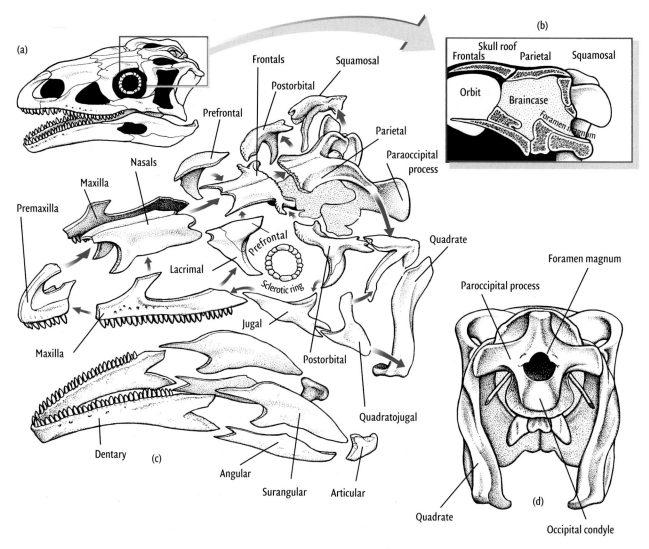

Figure 4.6. Skull and mandible of the primitive saurischian dinosaur *Plateosaurus*, exemplifying the general arrangement of bones in the skull and mandible. (a) Skull elements "exploded"; (b) cross-section through braincase; (c) "exploded" elements of mandible (lower jaw); (d) rear view of skull.

of openings – the **nares** (singular, naris), or nostril openings. Finally, flooring the skull, above the mandible, is a paired series of bones, organized in a flat sheet, which forms the **palate**.[4]

Within Tetrapoda

Tetrapods share a variety of derived features (Figure 4.7). We have seen many of these in the tetrapod skeleton: the distinctive morphologies of the girdles and limbs, as well as the fixed patterns of skull roofing bones. The hypothesis that all of these shared similarities evolved

---

[4] In mammals, a passage forms between the floor of the nasal cavity and the roof of the oral cavity (mouth), so that air breathed in through the nostrils is guided to the back of the throat, bypassing the mouth. As a result, it is possible for chewing and breathing to occur at the same time. Similar kinds of palates (called secondary palates) are known in other tetrapods besides mammals, but primitively the nostrils lead directly to the oral cavity. So, if food were to be extensively chewed in the mouth, it would quickly get mixed up with the air that is breathed in. For this reason, chewing is not a behavior of primitive tetrapods.

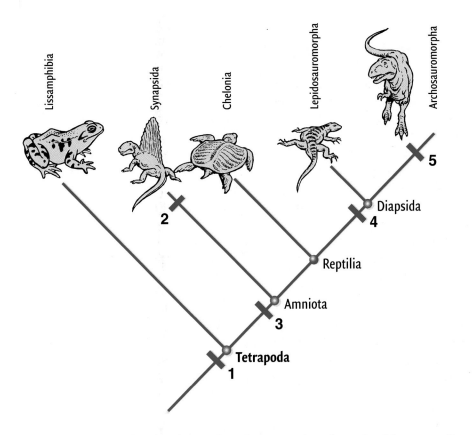

Figure 4.7. Cladogram of Tetrapoda. Derived characters include: at **1**, the tetrapod skeleton (see Figure 4.5); at **2**, a lower temporal fenestra (see Figure 4.9); at **3**, the presence of an amnion (see Figure 4.8); at **4**, lower and upper temporal fenestrae (see Figure 4.9); and at **5**, an antorbital fenestra (see Figure 4.12). Lepidosauromorpha is a monophyletic group, the living members of which are snakes, lizards, and the tuatara. Chelonia – turtles – are reptiles whose primitive, completely roofed skulls place them near the base of Amniota.

separately in distantly related organisms is not parsimonious; for this reason, these characters reaffirm Tetrapoda as a monophyletic group. Now we continue our journey to find out exactly what a dinosaur is.

Amniota

A subset of the tetrapods, Amniota, is characterized by the invention of a special membrane for the egg-bound, developing embryo called an amnion (Figure 4.8, and see below). The tetrapods without an amnion – anamniotes – are today represented only by frogs, salamanders, and a

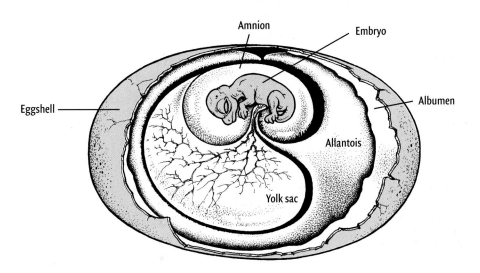

Figure 4.8. Amniote egg.

Box 4.2 **What, if anything, is a "reptile?"**

Organisms are commonly classified according to the biological classification system, first developed by the Swedish naturalist Carolus Linnaeus (1707–1778). His hierarchical system is the very famous (or infamous!) ranking of groups of organisms in groups of decreasing size: kingdom, phylum, class, order, family, genus, species.

Individuals are generally referred to by italicized **generic** (genus) and **specific** (species) names, for example in the case of a famous large dinosaur *Tyrannosaurus rex*. Any name in the hierarchy – representing a group of organisms – is considered a **taxon** (plural **taxa**).

All classifications have a purpose, and the biological classification is no exception. We classify for many purposes; for example, our movies are classified both by subject (Drama, Horror, Comedy, etc.) as well as by suitability for viewing (PG-13, R, etc.). In the case of the biota, implicit in the classification is the degree of relatedness. Thus all members of a taxon – at any level in the hierarchy – are said to be more closely related to each other than any one is to anything else. And that's where the term "Reptilia," as it's conventionally understood, gets into all kinds of trouble.

Linnaeus developed his classification long before evolution was proposed, so it really wasn't all about degree of relatedness; it was about grouping similar-looking things. His Reptilia (*reptere* – to crawl) denoted a group of scaly, four-legged creatures crawling around on their bellies. And a look at the living representatives of "Reptilia" suggests a certain superficial similarity among the living reptiles: snakes, lizards, crocodilians, and the tuatara.

But are snakes, lizards, crocodilians, and the tuatara really more closely related to each other than they are to anything else? The cladogram in Figure 4.11 and those in Chapter 10 demonstrate that, on the basis of shared derived characters, birds are more closely related to crocodiles than crocodiles are to lizards. And that's just not possible unless a bird is a reptile. But how can a bird be a reptile?

The simplest answer is that clearly we have a decidedly different Reptilia from your parents' (and Linnaeus's) traditional motley crew of crawling, scaly, non-mammalian, non-bird, non-amphibian creatures that were once tossed together as reptiles. If it is true that crocodiles and birds are more closely related to each other than either is to snakes and lizards, a monophyletic group that includes snakes, lizards, and crocodiles must also include birds.

Birds are reptiles because birds share the derived characters of Reptilia, as well as having unique characters of their own (see also Chapter 10). The inclusion (above) of birds among the living members of Reptilia is contrary to the conventional way of classifying birds, but more accurately reflects *who* they are (and where they come from).

So what, finally, is a reptile? The living reptiles = turtles + diapsids (including birds). Figure 4.7 shows the position of Reptilia on the cladogram and includes some diagnostic characters for the group.

rare, limbless, tropical amphibian known as a caecilian. If the living amphibians are any guide, the life cycles of anamniotes are intimately associated with water, as the eggs require, and likely required, an external source of moisture.

Amniotes, by contrast, are fully terrestrial, and need a means of retaining moisture within the egg. The semi-permeable amnion allows gas exchange but retains water, which permits the embryo to be continuously bathed in liquid. The evolutionary appearance of the amnion occurred in conjunction with several other features including a calcified shell, a large yolk for the nutrition of the developing embryo, and a special bladder for the management of embryonic waste. Amniotic eggs can thus be laid on land without drying out, which allowed amniotes to sever all ties with water (other than for drinking). This was a key step in the evolution of a completely terrestrial lifestyle, and is commonly associated with the advent of reptiles (Box 4.2).

There are three great groups of amniotes – primitive amniotes, sometimes termed **anapsids** (*a* – without; *apsid* – arch), **Synapsida** (*syn* – with), and **Diapsida** (*di* – two). They're most easily distinguished by the number and position of the openings in the skull roof behind the eyes, called **temporal fenestrae** (*fenestra* – window; Figure 4.9). Our main interest is in diapsids, but we'll detour briefly to look at some basal amniotes and Synapsida.

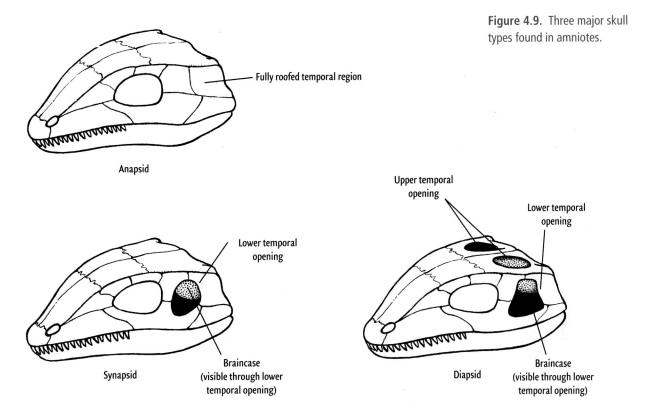

Figure 4.9. Three major skull types found in amniotes.

Anapsids and Synapsida

The anapsid condition represents what is thought to have been the original morphology of the skull roof in amniotes. In these amniotes, the skull behind the eyes is completely roofed; they therefore have no temporal fenestrae. The anapsid condition is seen in some long-extinct, bulky quadrupeds that do not concern us here, and may persist today only in turtles (Chelonia). Legendary stalwarts of the world, turtles are unique: these venerable creatures with their portable houses, in existence since the Late Triassic (210 million years ago), will surely survive another 200 million years at least if we let them.

Synapsida is one of two great lineages of amniotic tetrapods. All mammals (including ourselves) are synapsids, as are a host of extinct forms, traditionally called "mammal-like reptiles" (Figure 4.10). The split between the earliest synapsids and the earliest representatives of the other great lineage, Diapsida (including dinosaurs), likely occurred between 310 and 320 Ma. Since then, therefore, the synapsid lineage has been evolving independently, genetically unconnected to any other group.

Synapsids are united by a skull roof that is a departure from primitive tetrapods: the skull roof has developed a low opening behind the eye – the lower temporal fenestra (see Figure 4.9). Jaw muscles pass through this opening and attach to the upper part of the skull roof. Synapsids are a remarkable and diverse group of amniotes, and could easily fill a book just like this; but because we're interested in dinosaurs, we'll regretfully move right on past them.

Diapsida

The other great clade of amniotes is Diapsida (see Figures 4.7 and 4.9). The living diapsids include about 15,000 total species including snakes, lizards, crocodiles, the tuatara (a reptile

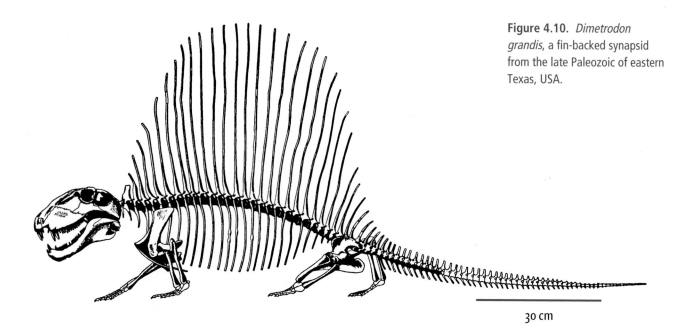

Figure 4.10. *Dimetrodon grandis*, a fin-backed synapsid from the late Paleozoic of eastern Texas, USA.

30 cm

found only in New Zealand), and birds; extinct diapsids include dinosaurs as well as many other forms. Nobody really knows how many members of this clade have come and gone.

Diapsida is united by a suite of shared, derived features, including having two temporal openings in the skull roof, and an upper and a lower temporal fenestra. The upper and lower temporal fenestrae are thought to have provided accommodation for the bulging of contracted jaw muscles, as well as increased the surface area for the attachment of these muscles.

Moving to the ultimate node in Figure 4.7, there are two major clades of diapsids. The first, Lepidosauromorpha (*lepido* – scaly; *morpho* – shape), is composed of snakes and lizards and the tuatara (among the living), as well as a number of extinct lizard-like diapsids;[5] the second, Archosauromorpha (*archo* – ruling), brings us within striking distance of dinosaurs.

Archosauromorpha

Archosauromorpha is supported by many important, shared, derived characters (Figure 4.11). Within archosauromorphs are a series of basal members that are known mostly from the Triassic. Some bear a superficial resemblance to large lizards; others look like beefed up crocodiles; a few even look like reptilian pigs (see Figure 13.4).

A subset of archosauromorphs possesses a number of significant evolutionary innovations (Figure 4.11), most notably an opening on the side of the snout, just ahead of the eye, called the antorbital fenestra (Figure 4.12). This is the key character that unites Archosauria, the group that contains crocodilians, birds, and dinosaurs. It is ironic that, for all its phylogenetic importance, the function of the antorbital fenestra is still uncertain, it may have contained a large air sac, or a salt gland.

Crocodilians and their close relatives belong to a clade called Pseudosuchia (*pseudo* – false; *suchia* – crocodile; ironically named because while some things that only *look* like crocodiles belong to this group, true crocodiles belong as well!) about which we won't be too concerned here; dinosaurs and their close relatives constitute a clade called Ornithodira (*ornith* – bird; *dira* – neck; see Figure 4.11).

---

[5] Two marine groups, ichthyosaurs and plesiosaurs (see Figure 16.9) have also been placed within Diapsida. Recently, it has been proposed that turtles, long thought to represent the anapsid condition, are also diapsids.

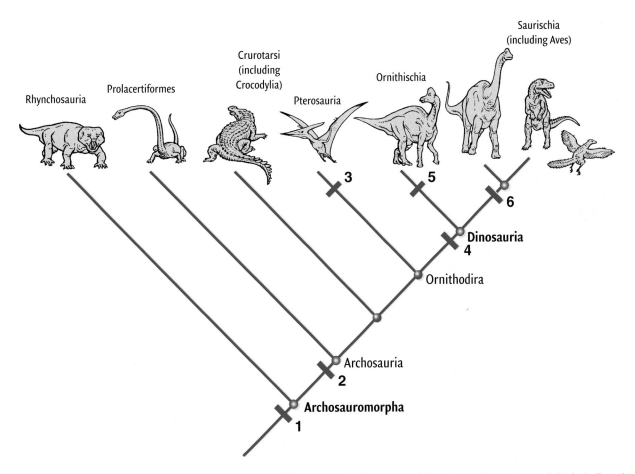

Figure 4.11. Cladogram of Archosauromorpha. Derived characters include: at **1**, teeth in sockets, elongate nostril, high skull, and vertebrae not showing evidence of embryonic notochord; at **2**, antorbital fenestra (see Figure 4.12), loss of teeth on palate, and new shape of articulating surface of ankle (calcaneum); at **3**, a variety of extraordinary specializations for flight, including an elongate digit IV; at **4**, erect stance (shaft of femur is perpendicular to head, upper part of hip socket is thickened or has a ridge, ankle has a modified mesotarsal joint), perforate acetabulum; at **5**, predentary and rearward projection of pubic processes (see introductory text for Part II: Ornithischia; at **6**, asymmetrical hand with distinctive thumb, elongation of neck vertebrae, and changes in chewing musculature (see introductory text for Part III: Saurischia; see also Chapter 15).

Ornithodira brings us quite close to the ancestry of dinosaurs. This group is composed of two monophyletic groups, Dinosauria (*deinos* – terrible) and Pterosauria (*ptero* – winged; Figure 4.13). That pterosaurs are unapologetically Mesozoic archosaurs has led to their being called "dinosaurs"; that they had wings and flew has led some to mistake them for birds; but in fact they were something utterly different from either dinosaurs or birds. They were unique, magnificent, and now, sadly, very extinct.

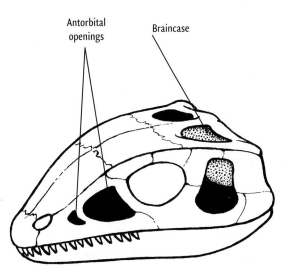

Figure 4.12. An archosaur skull with the diagnostic antorbital fenestra.

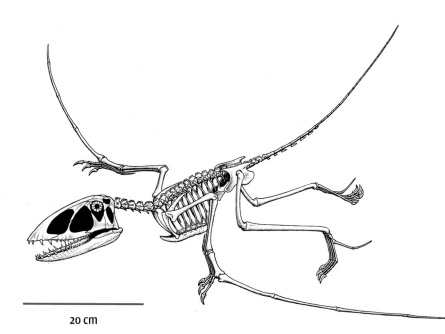

Figure 4.13. A close relative of Dinosauria: Pterosauria as represented by *Dimorphodon*, from the Lower Jurassic of Europe.

20 cm

Dinosaurs

This climb up the cladogram leaves us wheezing and gasping for air, but at long last situated at the subject of our book: Dinosauria. Dinosaurs can be diagnosed by a host of shared, derived characters (Figures 4.14 and 4.15). Most strikingly, dinosaurs are united by the fact that, within archosaurs, they possess an erect, or parasagittal stance; that is, a stance in which the plane of the legs is parallel to the vertical plane of the torso (see Figure 4.11). In dinosaurs, an erect stance consists of a suite of anatomical features with important behavioral implications (Box 4.3). The head of the femur (thigh bone) is oriented at approximately 90° to the shaft. The head of the femur itself is barrel-shaped (unlike the familiar ball shape seen in a human femur), so that motion in the thigh is restricted largely to forward and backward, within, as we've seen, a plane parallel to that of the body (see Figure

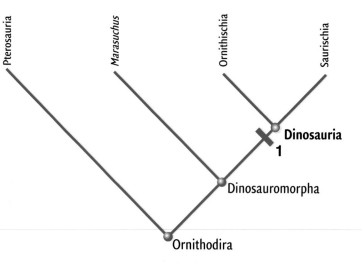

Figure 4.14. Cladogram of Ornithodira showing the monophyly of Dinosauria. Derived characters at **1**, elongate deltapectoral crest on humerus, extensively perforated acetabulum, tibia with transversely expanded subrectangular distal end, temporal musculature that extends anteriorly on to the skull roof, and epopophyses on the cervical vertebrates (see also Figure 4.15).

4.16). The ankle joint is modified to become a single, linear articulation. This type of joint, termed a modified mesotarsal joint, allows movement of the foot only in a plane parallel to that of the body: forward and backward (Figure 4.16). Note that again this situation differs from that in humans, in which the foot is capable of rotating. The upshot of these adaptations of stance is that all dinosaurs are highly specialized for cursorial locomotion (that is, running, as in the "cursor" on a computer screen). Dinosaurs are terrestrial beasts through and through (see Box 4.3).

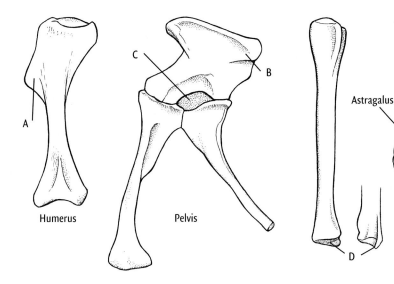

Figure 4.15. Some of the derived characters uniting Dinosauria. (A) Elongate deltopectoral crest on humerus; (B) brevis shelf on ventral surface of postacetabular part of ilium; (C) extensively perforated acetabulum; (D) tibia with transversely expanded subrectangular distal end, and (E) ascending astragalar process on front surface of tibia.

Beyond their fully erect stance, dinosaurs are diagnosed by a host of derived features (see Figure 4.14). These include loss of a skull roofing bone – the postfrontal – that lies on the top of the head along the front margin of the upper temporal fenestra, an elongate deltopectoral crest on the humerus, an extensively perforated acetabulum, a tibia with a transversely expanded subrectangular lower end, and an ascending process of the astragalus on the front surface of the tibia (see Figure 4.15).

Ornithischia and Saurischia

In 1887, the English paleontologist Harry Seeley first recognized a fundamental division among dinosaurs. Ornithischia (*orni* – bird; *ischia* – hip) were all those dinosaurs that had a bird-like pelvis, in which at least a part of the pubis runs posteriorly, along the lower rim of the ischium (Figure 4.17). Saurischia (*saur* – lizard) were those that had a pelvis more like a lizard, in which the pubis is directed anteriorly, and slightly downward (Figure 4.18). This pelvic distinction has held sway ever since.

That dinosaurs had one or the other kind of pelvis was of great importance to understanding the evolution of these animals, and, in Seeley's hands, it went considerably farther. For it implied

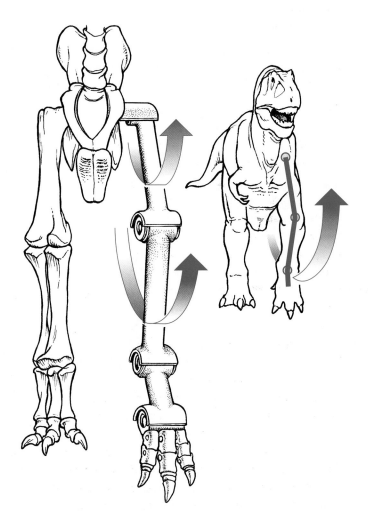

Figure 4.16. The fully erect posture in dinosaurs. Unlike in, for example, a human, the bones of the leg restrict movement to only one plane: forward and backward.

Box 4.3 Stance: it's both who you are and what you do

Tetrapods that are most highly adapted for land locomotion tend to have an erect stance. This clearly maximizes the efficiency of the animal's movements on land, and it is not surprising that, for example, all mammals are characterized by an erect stance. Tetrapods such as salamanders (which are adapted for aquatic life) display a sprawling stance, in which the legs splay out from the body nearly horizontally. The sprawling stance seems to have been inherited from the original position of the limbs in early tetrapods, whose sinuous trunk movements (presumably inherited from swimming locomotion) aided the limbs in land locomotion.

Some tetrapods, such as crocodiles, have a semi-erect stance, in which the legs are directed at something like 45° downward from horizontal (Figure B4.3.1). Does this mean that the semi-erect stance is an adaptation for a combined aquatic and terrestrial existence? Clearly not, because a semi-erect stance is present in the large, fully terrestrial monitor lizards of Australia (goanna) and Indonesia (Komodo dragon). If adaptation is the only factor driving the evolution of features, why don't completely terrestrial lizards have a fully erect stance, and why don't aquatic crocodiles have a fully sprawling stance? The issue is more complex and is best understood through adaptation to a particular environment or behavior, *as well as* through inheritance. If we consider stance simply in terms of ancestral and derived characters, the ancestral condition in tetrapods is sprawling. An erect stance represents the most highly derived state of this character, but are animals with sprawling stances not as well designed as those with erect stances? In 1987, D. R. Carrier of Brown University, Rhode Island, USA, hypothesized that the adoption of an erect stance represents the commitment to an entirely different mode of respiration (breathing) as well as locomotion (see Chapter 12 on "warm-bloodedness" in dinosaurs). Those organisms that possess a semi-erect stance may reflect the modification of a primitive character (sprawling) for greater efficiency on land, but they may also retain the less-derived type of respiration. Dinosaurs (see Figure 4.11) and mammals both have fully erect stances, which represent a full commitment to a terrestrial existence as well as to a more derived type of respiration. The designs of all these organisms are thus compromises among inheritance, habits, and mode of respiration. Who can say what other influences are controlling morphology?

Interestingly, the cladogram (see Figure 4.7) shows that the most recent common ancestor of dinosaurs and mammals – some primitive amniote – was itself an organism with a sprawling stance. Because dinosaurs and mammals (or their precursors) have been evolving independently since their most recent common ancestor, an erect stance must have evolved at least twice in Amniota: once among the synapsids and once in dinosaurs.

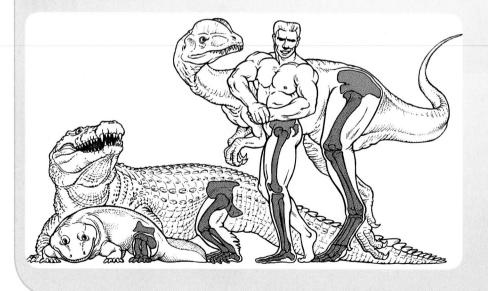

Figure B4.3.1. Stance in four vertebrates. To the left, the primitive amphibian and crocodile (behind) have sprawling and semi-erect stances, respectively. To the right, the human and the dinosaur (behind) both have fully erect stances.

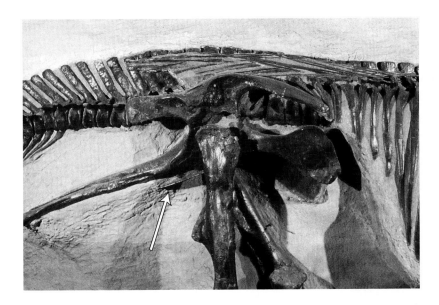

Figure 4.17. The pelvis of the hadrosaurid *Prosaurolophus*. A splint of the pubis points posterior, along the base of the ischium (see arrow), exemplifying the ornithischian condition.

to him that the ancestry of Ornithischia and Saurischia was to be found separately and more deeply embedded in a heterogeneous group of primitive archosaurs once called "Thecodontia."[6] To Seeley, therefore, Dinosauria was not monophyletic.

All that changed in 1986, when cladistic analysis, in the skilled hands of J. A. Gauthier, provided powerful evidence for a monophyletic Dinosauria. And, since Gauthier's studies, numerous other analyses by other paleontologists, using both newly discovered and familiar taxa, have confirmed that dinosaurs are monophyletic (see Chapter 15).

Is Saurischia more primitive than Ornithsichia?

The distinction between Ornithischia and Saurischia is valid and, as we'll see later in this book, both groups are diagnosed by suites of well-established characters. Indeed, the split between ornithischians and saurischians is the fundamental division within Dinosauria. But which is more primitive? Saurischians, with their claws, and teeth, appear to be a lot like their archosaurian forebearers – claws, teeth, etc. (see Figure 4.18) – and in many books, particularly the older ones, they are treated

Figure 4.18. The pelvis of the ornithomimosaur *Ornithomimus*. The pubis is directed anterior only (see arrow), exemplifying the saurischian condition.

---

[6] The group "Thecodontia" is based on the same characters that diagnose all archosaurs. If we're discussing archosaurs, we can't cherry- pick a few basal ones; we need to include *all* members of the group (including dinosaurs and pterosaurs) that bear the diagnostic characters. In short, the term "Thecodontia," though venerable, has not withstood cladistic scrutiny (see Chapters 10 and 14).

first, reflecting the intuitive notion that they are more primitive, and that somehow ornithischians must have evolved from a saurischian ancestor. But is this true?

The cladistic answer to this question is, in so far as we yet know, "No": Figure 4.14 shows that Saurischia is no less derived than Ornithischia; thus, we really don't know which came first. To underscore this important evolutionary point, we'll begin by studying Ornithischia first. We return to some of these ideas in Chapter 15, when we explore the origin of Dinosauria.

Summary

In this chapter cladograms are used to map the relationship of Dinosauria to the rest of the biota. Dinosaurs fall within Vertebrata, backbone-bearing animals within the larger group Chordata (bilateral creatures bearing a dorsal nerve cord and a notochord). Within vertebrates, dinosaurs are tetrapods, or animals with four limbs. Dinosaurs are amniotic tetrapods (or amniotes), which means that, like ourselves, they possess an amnion.

Amniotes are generally identified by the number and type of temporal fenestrae. Synapsida (the group that includes mammals) is united by the possession of lower temporal fenestrae, while Diapsida (the group that includes snakes, lizards, crocodilians, and birds) all possess lower and upper temporal fenestrae. The split between these two major groups of reptiles took place about 320 Ma.

Dinosaurs are diapsids and among living diapsids, dinosaurs are most closely related to crocodiles and birds, both of whom are archosaurs, a group of reptiles united by the presence of an antorbital opening. Calling birds reptiles is, of course, contrary to conventional Linnaean classification, which in this case fails to accurately depict their evolutionary relationships as revealed by cladistic analysis.

There are two major groups of dinosaurs: Ornithischia and Saurischia. These are identified by a variety of features, in particular the orientation of the anterior portion of the pubis, and the presence in ornithischians of a new element of the lower jaw, the predentary. As the phylogeny is currently understood, ornithischians and saurischians are equally derived; thus it is not possible to say whether one is more primitive than the other.

Finally, this chapter contains a brief summary of basic diapsid bone morphology. Elements of the axial skeleton are presented, including centra, neural arches, hemal arches, cervical vertebrae, caudal vertebrae, ribs, gastralia, pelvic and pectoral girdles. Limb skeletal structure is presented, including humerus, radius, and ulna; femur, tibia, and fibula, carpels, tarsals, metapodials, and ungual phalanges. The elements of the skull and mandible are also presented.

Selected readings

Bakker, R. T. and Galton, P. M. 1974. Dinosaur monophyly and a new class of vertebrates. *Nature*, **248**, 168–172.

Benton, M. J. (ed.). 1988. *The Phylogeny and Classification of the Tetrapods*, vol I. Systematic Association Special Volume no. 35A, Oxford, 377 pp.

Benton, M. J. 1997. Origin and early evolution of the dinosaurs. In Farlow, J. O. and Brett-Surman, M. K. (eds.), *The Complete Dinosaur*. Indiana University Press, Bloomington, IN, pp. 204–214.

Benton, M. J. 2004. Origin and relationships of Dinosauria. In Weishampel, D. B., Dodson, P. and Osmólska, H. (eds.), *The Dinosauria,* 2nd edn. University of California Press, Berkeley, pp. 7–20.

Carrier, D. R. 1987. The evolution of locomotor stamina in tetrapods: circumventing a mechanical constraint. *Paleobiology,* **13,** 326–341.

Gauthier, J. A., Kluge, A. G. and Rowe, T. 1988. Amniote phylogeny and the importance of fossils. *Cladistics,* **4,** 105–209.

Parrish, M. J. 1997. Evolution of the archosaurs. In Farlow, J. O. and Brett-Surman, M. K. (eds.), *The Complete Dinosaur. Indiana University Press,* Bloomington, IN, pp. 191–203.

Sereno, P. C. 1991. Basal archosaurs: phylogenetic relationships and functional implications. *Journal of Vertebrate Paleontology,* **11** (suppl.), 1–53.

Sereno, P. C., Forster, C. A., Rogers, R. R. and Monetta, A. F. 1993. Primitive dinosaur skeleton from Argentina and the early evolution of Dinosauria. *Nature,* **361,** 64–66.

Topic questions

1 Define: Chordata, notochord, gnathomstome, Sarcopterygii, skull, mandible, girdles, neural arch, centrum, process, ilium, ischium, pubis, acetabulum, sternum, humerus, femur, radius, ulna, tibia, fibula, phalanx, ungual, metacarpals, metatarsals, metapodials, occipital condyle, foramen magnum, stapes, skull roof, tympanic membrane, nares, orbit, palate, amniote, anamniote, amnion, anapsid, synapsid, diapsid, upper temporal fenestra, lower temporal fenestra, archosaur, Archosauromorpha, lepidosaur, Crurotarsi, Ornithodira, Dinosauria, pterosaur, mesotarsal, Ornithischia, Saurischia.

2 Explain the importance of the amnion in the evolution of terrestrial tetrapods.

3 Describe the basic structure of the vertebrate limb.

4 Draw a skull and lower jaws, indicating the skull roof, braincase, temporal region, orbit, nares, and snout.

5 For how long have the mammal and bird lineages been evolving separately?

6 Why is it that we said that a bird was more closely related to a crocodile than a crocodile is to a lizard?

7 Construct a cladogram with only Vertebrata, Diapsida, Dinosauria, Synapsida, and Lepidosauria marked on it.

8 Construct a cladogram with just Dinosauria, Ornithischia, Archosauria, and Crocodylia marked on it.

9 How can birds be reptiles?

10 Within Amniota, warm-bloodedness and flight occur in bats, in birds, and in pterosaurs. Use a cladogram to show how many separate evolutionary events this required.

PART II

ORNITHISCHIA:
ARMORED, HORNED, AND DUCK-BILLED DINOSAURS

In the preceding chapter we met Ornithischia, one of the two great branches of dinosaurs. Although it was identified as early as 1887, nobody at that time had an inkling how diverse the group really was. Since then, we have learned about the richness of ornithischians as well as about the anatomy, evolution, and even *behavior* of these dinosaurs. We'll check out ornithischians in Chapters 5–7. But first, let's introduce Ornithischia a bit more completely.

What makes an ornithischian an ornithischian?

Diagnostic features for Ornithischia abound. Among these, two stand out:

- In all ornthsichians, at least a part of the pubis has rotated backward to lie close to and parallel with the ischium;[1] this orientation is called opisthopubic (Figure II.1).
- All ornithischians had a unique bone, the predentary, an unpaired, scoop-shaped element that capped the front of the lower jaws (Figure II.2).

Figure II.1. Left lateral view of the ornithischian pelvis as exemplified by Stegosaurus. Note that the pubic bone is rotated backward to lie under the ischium (see arrow) in what is known as the opisthopubic condition.

Both of these adaptations were associated with food consumption and processing. The evolution of the rearward-directed pubis (recall that, primitively, the pubis points forward; see Figure 4.5) is believed to be associated with the development of a large stomach (or stomachs?) and intestinal region (gut), the better for extracting nutrients from plants. Accommodating the large gut was a barrel-shaped torso, recognizable from the shape of the ribs. The predentary supported the lower portion of a beak, a characteristic feature of all ornithischians (see below).

Other ornithischian diagnostic characters include: a toothless, roughened front tip of the snout; a narrow bone (the palpebral) that crossed the outside of the eye socket; a jaw joint set

---

[1] Ornithischian dinosaurs are called, confusingly, "bird-hipped"; but birds themselves belong to Saurischia – the "lizard-hipped" clade (Saurischia) of dinosaurs (see Chapter 10).

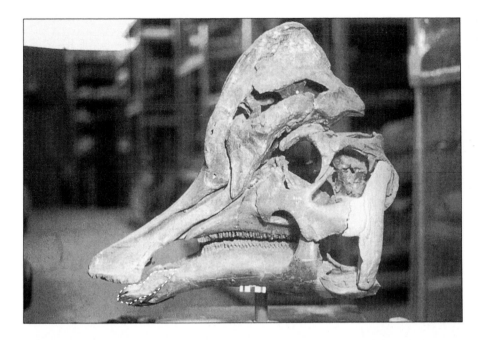

Figure II.2. Left lateral view of the skull of the lambeosaurine hadrosaurid Corythosaurus. Predentary capping the front of the lower jaw is outlined in white.

below the level of the upper tooth row; cheek teeth with low crowns somewhat triangular in shape; five or more vertebrae in the sacrum; and ossified tendons above the sacral region (and probably further along the vertebral column as well), for stiffening the backbone at the pelvis. The monophyly of Ornithischia is *extremely* well supported.

Chew on this!

One essential quality of ornithischians is that, to a greater or lesser extent, all apparently chewed their food. Because humans and many other mammals chew, it can be surprising to learn that most vertebrates don't chew; that teeth are really most commonly used only for biting off chunks of whatever is being eaten. The fundamental act of *chewing* – of grinding food down to a paste that can be digested relatively efficiently – is, as we shall see in the following chapters, done by other organs in most vertebrates. But ornithischian dinosaurs got into the chewing game, so a look at what chewing's all about is useful for understanding these dinosaurs.

Chewing in mammals

We start by looking at the basics of chewing in a familiar living group: mammals. Among herbivorous mammals, no matter what the size, the skull is generally divided into three sections (Figure II.3): at the front is the cropping part, where blade-like teeth (generally incisors) bite off chunks of food. Behind the cropping section is the diastem, a gap that is toothless (or nearly so), likely used for the manipulation of the food by the tongue. Finally, further back in the mouth are the cheek teeth (molars in mammals) – most commonly a block of teeth, relatively tightly fitted against one another, which are used to grind or shear plant material. In mammals the upper and lower cheek teeth occlude – or fit tightly against each other when the jaw is shut – which ensures that, as the chewing takes place, the grinding is efficient.

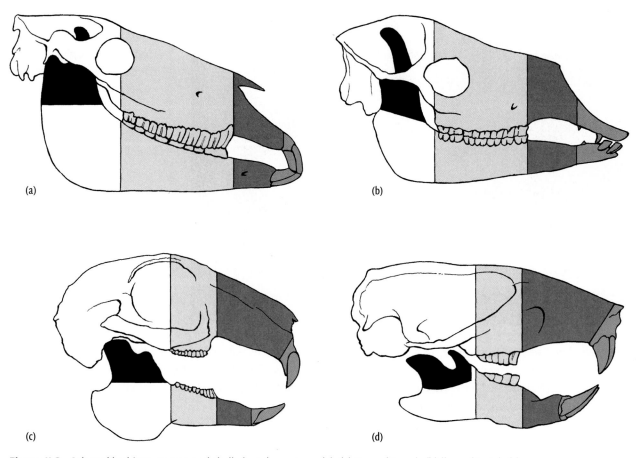

Figure II.3. Selected herbivorous mammal skulls (not drawn to scale). (a) Horse (*Equus*), (b) llama (*Lama*), (c) rabbit (*Lepus*), and (d) rat (*Rattus*). Divisions of skulls indicate: anterior cropping section (dark blue), diastem (gray), block of grinding cheek teeth (light blue), and coronoid process (black). Despite the range of sizes and herbivorous behaviors, all skulls show the same basic organization.

Two other features are associated with chewing in mammals. Firstly, toward the back of the lower jaw is a large expansion of bone, the coronoid process, which serves as an attachment site for strong jaw-closing muscles (Figure II.3). Secondly, the tooth row is deeply inset toward the midline of the skull. This makes room for cheeks, muscular tissues that play the obviously essential role of keeping food in the mouth while it is being chewed.

In mammals, then, chewing leaves a recognizable imprint on the design of the skull, the lower jaw, and the teeth. It's striking that in almost all ornithischian dinosaurs, many of these same adaptations for chewing can be found.

Chewing in dinosaurs

The primitive tetrapod condition is that in which the jaw joint is right at the same level as the tooth row (Figure II.4a). The jaw thus functions a bit like scissors: the bite slices sequentially, from the back of the jaw forward. This means that the teeth in the upper jaw move past those in the mandible as the jaw closes.

By contrast, when the jaw joint is below the level of the tooth row, as it is in all orni-thischians, the jaw functions a bit like water-pump pliers, in which the jaws close simultane-ously along their entire length (Figure II.4b). The blocks of cheek teeth grind against each other simultaneously instead of shearing past each other sequentially. As it turns out, even

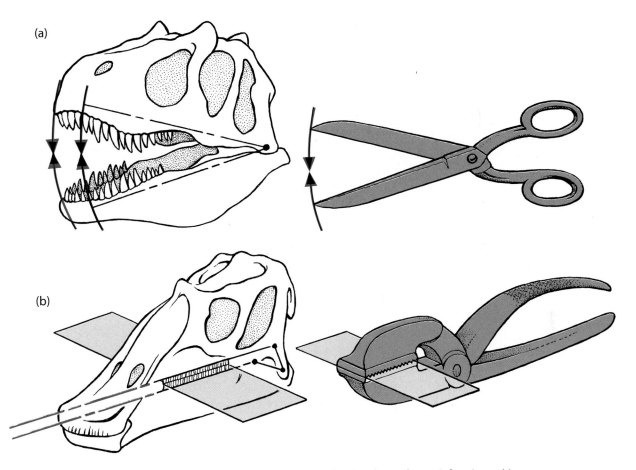

Figure II.4. Positions of jaw joints. When the jaw joint is at the same level as the tooth row, it functions a bit like a pair of scissors, slicing sequentially from the back of the jaw forward (a). By contrast, when the jaw joint is below the level of the tooth row, as it is in all ornithischians, the blocks of cheek teeth grind against each other simultaneously, much like the water-pump pliers (b).

the most primitive ornithsichians had the required goods for serious chewing. And the many ornithischians that followed never kicked the habit.

In many dinosaurs, including all ornithischians, the cropping function of the mouth was carried out not by teeth, but by a beak, or **rhamphotheca**. Rhamphothecae were made of **keratin**, a protein-based substance that makes up horns, nails, hooves, and claws. It is also a key ingredient of hair, although you'll never find *that* in a dinosaur!

Ornithischia: the big picture

The fundamental split of Ornithischia is between the very primitive ornithischian *Lesothosaurus* and everything else ornithischian (Figure II.5). *Lesothosaurus* was a small, long-limbed Early Jurassic herbivore from South Africa (Figure II.6). It had a typical suite of diagnostic ornithischian characters including a jaw joint lower than the tooth row (see Figure II.4). That character hints at chewing; but mere hints won't be necessary for the rest of Ornithischia.

In contrast, "everything else ornithischian" is within the clade Genasauria (*gena* – cheek; see Figure II.5). They all share the derived characters of muscular cheeks, indicated by the deep-set position of the tooth rows, away from the sides of the face, a spout-shaped front to

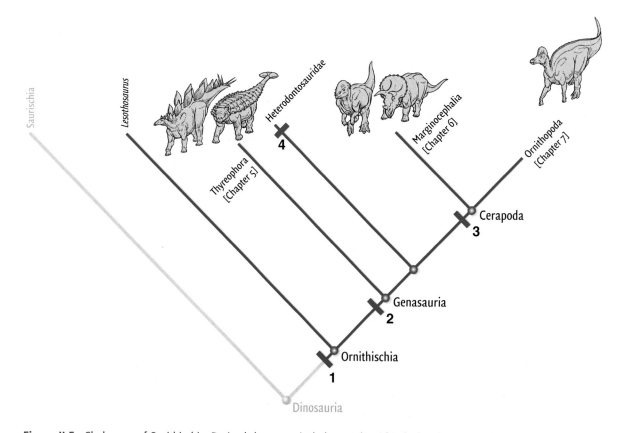

Figure II.5. Cladogram of Ornithischia. Derived characters include: at 1 (Ornithischia), opisthopubic pelvis, predentary bone, toothless and roughened tip of snout, reduced antorbital opening, palpebral bone, jaw joint set below level of the upper tooth row, cheek teeth with low subtriangular crowns, at least five sacral vertebrae, ossified tendons above the sacral region, small prepubic process along the pubis, long and thin pre-acetabular process on the ilium; at 2 (Genasauria), emarginated dentition (indicating large cheek cavities, and reduction in the size of the opening on the outside of the lower jaw (the external mandibular foramen); at 3 (Cerapoda), gap between the teeth of the premaxilla and maxilla, five or fewer premaxillary teeth, finger-like anterior trochanter; at 4, high-crowned cheek teeth, denticles on the margins restricted to the terminal third of the tooth crown, canine-like tooth in both the premaxilla and dentary.

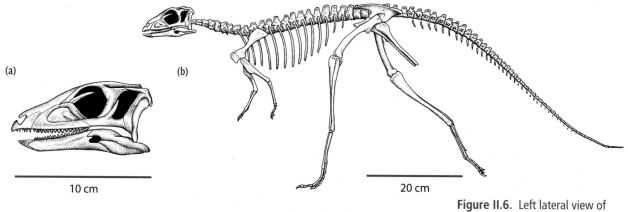

(a)

(b)

10 cm

20 cm

Figure II.6. Left lateral view of the skull (a) and skeleton (b) of the basal ornithischian *Lesothosaurus.*

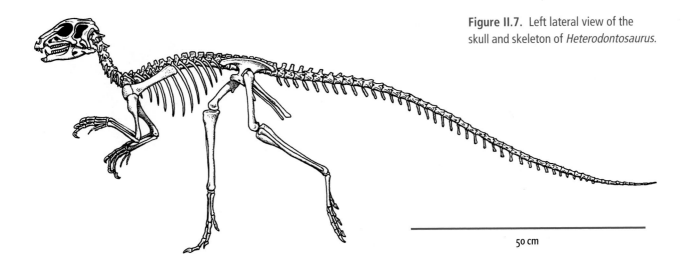

Figure II.7. Left lateral view of the skull and skeleton of *Heterodontosaurus*.

50 cm

the mandibles, and reduction in the size of the opening on the outside of the lower jaw (the external mandibular foramen), among others. Because it's hard to understand cheeks without chewing, chewing should be thought of as a fundamental genasaur behavior. Then it only becomes a matter of how efficiently the various groups of genasaurs chewed. Genasaurs include the basal heterodontosaurids,[2] and then the two great groups of ornithsichians, Thyreophora (Chapter 5), and Cerapoda (Marginocephalia + Ornithopoda; Chapters 6 and 7, respectively), as well as a few assorted forms with which we won't concern ourselves here.

Heterodontosaurids were small, bipedal ornithischians (Figure II.7). Despite their low position on the cladogram, however, in many respects they were not exactly primitive; for example, they evolved teeth bearing a high, chisel-shaped crown ornamented with tiny bumps (Figure II.8a; correctly termed denticles) as well as a large canine-like tooth on both upper and lower jaws (the basis for the name "heterodontosaurid"; Figure II.8b). Moreover, they chewed distinctively: they amplified the familiar vertical movement of the lower jaws with slight outward rotations of each side of the mandible about its long axis (Figure II.9). This allowed them to get a bit more grind out of each bite. Powerful forelimbs and clawed hands likely were used to grab vegetation or to dig up roots and tubers.

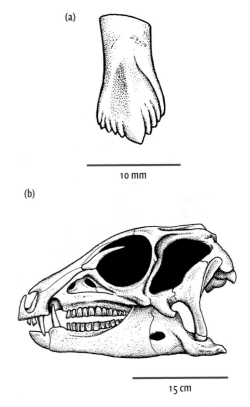

(a)

10 mm

(b)

15 cm

Figure II.8 (a) Upper tooth of the heterodontosaurid *Lycorhinus*. (b) Left lateral view of the skull of *Heterodontosaurus*.

---

[2] A recent, important reanalysis of Ornithischia (Butler et al., 2008) removed *Lesothosaurus* from its conventional basal position in Ornithischia and made it a basal thyreophoran (see Chapter 5), closely related to stegosaurs and ankylosaurs. Until this idea is generally accepted, however, we continue to regard it as among the most primitive of ornithischians. In the same analysis, heterodonotosaurs, traditionally considered ornithipods (or very close to them; see Chapter 7), were reassigned to an extremely basal position in Ornithsichia; we follow this analysis here. But because heterodontosaurids bear the inset tooth rows (and thus cheeks) diagnostic of genasaurs, we consider them to be basal genasaurs.

Like many ornithsichians, the evolution of canine-like teeth of heterodontosaurids likely was related to combat between animals of the same species (males?), ritualized display, social ranking, and possibly even courtship (see Chapters 5 and 6). A modern analog is found in tusked tragulids, living mammals from southeastern Asia and Africa, closely related to deer. In these mammals, tusk development is tied to sexual maturity, and is a dimorphic feature that is, as we propose for heterodontosaurids, used for intraspecific combat, ritualized display, and social ranking. Similarly, the development of a bony boss in the cheek region (the jugal boss) in heterodontosaurids might also be interpreted as a form of visual display.

Display in even the most primitive ornithischians may have come in another form as well. The 2009 discovery of *Tianyulong*, a heterodontosaur from western Liaoning Province, China, stunned researchers: the thing was evidently – and certainly unexpectedly – covered with a fine carpet of primitive feather-like filaments (see Figure 10.14a)! Was having feathers – something that 20 years ago people thought only birds possess – a primitive condition pertaining to all Dinosauria?!

Depending upon the colors and patterns, these primitive feathers could have served as display function and may have been intimately associated with behaviors ranging from choosing mates (see Chapter 6) to marking territory, to camouflage, to keeping warm (Chapter 12). The implications of *Tianyulong* and its feathers is discussed in Chapter 10 ("Feathers without flight").

Thyreophora (*thyreo* – shield; *phora* – bearer; a reference to the fact that these animals have dermal armor) consists of genasaurs in which there are parallel rows of keeled dermal armor scutes (or bony plates) on the back surface of the body. The most familiar thyreophorans are stegosaurs and ankylosaurs, but we'll encounter others along the way.

Cerapoda (*cera* – horn) are those genasaurs that share a pronounced diastem between the teeth of the premaxilla and maxilla among other

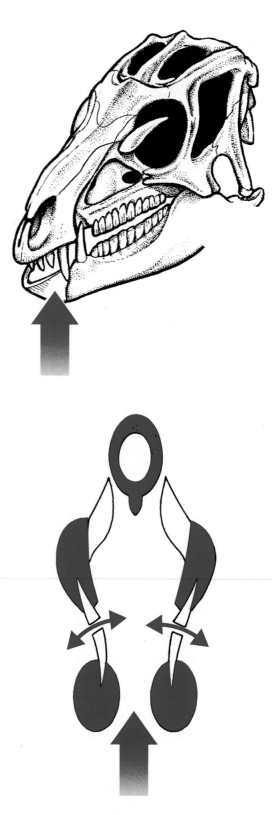

Figure II.9. Jaw mechanics in Heterodontosauridae, showing mobility of the lower jaws. As the lower jaw closed, each side rotated slightly along its long axis.

derived characters (see *Figure II.3*). Marginocephalians (*margin* – margin; *cephal* – head), a group united by having, primitively, a distinctive, narrow shelf that extended over the back of the skull (see Figure II.5), consists of two well-known ornithischian taxa, the dome-headed pachycephalosaurs and ceratopsians (Chapter 6), the latter being the horned dinosaurs most famously known from the Late Cretaceous of North America.

Ornithopods took chewing to new levels, possibly unmatched in the history of life. Within this group are found the familiar duck-billed dinosaurs, as well as one of the very first dinosaurs ever recorded, *Iguanodon*. We'll visit these animals in Chapter 7.

Selected readings

Butler, R. J., Upchurch, P. and **Norman, D. B.** 2008. The phylogeny of the ornithsichian dinosaurs. *Journal of Systematic Paleontology*, **6**, 1–40.

Norman, D. B., Witmer, L. M. and **Weishampel, D. B.** 2004. Basal Ornithischia. In Weishampel, D. B. and Dodson, P. (eds.), *The Dinosauria*, 2nd edn. University of California Press, Berkeley, pp. 325–334.

Sereno, P. C. 1986. Phylogeny of the bird-hipped dinosaurs (Order Ornithischia). *National Geographical Research*, **2**, 234–256.

CHAPTER 5

THYREOPHORANS:
THE ARMOR-BEARERS

CHAPTER OBJECTIVES

- Introduce Thyreophora, particularly its two large constituent groups, Stegosauria and Ankylosauria

- Develop familiarity with current thinking about the lifestyles and behaviors of thyreophorans

- Develop an understanding of thyreophoran evolution using cladograms, and an understanding of the place of Thyreophora within Dinosauria

Thyreophora

In life as in games, offense and defense are strategies, each with its own advantages and disadvantages. Thyreophora (*thyr* – shield; *phora* – bearing or carrying; literally, "armor-bearers") went with defense, evolutionarily opting for fortress-like protection and armor. And the strategy paid off: these dinosaurs did very well during their approximately 100 million years on Earth, spawning upward of 50 species currently known.

Who are thyreophorans?

All thyreophorans are characterized by parallel rows of special bones, embedded in the skin, called osteoderms (*osteo* – bone; *derm* – skin), that run down the necks, backs, and tails. The group is dominated by two great clades: Stegosauria (*stego* – roof); and Ankylosauria (*ankylo* – fused). Together, stegosaurs and ankylosaurs make up a monophyletic clade known as Eurypoda (*eury* – broad; *poda* – feet). Along with these two big groups, a few other miscellaneous, primitive Early Jurassic thyreophorans round out our story. Figure 5.1 lays out basic thyreophoran relationships.

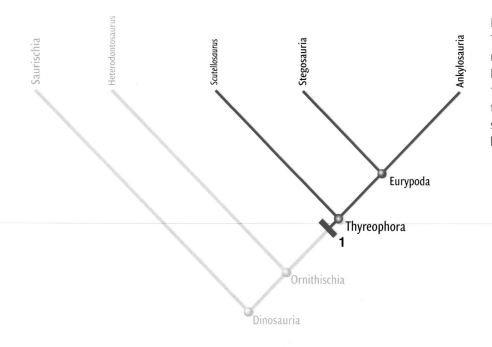

Figure 5.1. Cladogram of Thyreophora, emphasizing relationships within Ornithischia. Derived characters include: at **1**, transversely broad process of the jugal, parallel rows of keeled scutes on the back surface of the body.

Primitive Thyreophora

Primitive thyreophorans, outside of Ankylosauria and Eurypoda, are represented by three forms: *Scutellosaurus*, *Emausaurus*, and *Scelidosaurus* (Figure 5.2). Although primitive ornithischians in most respects, all have the diagnostic thyreophoran character of rows of osteoderms down the back and tail. Two were bipedal, reflecting the primitive condition in Dinosauria; *Scelidosaurus* was quadrupedal, foreshadowing the trend in the rest of Thyreophora.

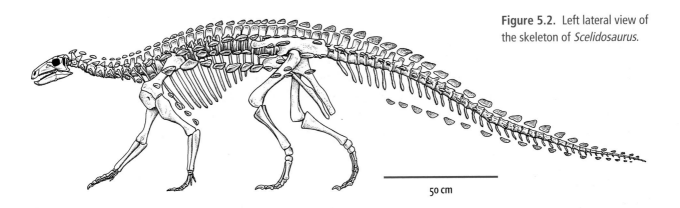

Figure 5.2. Left lateral view of the skeleton of *Scelidosaurus*.

50 cm

Eurypoda: Stegosauria – hot plates

Stegosaurs were medium-sized dinosaurs, 3–9 m in length and weighing 300–1500 kg, characterized by osteoderms that developed into spines and plates, as well as by their quadrupedal stance (Figure 5.3). Their profiles sloped strongly forward and downward toward the ground as a result of the hindlimbs being substantially longer than the forelimbs (Figure 5.4). All toes had broad hooves. They seem to have been relatively uncommon dinosaurs, yet clearly had a global distribution (Figure 5.5).

Figure 5.3. *Tuojiangosaurus*, a stegosaur from the Late Jurassic of Sichuan Province, China.

Figure 5.4. The best known of all plated dinosaurs, the North American *Stegosaurus* from the Late Jurassic.

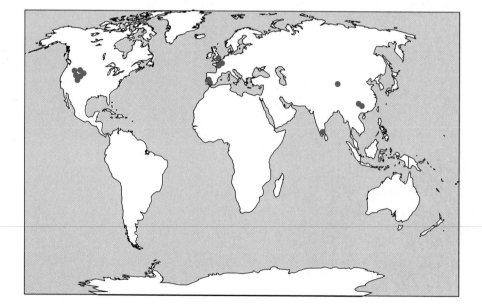

Figure 5.5. Global distribution of Stegosauria.

Stegosaurian lives and lifestyles

Locomotion The general body plan of stegosaurs does not suggest life in the fast lane (Figure 5.6). Indeed, with their long back legs and short front legs, stegosaurs must have had a locomotor conundrum: at the same cadence (the rate of feet hitting the ground), the hindlimbs would have covered much more distance than the forelimbs. At high speeds, therefore, the rear of the animal would have overtaken its head! This problem could be avoided in two ways: (1) by drawing the forelimbs up from the ground (that is, temporarily being bipedal while running) or (2) by limiting movement to a slow walking gait. Because of the mass distribution of stegosaurs,

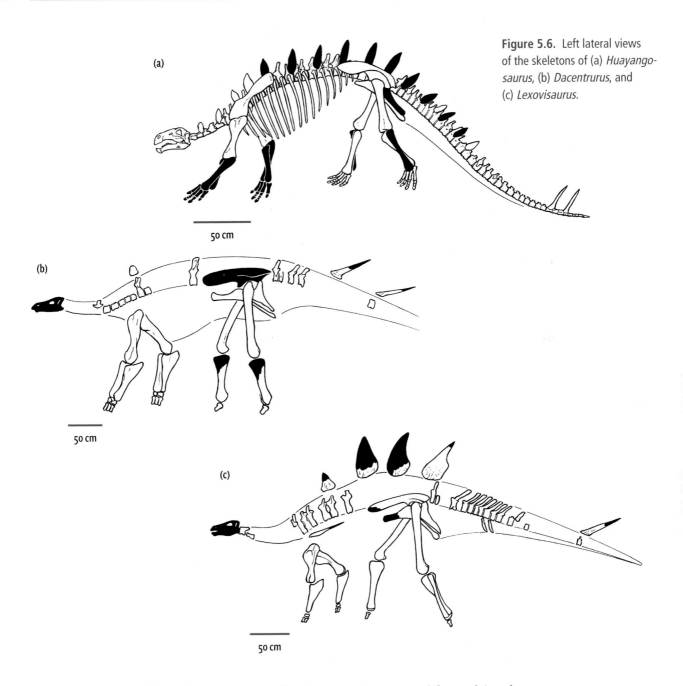

Figure 5.6. Left lateral views of the skeletons of (a) *Huayango-saurus*, (b) *Dacentrurus*, and (c) *Lexovisaurus*.

(a)

50 cm

(b)

50 cm

(c)

50 cm

the first option is unlikely. Our best guess is that the pace of stegosaur life was leisurely, on the order of 6.5 to 7.0 km/h maximum speed (see Box 12.3). Chasing fleet prey was not too important to a hungry herbivore.

Neither, it would seem, was growing up quickly. Paleontologists can gain insights about the speed with which a dinosaur grew from juvenile to adult by the type of bone tissue it produces (Chapter 12). Recent studies on stegosaur development, based upon the type of bone tissue they have, suggest that by the standards of most other dinosaur groups, stegosaurs generally took their time to grow, not getting too large too quickly. This has implications for how stegosaurs used their distinctive plates (see "Spines and plates" below).

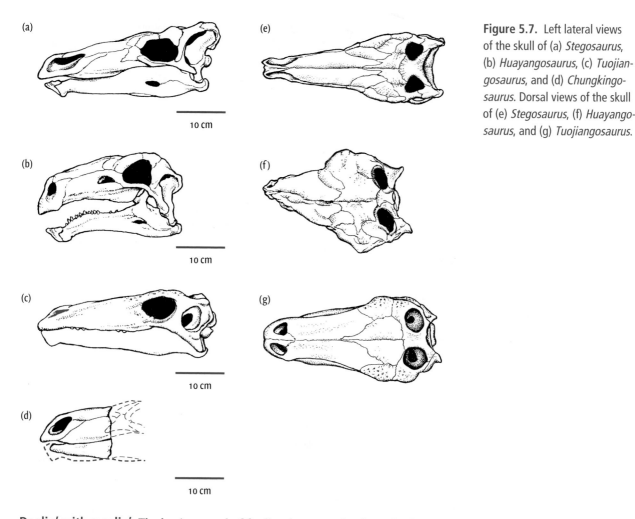

Figure 5.7. Left lateral views of the skull of (a) *Stegosaurus*, (b) *Huayangosaurus*, (c) *Tuojiangosaurus*, and (d) *Chungkingosaurus*. Dorsal views of the skull of (e) *Stegosaurus*, (f) *Huayangosaurus*, and (g) *Tuojiangosaurus*.

Dealin' with mealin' The business end of feeding began at the rhamphothecae, similar to those seen in modern turtles and birds, which covered the fronts of both the upper and lower jaws (Figure 5.7). The rhamphothecae were probably sharp-edged, and were used to crop and strip foliage.

Like all genasaurs, stegosaurs had an inset tooth row, implying cheeks, which in turn suggest chewing; however, exactly how that must have worked is baffling. The cheek teeth of stegosaurs were relatively small, simple, and triangular (Figure 5.8), and not tightly pressed together in a block for efficient grinding. Moreover, the teeth lack regularly placed, well-developed worn surfaces, features present in herbivores that chew by grinding. Furthermore, the coronoid process was low, lending little mechanical advantage to the jaw musculature. Chewing? Perhaps, but not particularly efficient when compared with other chewing vertebrates.

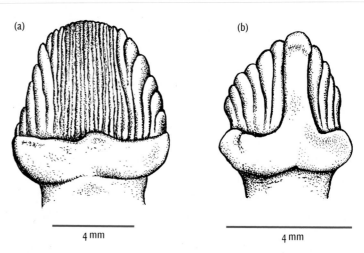

Figure 5.8. Inner views of an upper tooth of (a) *Stegosaurus* and (b) *Paranthodon*.

So how else might stegosaurs have ground their food? Birds use gastroliths (stones within the muscular part of the stomach; see Chapter 8) to grind food. The problem is that gastroliths have never been found with stegosaur remains, as they have with other dinosaurs (prosauropods, sauropods, psittacosaurs, and ornithomimosaurs). Ultimately, however, the co-existence in stegosaurs of simple, irregularly worn teeth, large gut capacity, cropping rhamphothecae, weak jaw musculature, and cheeks all conspire to make the business of dealin' with mealin' poorly understood in these dinosaurs.

What might stegosaurs have eaten? In most, the head was held near the 1 m level. Thus stegosaurs were likely low-browsers, consuming ground-level plants such as ferns, cycads, and other herbaceous gymnosperms (see Chapter 13).

The Mesozoic world of the low-browsers was not filled only with stegosaurs. It is very likely that stegosaurs competed with a variety of other dinosaurs, many of whom appear to have been more efficient chewers. Could stegosaurs have used their narrow skulls to select only the most nutritious parts of the plant, while everybody else dined less discriminately?

On the other hand, maybe stegosaurs weren't confined to low-browsing. Some paleontologists have argued that stegosaurs could have reared up on their hindlimbs in order to forage. Then the strong, flexible tail might have acted as a third "leg" to form a tripod. If so, these animals could have reached as high as 6 m in the largest forms.

No brains, one brain, or two brains? It is as clear as most anything can be at a distance of 100 million years that stegosaurs were just not all that bright. Their brains were an estimated 0.001% of the adult stegosaur body weight, putting them near the bottom of the dinosaur – for that matter, vertebrate – gray-matter scale (Figure 5.9). Brainy-ness must not have been

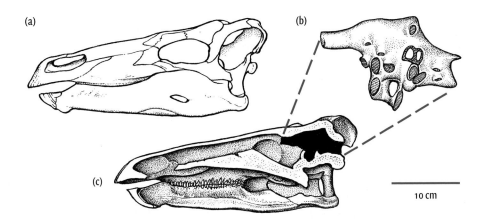

(a)

(b)

(c)

10 cm

Figure 5.9. Mold of the inside of the braincase of *Stegosaurus* and its silhouette imposed on the skull: (a) left profile; (b) brain endocast; (c) section through skull with braincase *in situ*.

part of the stegosaur life strategy, as indeed they were so small-brained that early workers felt compelled to assign them an extra brain: based upon an enlargement of the canal in the centra of the vertebrae (see inset to Figure 4.5) in the hip region, in which the spinal cord rests. Here began the legend of the dinosaur with two brains: a small one in the head and another in the pelvis, presumably to pick up the slack left by the first. All of this inspired literary effulgence, two samples of which we offer in Box 5.1.

The enlargement of the stegosaur spinal canal in the pelvic region, the putative rear "brain," is yet another stegosaur mystery. Many vertebrates have enlargements in the sacrum for nerves going to the hind legs, but the neural canal at the front of the stegosaur pelvis is

Box 5.1 **The poetry of dinosaurs**

Dinosaurs have been subjects of doggerel virtually since the time of their earliest discovery. Most have contrasted their enormity with their putative lack of brain power (and, no doubt, social graces), with few recent efforts to balance such dismal views.

The most famous dinosaurian poem celebrates the mental achievements of *Stegosaurus*, in particular the cerebral gymnastics powered by its double brains. The piece, by Bert L. Taylor, a columnist in the 1930s and 1940s for the *Chicago Tribune*, goes like this:

> Behold the mighty dinosaur,
> Famous in prehistoric lore,
> Not only for his power and strength
> But for his intellectual length.
> You will observe by these remains
> The creature had two sets of brains –
> One in his head (the usual place),
> The other at his spinal base.
> Thus he could reason a priori
> As well as a posteriori
> No problem bothered him a bit
> He made both head and tail of it.
> So wise was he, so wise and solemn,
> Each thought filled just a spinal column.

> If one brain found the pressure strong
> It passed a few ideas along.
> If something slipped his forward mind
> 'Twas rescued by the one behind.
> And if in error he was caught
> He had a saving afterthought.
> As he thought twice before he spoke
> He had no judgement to revoke.
> Thus he could think without congestion
> Upon both sides of every question.
> Oh, gaze upon this model beast,
> Defunct ten million years at least.

As a poetic counterpoint to the range of Mesozoic intelligentsia, we also provide some thoughts, entitled *The Danger of Being too Clever*, by John Maynard Smith, English evolutionary biologist *extraordinaire*.

> The Dinosaurs, or so we're told
> Were far too imbecile to hold
> Their own against mammalian brains;
> Today not one of them remains.
> There is another school of thought,
> Which says they suffered from a sort
> Of constipation from the loss
> Of adequate supplies of moss.

upward of 20 times the volume of the brain. Some living birds have a similar enlargement that houses an organ whose function is thought to supply glycogen (a complex sugar-based molecule which the body stores, but can break down to obtain energy) to the nervous system. Could the enlargement in the stegosaur sacrum have housed a glycogen body?

Speculation aside, the two stegosaurs in which the brain cavities are known – *Kentrosaurus* and *Stegosaurus* – suggest that stegosaur brains were relatively long, slightly flexed, and *small* (Figure 5.9; Box 5.2). Only the olfactory bulbs, the portions of the brain that provide the animal with its sense of smell, are somewhat enlarged. Clearly stegosaurs, animals that had an unhurried lifestyle and possibly a relatively uncomplicated range of behaviors, would have had time to stop and smell the roses ... had there *been* any roses![1]

Social lives of the enigmatic We don't have much of an idea about the social behavior of stegosaurs or know much about their life histories. No nests, isolated eggs, eggshell fragments, or hatchling material is yet known for any stegosaur. In fact, only a few juvenile and adolescent stegosaur specimens can tell us anything about the lives of subadult stegosaurs.

---

[1] Roses didn't appear until long after the last stegosaur went extinct – see Chapter 13.

But Science now can put before us
The reason true why Brontosaurus
Became extinct. In the Cretaceous
A beast incredibly sagacious
Lived & loved & ate his fill;
Long were his legs, & sharp his bill,
Cunning its hands, to steal the eggs
Of beasts as clumsy in the legs
As Proto- & Triceratops
And run, like gangster from the cops,

To some safe vantage-point from which
It could enjoy its plunder rich.
Cleverer far than any fox
Or Stanley in the witness box
It was a VERY GREAT SUCCESS.
No egg was safe from it unless
Retained within its mother's womb
And so the reptiles met their doom.

The Dinosaurs were most put out
And bitterly complained about
The way their eggs, of giant size,
Were eaten up before their eyes,
Before they had a chance to hatch,
By a beast they couldn't catch.

This awful carnage could not last;
The age of Archosaurs was past.
They went as broody as a hen
When all their eggs were pinched by men.
Older they grew, and sadder yet,
But still no offspring could they get.
Until at last the fearful time, as
Yet unguessed by Struthiomimus
Arrived, when no more eggs were laid,
And then at last he was afraid.
He could not learn to climb with ease
To reach the birds' nests in the trees,
And though he followed round and round
Some funny furry things he found,
They never laid an egg – not once.
It made him feel an awful dunce.
So, thin beyond all recognition,
He died at last of inanition.

MORAL
This story has a simple moral
With which the wise will hardly quarrel;
Remember that it scarcely ever
Pays to be too bloody clever.

Among fully adult individuals, some workers have proposed that there is some sexual dimorphism; that is, differences between the sexes. It has been claimed that this shows up in, of all places, the number of ribs that contribute to the formation of the pelvis. If so, whether it is the male or the female that has the greater number of ribs is anybody's guess. Sex-based differences in the size and shape of the spines and/or plates might be predicted if only we had larger and better samples.

Little is also known about the degree of sociality among stegosaurs. The mass accumulation of disarticulated, yet associated, *Kentrosaurus* material from Tendaguru in Tanzania (see Chapter 14) provides us with a hint that *Kentrosaurus* was gregarious (exhibited herding and other social behaviors). In other genera, however, we have no such information. The fossil record is simply silent on this issue – so far.

Spines and plates Whether or not stegosaurs were gregarious, there are still some features of these animals that give clues about behavior: the spines and plates. As we have learned, the majority of stegosaurs had at least one row of osteoderms along the dorsal margin of each side

Box 5.2 **Dino brains**

Brains, of course, are soft anatomy, and so finding one preserved is next to impossible. Yet, we have good evidence for how brains looked in many dinosaurs and, considering areas of enlargement, we can suggest that a particular dinosaur had a well-developed sense of smell, or perhaps good vision. How can this be?

If the bones of the braincase were well preserved, paleontologists initially obtained brain **endocasts**, three-dimensional models of the brain. This was not so easy: all of the interior rock needed to be prepared out of the braincase, a process that generally involved breaking the braincase, and then the thing in effect reassembled, so that its inside could be cast. Reassembly took a *very* skilled preparator. Casting was accomplished by painting several layers of latex on the interior of the braincase, and then, once it had dried, pulling the flexible material through the foramen magnum (the opening through which the spinal cord contacts the brain) if the braincase was intact; peeling it off of the exposed interior if it was not. Latex cast in hand, a mold could then be prepared that faithfully recorded the contours of the inside of the braincase. From this mold were cast the brain endocasts; in plaster, or perhaps resin. Figure 5.9b shows such a brain endocast for *Stegosaurus*.

Very much more recently, paleontologists have begun to use computer-aided tomography (CT) scanning techniques to image the brain. As in humans, the technique is non-invasive, and responds to density differences between the fossil bone and the rock that encases it. Preparation is therefore not necessary. As in medical imaging, the machine obtains a

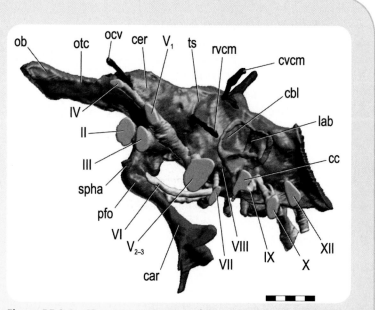

Figure B5.2.1. CT scan reconstruction of the brain of the ceratopsian genus *Pachyrhinosaurus* (see Figure 6.20f). Scale bar = 4 cm. For abbreviations see Figure 9.20 on page 202.

sequence of two-dimensional "slices" – images, in fact – which can then be assembled into a computer-generated composite image, and be rotated and visualized from any angle. The advantages of this are multiple: aside from obviating the need for preparation – which, as we have seen, is costly, time-consuming, and potentially destructive – the CT scan provides as much detail as is preserved – the equal of, or better than, the very finest preparation jobs. For example, the CT-scanned image presented here (Figure B5.2.1) shows details of the cranial nerves (identified by Roman numerals) unavailable by conventional means.

of the body. And these osteoderms generally take the form of spines, spikes (Figure 5.10), blunt cones, or plates. In all cases, at the end of the tail were pairs of long spines. All of these, like all osteoderms, were embedded in the skin (Figure 5.11). What purpose might they have served? Originally, the idea was that they were all about protection and defense. But defense, if it is any part of the story, isn't the whole story.

The shapes and patterns of plates and spines in stegosaurs are nearly always species specific; that is, diagnostic for a particular species (in this case stegosaurs). Moreover, they have their maximum visual effect when viewed from the side. The osteoderms, therefore, might have served a display function – both for predators *and* for other stegosaurs. If intraspecific display (display among members of a species) was involved, it is likely that individual stegosaurs would have used these structures not only to tell each other apart, but also to gain dominance in territorial disputes and/or as libido-enhancers during the breeding season. But that's yet another question we can't answer: the stegosaur fossil record just isn't rich enough.

Figure 5.10. The skeleton of *Kentrosaurus*, a spiny stegosaur from the Late Jurassic of Tanzania.

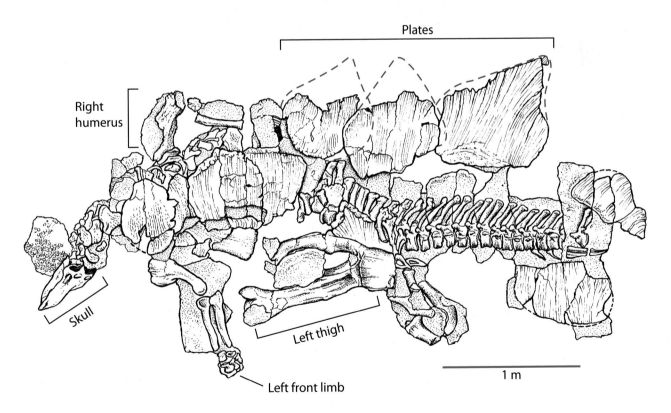

Figure 5.11. Diagram of one of the best skeletons of *Stegosaurus* as it was found in the field. Note that the plates do not articulate directly with the vertebrae.

Hot plates The surfaces of the plates of *Stegosaurus* are covered with an extensive pattern of grooves, while the insides are filled with a honeycomb of channels (Figure 5.12). These external grooves and internal channels most likely formed the bony walls for an elaborate network of blood vessels. With such a rich supply of blood from adjacent regions of the body, could the plates have been used to cool the body by dissipating heat as air passed over them, or to warm the body by absorbing solar energy? In short, could the plates have been used for thermoregulation (temperature control)? As a test of this idea, paleontologist J. O. Farlow and colleagues

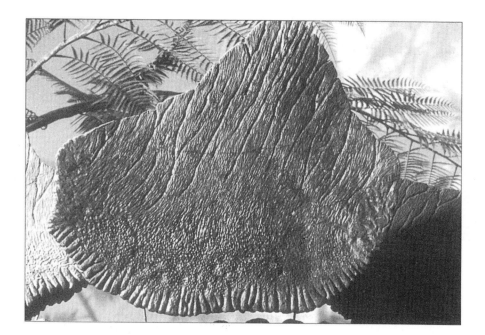

Figure 5.12. Lateral view of one of the dermal plates of *Stegosaurus*. Note the great number of parallel grooves, presumably conveying blood vessels across the outer surface of the plate.

tested the ability of the plates to radiate (or absorb) heat. Arranged in symmetrical pairs, as they are in life, the plates provided significant heat dissipation, suggesting that thermoregulation may also have been a role of the plates.

The slow growth rates that have been hypothesized for stegosaurs may not completely support the idea that the plates were entirely for thermoregulation. If stegosaurs took a long time to mature, as has been hypothesized, then how did the juveniles defend themselves? Size, then as now, can be an important defense mechanism: if you grow big, they can't touch you. But because stegosaurs are thought to have taken a longer time to mature than most other dinosaurs, size did not come into play as quickly as it did for other dinosaurs. Therefore, some paleontologists have argued that the plates were necessary for defense – particularly in a smaller-sized animal.

Unfortunately for this line of reasoning, the few juvenile stegosaurs known appear not to have had large spines or plates on their backs. The absence of these features in small, sexually immature individuals suggests that only when maturity was reached did looking big and sexy acquire importance. Likewise, thermoregulation may have only been important to adult stegosaurs; ironically, as we shall see (Chapter 12), the increased metabolic rates generally associated with thermoregulation – e.g., "warm-bloodedness" – are commonly most important to fast-growing juveniles. Interpretations of stegosaurs are never obvious!

Or mostly so: the role of the long, pointed tail spikes is pretty obvious! These were likely slashed from side to side on the powerful tail. While older reconstructions show these spikes as pointing primarily upward, recent discoveries suggest that the spikes actually splayed out to the sides, producing a much more effective defensive weapon.

So stegosaurs appear as a mass of contradictions: chewers that may not have chewed that well; thermoregulators that appear to have moved slowly and lacked the sophisticated neural controls usually associated with thermoregulation; and animals in which supposed sexual display functions were prominent but in which there is little evidence for gregarious behavior. Until very recently, we didn't even know the positions of the plates in *Stegosaurus*, the best-known genus. So much of what is apparently contradictory about stegosaurs is likely due to how little we know about them – a situation that we'd like to see changed!

Eurypoda: Ankylosauria – mass and gas

Ankylosaurs were masters of the art of defense-by-hunkering. As their name implies, ankylosaurs were encased in a pavement of bony plates and spines – each embedded in skin and interlocked with adjacent plates – that formed a continuous shield across the neck, throat, back, and tail (Figures 5.13, 5.14, and 5.15). In many cases, it covered the top of the head, cheeks, and even eyelids.

Under all that armor, ankylosaurs were round and very broad (see Figure 5.13); clearly their girth accommodated a large gut. The head was low and broad (Figure 5.16), and equipped with simple, leaf-shaped teeth for pulverizing whichever plants an ankylosaur chose to eat (Figure 5.17).

Among the most striking features of ankylosaurs are the suite of complex nasal passageways. These have been revealed by CT scans (see Box 5.2) of the skulls. These passageways were elaborate, and in some cases, convolute (see below, Figure 5.19) Their precise function is

Figure 5.13. *Euoplocephalus*, the armored, club-tailed ankylosaur.

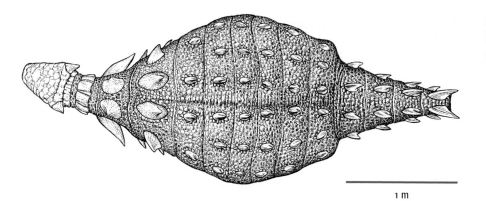

Figure 5.14. Dorsal view of the body armor of *Euoplocephalus;* reconstruction.

1 m

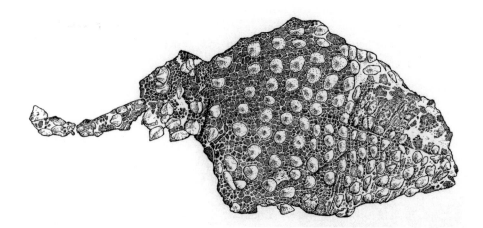

Figure 5.15. Dorsal view of the body armor of *Sauropelta*.

(a)

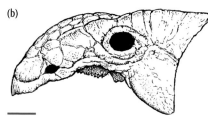

10 cm

(d)

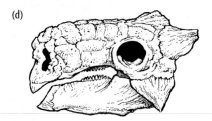

10 cm

(b)

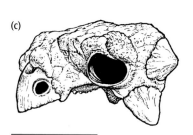

10 cm

(e)

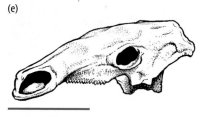

10 cm

(c)

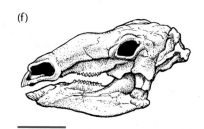

10 cm

(f)

10 cm

Figure 5.16. Left lateral view of the skulls of (a) *Shamosaurus,* (b) *Ankylosaurus,* (c) *Pinacosaurus,* (d) *Tarchia,* (e) *Silvisaurus,* and (f) *Panoplosaurus.*

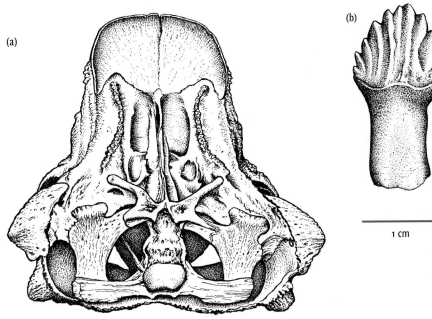

(a)

(b)

Figure 5.17. Palatal view of the skull of (a) *Euoplocephalus* and (b) a tooth of *Edmontonia*.

1 cm

10 cm

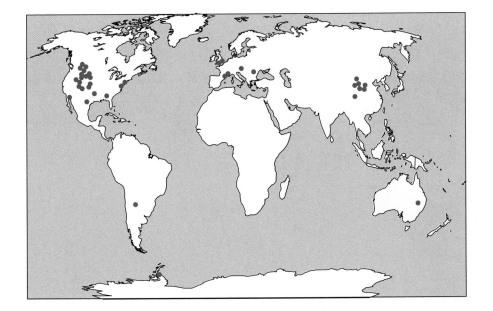

Figure 5.18. Global distribution of Ankylosauria.

uncertain, although they suggest that olfaction, the sense of smell, was an important sense in this group of dinosaurs.

Ankylosaurs were mid-sized dinosaurs, rarely exceeding 5 m in length, although some (such as *Ankylosaurus*) ranged upward of 9 m. As in stegosaurs, the limbs were of different lengths, with the hindlimb exceeding the length of the forelimb by as much as 150%.

Like stegosaurs, ankylosaurs also had a global distribution, coming predominantly from North America and Asia, but also from Europe, Australia, South America, and Antarctica (Figure 5.18). As a group, they reached their peak diversity during the Cretaceous.

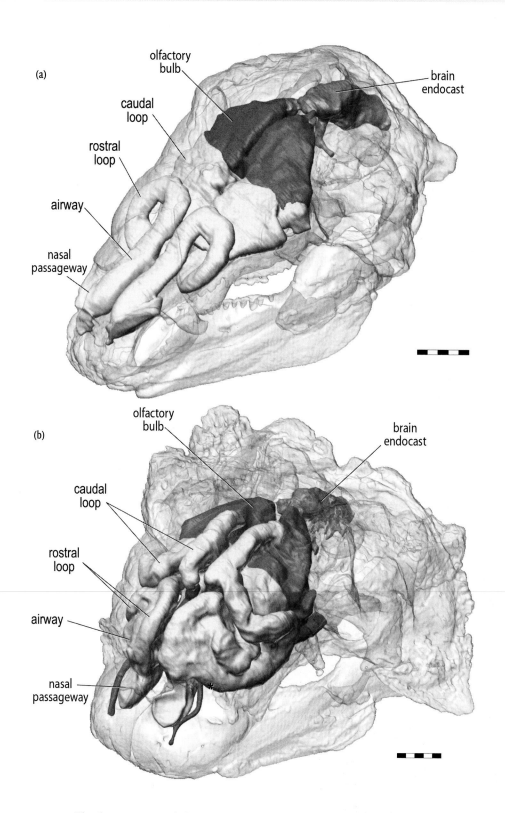

(a)

olfactory bulb

caudal loop

rostral loop

airway

nasal passageway

brain endocast

(b)

olfactory bulb

caudal loop

rostral loop

airway

nasal passageway

brain endocast

Figure 5.19. The skulls of nodosaurs (above) and ankylosaurs (below) compared. The skull of the ankylosaur is broader, and was held at a more downward angle than that of the nodosaur. Three-quarter view CT scan reconstructions of (a) the nodosaur *Panoplosaurus* and (b) the ankylosaur *Euoplocephalus*. In both cases, convolute nasal sinuses are revealed, running from the external nasal opening towards the braincase. These sinuses doubtless were involved in a heightened sense of smell; we don't know if ankylosaurs smelled good, but the evidence suggests that they smelled well! The small, darkened region to the rear of the skull is the brain. Scale bars = 5cm; for abbreviations see Figure 9.20 on page 202.

The best-preserved fossils come from Mongolia and China, where stunning specimens have been found inland, far from the ocean (either then or now), nearly complete and articulated, and in most cases preserved upright or on their sides. In contrast, in North America, only partial skeletons have been found, and these are often upside-down, sometimes in rocks deposited along the sea shore or even in rocks representing open marine environments. The North American forms evidently lived sufficiently close to the sea that their bloated carcasses might have been carried out with the tide, flipping upside-down because of their heavily armored backs.

Ankylosaur finds most commonly consist of individual skeletons or isolated partial remains; there is only one ankylosaur bonebed known. Perhaps this indicates that these animals had solitary habits or lived in very small groups. Even from our incomplete window on the past, it appears reasonably certain that ankylosaurs did not enjoy the company of huge herds.

Ankylosauria consists of two great clades: Nodosauridae (*nodo* – knot; referring to the rounded osteoderms) and Ankylosauridae (Figures 5.20 and 5.21).

Nodosauridae Nodosaurids had relatively long snouts and well-muscled shoulders, reflected by the presence of a large knob of bone on the shoulder blade, the acromial process (Figure 5.21), an attachment site for the heavy shoulder musculature that characterizes nodosaurs (Figure 5.21). Nodosaurs also had flaring hips, and pillar-like limbs. Many had tall spines at the shoulder (parascapular spines). Nodosaurids are known principally from the Northern Hemisphere (North America and Europe), although new discoveries in Australia and Antarctica have extended the geographical range of these animals deep into the Southern Hemisphere.

Ankylosauridae Members of Ankylosauridae give the impression of impregnable, mobile fortresses. All are well armored (see Figure 5.14), but there are fewer tall spines along the body than in nodosaurids. The tail ends in a massive bony club, in some instances with several paired knobs or triangular spikes along its length. The head is shorter and broader in ankylosaurids than in nodosaurids and there are large triangular plates attached to the rear corners of the skull (termed "squamosal horns").

(a)

(b)

Figure 5.20. (a) An ankylosaur (*Euoplocephalus*) and (b) a nodosaur (*Sauropelta*) compared. The ankylosaur is stockier, and has a tail-club. The nodosaur is more lightly built, and bears distinctive dermal armor.

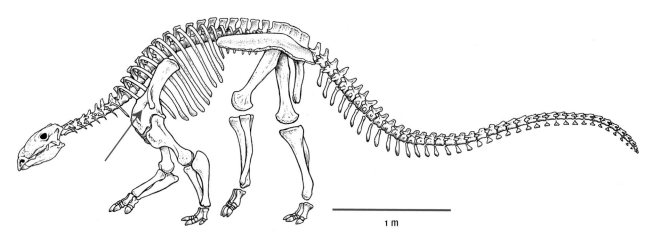

1 m

Figure 5.21. Left lateral view of the skeleton of *Sauropelta* without its armor shield. Note the projection on the scapula (indicated by arrow), known as the acromial process.

Ankylosaur lives and lifestyles

Mouths to feed Ankylosaurs doubtless had a very low browsing range, foraging for plants no more than a meter or so above the ground. The different beak shapes suggest different feeding preferences. The narrow beak of nodosaurids may suggest selective feeding, plucking or biting at particular kinds of foliage and fruits with the sharp edge of the rhamphotheca. In contrast, the very broad beak of ankylosaurids may imply less selective feeding, in which plant parts were indiscriminately bitten off from the bush or pulled from the ground.

How the food was then prepared for swallowing is a bit of a mystery. Like stegosaurs (and pachycephalosaurs; see Chapter 6), the triangular teeth of both nodosaurids and ankylosaurids are small, not particularly elaborate, and less tightly packed than animals with well-developed chewing behaviors (see Figure 5.17). However, the wear marks on the teeth indicate that grinding took place. In addition, it is likely that ankylosaurs had a long, flexible tongue (in their throats, they have large hyoid bones that support the base of the tongue) and an extensive secondary palate, which allowed them to chew and breathe at the same time. Moreover, deeply inset tooth rows suggest well-developed cheeks to keep whatever food was being chewed from falling out of the mouth. The jaw bones themselves were relatively large and strong (although lacking enlarged areas for muscle attachment). Indeed, most ankylosaur jaw features – except for tooth design and placement – suggest that ankylosaurs were reasonably adept chewers.

Perhaps the paradox of simple teeth in strong, cheek-bound jaws can be understood by looking not at how much chewing was done prior to swallowing (there obviously was some), but at the very substantial rear end of the animal, where the bones show a very deep rib cage circumscribing an enormously expanded abdominal region (see Figure 5.13). Commodious abdomens mean huge guts, and huge guts suggest that that digestion took place in a very large, perhaps highly differentiated, fermentation compartment(s) in these armored dinosaurs. Bacteria likely lived symbiotically within the stomach (or stomachs, because ankylosaurs may well have had a series of them). The stomach(s) must have served as a great fermentation vat(s), bacterially decomposing even the toughest woody plant material. Among modern mammals, this method of breaking down tough plant material is well known in ruminants such as cows.

The combination of browsing at low levels and having anatomy indicative of chewing and fermenting places limits on what kinds of plants ankylosaurs may have fed on. At the levels where ankylosaurs could forage, the undergrowth consisted of low-stature ferns, gymnosperms such as cycads, and shrubby angiosperms (flowering plants), a rich array of plants to choose from within the first few meters above the ground (see Chapter 13).

Brains and senses Ankylosaurs may not have been particularly adept at making foraging choices, however, because their brain power was close to the bottom of the dinosaur range (only sauropods had smaller brains for their size; see Chapter 8 and Box 12.5). Still, had there been roses (see footnote 1, this chapter), ankylosaurs, too, might have stopped to smell them; the skulls were equipped with enlarged, convolute nasal passages (Figure 5.19), suggesting that olfaction may have been an important sense for these dinosaurs. Hand in hand with slow thinking, ankylosaurs were among the slowest moving of all dinosaurs for their body weight. Estimates suggest that they were able to run no faster than 10 km/h and walked at a considerably more leisurely pace (about 3 km/h). Ankylosaurs were built for digestion, not for speed.

Defense Although ankylosaurs were slow on their feet, other aspects of the limb skeleton suggest that they could actively defend themselves against predators. Firstly, in all ankylosaurs the entire upper surface of the body – the head, neck, torso, and tail – was covered by a pavement of bony plates (see Figures 5.14 and 5.15).

In the case of nodosaurids, the shoulder region was exceedingly powerful, with tall spines. This powerful, well-defended front end, in conjunction with relatively long hindlimbs and wide stance, may have had the effect of dropping the center of gravity of the animal forward, providing nodosaurids with an aggressive frontal, spine-based defense.

Ankylosaurids went even further, augmenting their armor by a powerful tail-club (Figure 5.22a), constructed of paired masses of bone set at the end of a tail (itself about half the length of the body). In some cases the tail was also equipped with spikes along its length (Figure 5.22b).

Thanks to CT scans and careful muscle reconstructions, we now know a lot about the design of the ankylosaurid tail-club. The paired masses of bone varied in density, with the densest bone located outermost (Figure 5.23). While the tail was flexible at its base, the rear half was stiffened by modified, interlocking vertebrae, as well as by a series of tendons that had become ossified – turned to bone, and thus inflexible – running down its length. This part of

(a)   (b)

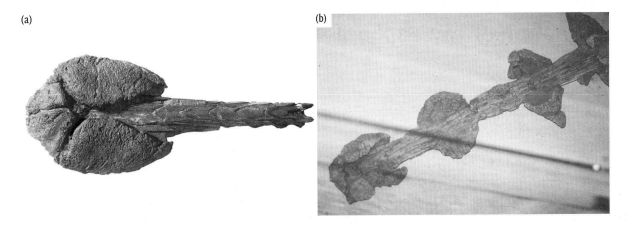

Figure 5.22. (a) The bony tail-club of *Euoplocephalus.* (b) Tail spikes in the tail of *Pinacosaurus.*

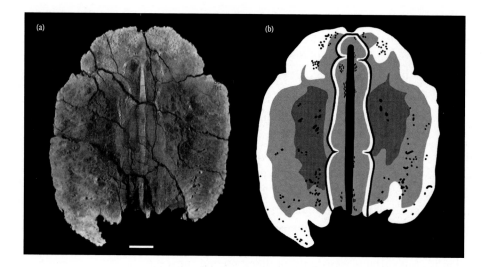

Figure 5.23. (a) CT scan image through an ankylosaur tail-club. (b) Interpretation of (a), showing densest bone (white), bone of medium density (light gray), and bone of lowest density (dark gray). Scale bar equals 5 cm.

the tail was not meant to bend, and it provided firm attachment for the powerful muscles. Using the flexibility of the tail vertebrae at the base and the stiffened tail farther out, the club could have been forcefully swung side to side.

Just how forcefully? Paleontologist Victoria Arbour theoretically calculated the velocity and thus the power of the swing; her estimates of the velocity, depending upon the type of muscle reconstruction used, are as high as 18.9 m/sec, producing bone crunching impulses of 376 kgm/sec. The idea, clearly, would be to avoid any such contact with an ankylosaur tail-club in the first place!

For nodosaurids, a proactive defense must have been a head-first (or shoulder-first) proposition, keeping the parascapular spines pointed toward the predator. Ankylosaurids, on the other hand, likely first planted their hindlimbs and then rotated their forequarters with their strong forelimb muscles, ever keeping watch on the threatening opponent. Then they would have wielded that massive club at their opponent's legs and feet.

In both ankylosaurids and nodosaurids, however, the last resort must have been to hunker down defensively. With its legs folded under its body, a 3,500 kg ankylosaur would have been formidably immobile. Safe under protective armor, both nodosaurids and ankylosaurids were the best-defended fortresses of the Mesozoic.

The evolution of Thyreophora

Thyreophora

We can be reasonably confident that the evolution of thyreophorans embodied increasing size and a return to quadrupedality, because the cladograms show that the quadrupedal eurypodan clades were all derived relative to these primitive forms. The primitive thyreophorans suggest an evolutionary sequence from gracile, small, bipedal creatures like *Scutellosaurus* to larger, quadrupedal dinosaurs like *Scelidosaurus* (see Figures 5.2 and II.5).

Eurypoda

It is not difficult to imagine the evolution of an ankylosaur from an armored, primitive quadruped like *Scelidosaurus*. The cladogram suggests that the basal eurypodan must have looked something like *Scelidosaurus*, and the step to a larger, more powerful, more heavily armored primitive ankylosaur or stegosaur is easy to conceive.

Stegosauria

Stegosauria is a monophyletic clade of ornithischian dinosaurs, diagnosed on the basis of a number of important features shown in Figure 5.24. The ancestral stegosaur must have been an animal with spine-shaped osteoderms and fore- and hindlimbs of not too dissimilar lengths. Within Stegosauria, the basal split is between *Huayangosaurus* on the one hand and remaining species on the other (Figure 5.25). This divergence took place sometime before the latter half of the Middle Jurassic. *Huayangosaurus* itself has a number of uniquely derived features shared by a more inclusive group of stegosaurs, Stegosauridae (Figure 5.25).

Within Stegosauridae, *Dacentrurus* represents the most basal form. The remainder of Stegosauridae includes *Stegosaurus*, *Wuerhosaurus*, *Kentrosaurus*, and *Tuojiangosaurus*. The evolution of this group was evidently characterized by an increase in the difference in length between the fore- and hindlimbs (Figure 5.25).

Finally, there is *Stegosaurus* itself, the best-known, most common stegosaur. *Stegosaurus* must have evolved its distinctive plates from the spiny, conical osteoderms present in its ancestry. Plates, however, are only known in *Stegosaurus*, and their evolution occurred sometime during the Middle or early Late Jurassic.

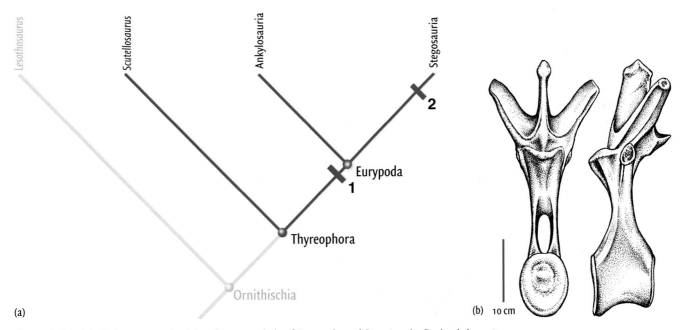

(a)

(b) 10 cm

Figure 5.24. (a) Cladogram emphasizing the monophyly of Eurypoda and Stegosauria. Derived characters include: at **1** (Eurypoda), bones that fuse to the margins of the eye sockets, loss of a notch between the quadrate (see Figure 4.6) and the back of the skull, and enlargement of the anterior part of the ilium; at **2** (Stegosauria), back vertebrae with very tall neural arches and highly angled transverse processes, loss of ossified tendons down the back and tail, a broad and plate-like acromial process, large and block-like wrist bones, elongation of the prepubic process, loss of the first pedal digit, and loss of one of the phalanges of the second pedal digit, and a great number of features relating to the development of osteoderms, and formation of long spines on plates from the shoulder toward the tip of the tail. (b) The front and left lateral view of one of the back vertebrae of *Stegosaurus*. Note the great height of the neural arch; a diagnostic stegosaurian character at **2**.

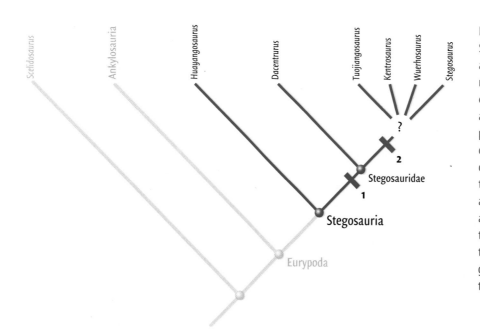

Figure 5.25. Cladogram of Stegosauria, with Ankylosauria and *Scelidosaurus* as successively more distant relatives. Derived characters include: at **1**, large antitrochanter, long prepubic process, long femur, absence of lateral rows of osteoderms on the trunk; at **2**, widening of the lower end of the humerus, an increase in femoral length, and an increase in the height of the neural arch of the back and tail vertebrae. Relationships of genera of the ultimate node on the cladogram remain uncertain.

Ankylosauria

Reflecting the importance of heavy armor to ankylosaurs and the ease of its preservation, it is not surprising that armor and/or its support comprise the majority of derived features uniting the clade Ankylosauria (Figure 5.26). Ankylosaur evolution followed two principal pathways since the origin of the group sometime in the Jurassic: Ankylosauridae and Nodosauridae.

Primitively in all ankylosaurs, the beak was scoop-shaped but relatively narrow (although slightly broader than in stegosaurs) and remained so in the nodosaurids and in the ankylosaurid *Shamosaurus*. In all other ankylosaurids, by contrast, the beak became very broad, which matched the general broadening of the animal and the development of tail-clubs.

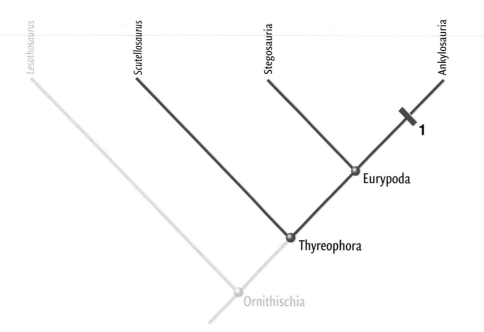

Figure 5.26. Cladogram of Eurypoda, emphasizing the monophyly of Ankylosauria. Derived characters include: at **1**, closure of antorbital and upper temporal fenestrae, ossification and fusion of keeled plate onto side of lower jaw, fusion of first tail vertebrae to sacral vertebrae and ilium, rotation of ilium to form flaring blades, closure of hip joint, development of dorsal shield of symmetrically placed bony plates and spines.

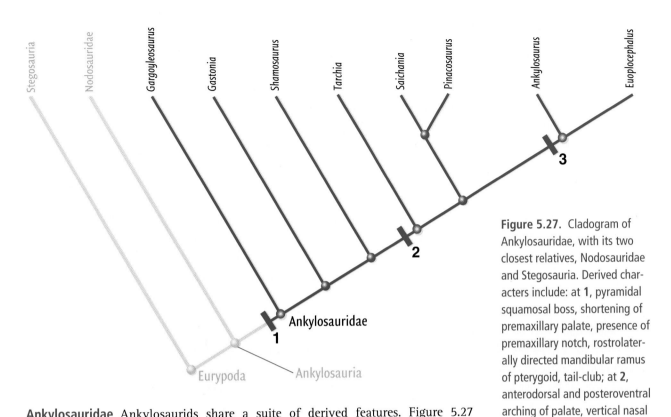

Figure 5.27. Cladogram of Ankylosauridae, with its two closest relatives, Nodosauridae and Stegosauria. Derived characters include: at **1**, pyramidal squamosal boss, shortening of premaxillary palate, presence of premaxillary notch, rostrolaterally directed mandibular ramus of pterygoid, tail-club; at **2**, anterodorsal and posteroventral arching of palate, vertical nasal septum, rugose and crested basal tubera; at **3**, caudal end of postaxial cervical vertebrae dorsal to cranial end, fusion of sternals.

Ankylosauridae Ankylosaurids share a suite of derived features. Figure 5.27 highlights the relationships among ankylosaur genera, as well as some of the key characters supporting these relationships.

Nodosauridae Turning to the other great clade of ankylosaurs, nodosaurids share a number of derived features, as shown in Figure 5.28, particularly the well-developed acromial process. This musculature, and the parascapular spines that accompanied it, may have played a role in their defensive behavior. Nodosaurids changed little during their tenure on Earth; however, various diagnostic characters allow us to learn something of their relationships (Figure 5.28).

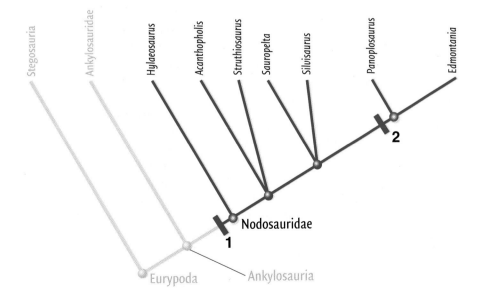

Figure 5.28. Cladogram of Nodosauridae, with its two closest relatives, Ankylosauridae and Stegosauria. Derived characters include: at **1**, knob-like acromion on scapula, occipital condyle derived from but one bone (the basioccipital), large knob (supraorbital boss) above the eye; at **2**, premaxillary teeth lacking, distinct rostro-dorsal feature of secondary palate.

Summary

Thyreophorans were small- to medium-sized quadrupedal ornithischians with rows of scutes down the back. Aside from some primitive forms, the two great groups of thyreophorans were Stegosauria and Ankylosauria, linked together within a monophyletic Eurypoda. Eurypodans, as a group, were not renowned for their intellects; some of the lowest brain:weight ratios known for dinosaurs come from Eurypoda.

Stegosaurs are relatively poorly known eurypodans with hindlimbs significantly longer than the forelimbs, and paired rows of plates or spines down the back, terminating in a tail with elongate spines: likely a defensive weapon. Their fossil record extends from the Middle Jurassic to the Early Cretaceous.

Stegosaurs are rare finds, and they likely functioned in isolation rather than gregariously. The plates in *Stegosaurus* may have been involved in thermoregulation. Stegosaurs – like all eurypodans – had cheeks, which suggest chewing. The teeth, however, occluded relatively poorly, suggesting inefficient grinding. Due in part to their rarity, the reproductive strategies and behavior of stegosaurs are still largely unknown.

Ankylosaurs were armored tank-like quadrupeds, coated with a pavement of osteoderms, and bony protection everywhere. They lived from the mid-Jurassic to latest Cretaceous. Two groups are known: Nodosauridae and Ankylosauridae. Nodosaurids were slightly more lightly built, with tall parascapular spines, while ankylosaurids were more tank-like, and equipped with a large tail-club.

Ankylosaurs were low-browsing animals. Their teeth and cheeks were rather like those of stegosaurs, but ankylosaurs had secondary palates, which may have aided in the efficiency of mastication. Unquestionably, though, much of their energy was obtained through gut fermentation, as suggested by the striking breadth of their girths. The presence of bonebeds suggests that, unlike stegosaurs, ankylosaurs may have been gregarious animals. Little is known of their reproductive strategies and, by morphology, they were evidently animals that relied heavily upon defense, either by simply hunkering down or, in the case of ankylosaurids, by wielding their tail-clubs.

Selected readings

Buffrenil, V. de, Farlow, J. O. and **de Ricqlès, A.** 1986. Growth and function of *Stegosaurus* plates: evidence from bone histology. *Paleobiology*, **12**, 459–473.

Carpenter, K. (ed.). 2001. *The Armored Dinosaurs*. Indiana University Press, Bloomington, IN, 512 pp.

Coombs, W. P., Jr. and **Maryańska, T.** 1990. Ankylosauria. In Weishampel, D. B., Dodson, P. and Osmólska, H. (eds.), *The Dinosauria*. University of California Press, Berkeley, pp. 456–483.

Coombs, W. P., Weishampel, D. B. and **Witmer, L. M.** 1990. Basal Thyreophora. In Weishampel, D. B., Dodson, P. and Osmólska, H. (eds.), *The Dinosauria*. University of California Press, Berkeley, pp. 427–434.

Galton, P. M. and **Upchurch, P. A.** 2004. Stegosauria. In Weishampel, D. B., Dodson, P. and Osmólska, H. (eds.), *The Dinosauria*, 2nd edn. University of California Press, Berkeley, pp. 343–362.

Giffin, E. B. 1991. Endosacral enlargements in dinosaurs. *Modern Geology*, **16**, 101–112.

Hopson, J. A. 1980. Relative brain size in dinosaurs – implications for dinosaurian endothermy. In Thomas, R. D. K. and Olson, E. C. (eds.), *A Cold Look at the Warm-Blooded Dinosaurs*. AAAS Selected Symposium no. 28, pp. 278–310.

Jerison, H. J. 1973. *Evolution of the Brain and Intelligence*. Academic Press, New York, 482pp.

Maryańska, T. 1977. Ankylosauridae (Dinosauria) from Mongolia. *Palaeontologia Polonica*, **37**, 85–151.

Norman, D. B., Witmer, L. M. and Weishampel, D. B. 2004. Basal Thyreophora. In Weishampel, D. B., Dodson, P. and Osmólska, H. (eds.), *The Dinosauria*, 2nd edn. University of California Press, Berkeley, pp. 335–342.

Sereno, P. C. and Dong, Z.-M. 1992. The skull of the basal stegosaur *Huayangosaurus taibaii* and a cladistic analysis of Stegosauria. *Journal of Vertebrate Paleontology*, **12**, 318–343.

Vicaryous, M., Maryańska, T. and Weishampel, D. B. 2004. Ankylosauria. In Weishampel, D. B., Dodson, P. and Osmólska, H. (eds.), *The Dinosauria*, 2nd edn. University of California Press, Berkeley, pp. 363–392.

Topic questions

1   What are the diagnostic characters of Thyreophora? How are thyreophorans related to other ornithischians?

2   What are the diagnostic characters of Stegosauria?

3   What are the diagnostic characters of Ankylosauria?

4   How do ankylosaurids differ from nodosaurids?

5   Describe the evidence for there being two brains in *Stegosaurus*. What is the prevailing view on this issue now?

6   Did ankylosaurs defend themselves passively, or could they actively defend themselves?

7   What are the apparent contradictions in the mouths of stegosaurs? That is, what features appear to be well designed for chewing and what features appear not to be so well designed for chewing?

8   Do ankylosaurs have the same apparent contradictions in their chewing mechanisms as stegosaurs? Elaborate.

9   What might the plates of *Stegosaurus* have been used for? Would this use apply to all stegosaurs? Why?

10   Do the small brains in stegosaurs appear to have hindered their evolutionary success? Why?

11   What is the evidence for intraspecific competition in stegosaurs?

12   What are some of the differences between ankylosaurid defense and nodosaurid defense?

13   Why do we think ankylosaurs likely ruminated?

14   Summarize what is known about herding behaviors in thyreophorans.

15   How does the interpretation of thermoregulation in stegosaurs correlate with their inferred activity levels?

16   What is known about sexual dimorphism in thyreophorans?

CHAPTER OBJECTIVES

- Introduce Marginocephalia, particularly its two large constituent groups, Pachycephalosauria and Ceratopsia

- Develop familiarity with current thinking about lifestyles and behaviors of marginocephalians

- Develop an understanding of marginocephalian evolution using cladograms, and an understanding of the place of Marginocephalia within Dinosauria

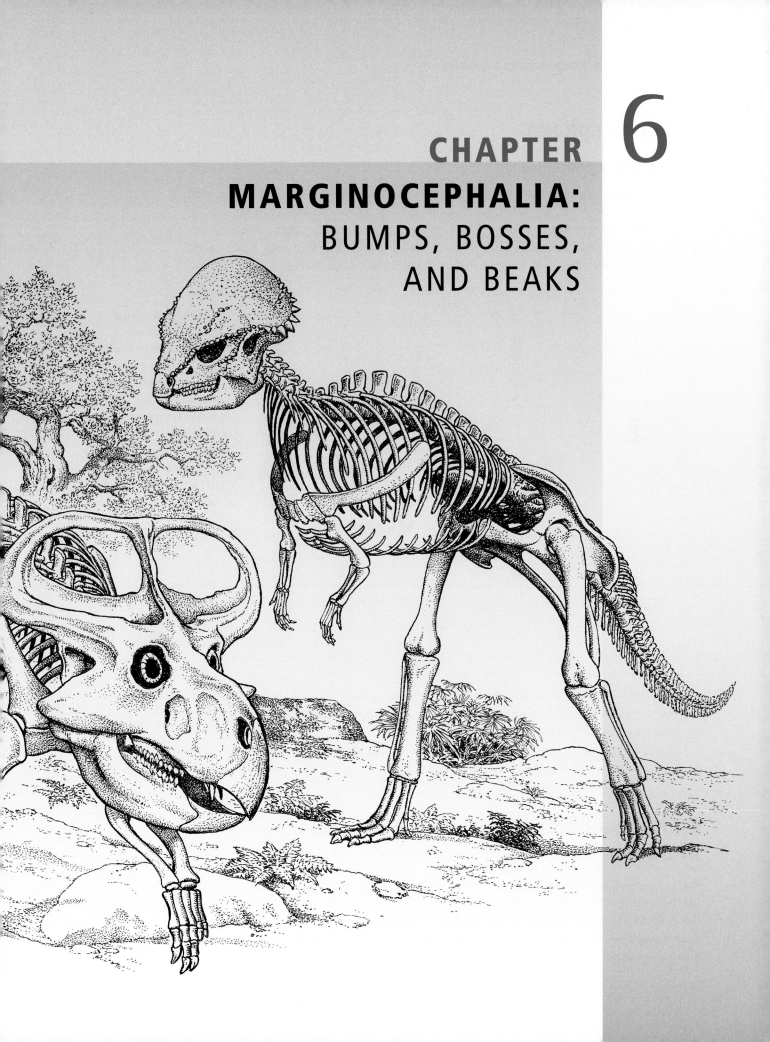

CHAPTER 6

MARGINOCEPHALIA:
BUMPS, BOSSES, AND BEAKS

Marginocephalia

Who were marginocephalians?

Marginocephalia (margin – edge; *cephal* – head). It's not a name you'll hear from the local 5-year-old diknow-it-all. Yet, the name Marginocephalia reflects an important connection between two major, somewhat different-looking, groups of dinosaurs: Pachycephalosauria (*pachy* – thick;) and Ceratopsia (*kera* – horn; *tops* – face). Together with Ornithopoda (Chapter 7), marginocephalians make up the taxon known as Cerapoda (Figure 6.1).

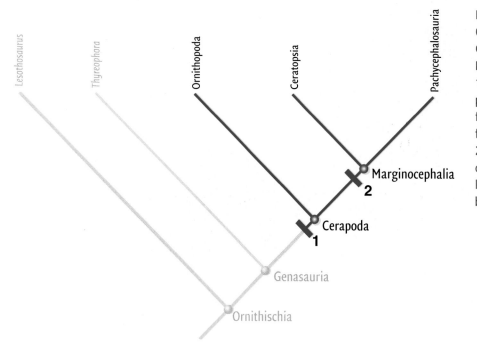

Figure 6.1. Cladogram of Ornithischia, emphasizing Cerapoda and Marginocephalia. Derived characters include: at **1**, significant diastem between premaxillary and maxillary teeth, five or fewer maxillary teeth, finger-like anterior trochanter; at **2**, narrow parietal shelf obscuring occipital elements in dorsal view, lateral portions of shelf formed by squamosal.

Marginocephalians all bear a ridge, or shelf, of bone running across the back of the skull. The size of this feature varies greatly, but in all cases, when viewed from above, it blocks from sight the bones at the back of the skull.

Although marginocephalians come in many shapes and sizes, they were restricted to the Northern Hemisphere during the Cretaceous Period.

Marginocephalia: Pachycephalosauria – In Domes We Trust

Pachycephalosaurs were bipedal ornithischians with thickened skull roofs (Figure 6.2). In the North American forms, this took the form of high domes, but several Asian varieties had flattened, thickened skulls (Figure 6.3); some, however, are considered to be juvenile forms of fully adult dome-headed pachycephalosaurs. Figure 6.4 shows the distinctively Northern Hemisphere distribution of pachycephalosaur sites.

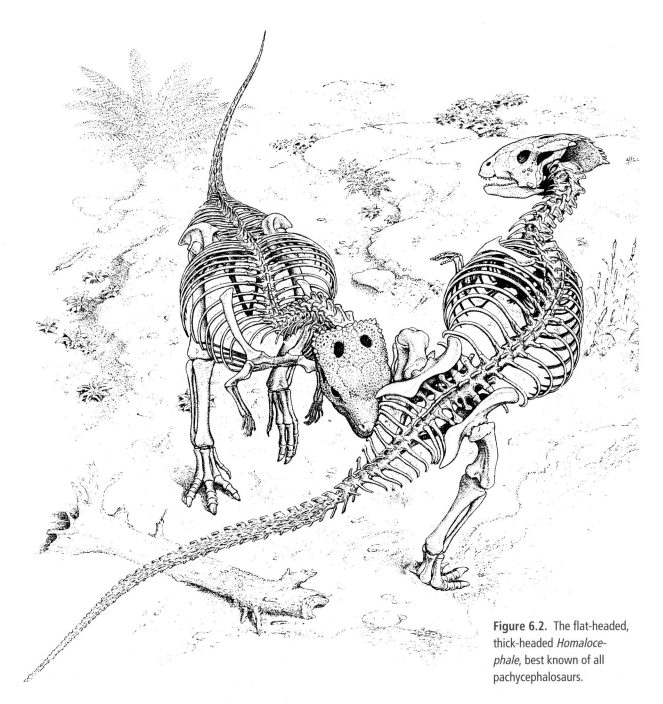

Figure 6.2. The flat-headed, thick-headed *Homalocephale*, best known of all pachycephalosaurs.

Pachycephalosaur lives and lifestyles

Where did a pachycephalosaur call home? In Asia, pachycephalosaurs apparently lived in a Sahara-like desert, punctuated by ephemeral streams in small drainage basins. Their remains are commonly found as nearly complete skulls and beautifully articulated skeletons. These fossils show little evidence of transport, and were apparently fossilized close to where the living animal died.

In North America, by contrast, the environments were very different. The rocks where marginocephalian remains are found represent a broad, Cretaceous coastal plain in a then-temperate climate – built from sediment eroded as the Rockies mountain range rose to the west

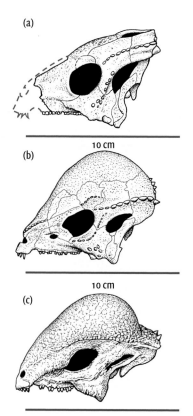

(a)

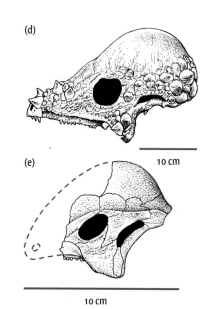

(d)

(b)

(e)

(c)

Figure 6.3. Left lateral view of (a) *Homalocephale*, (b) *Prenocephale*, (c) *Stegoceras*, (d) *Pachycephalosaurus*, and (e) *Tylocephale*.

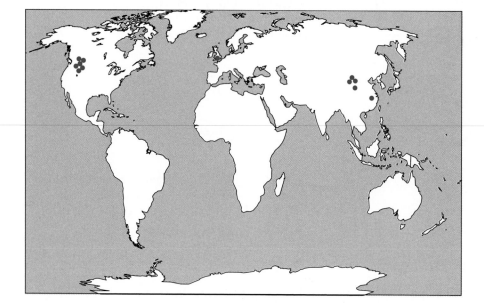

Figure 6.4. Global distribution of Pachycephalosauria.

(Figure 6.5). The most common finds are isolated, thickened skull caps, whose water-worn appearance suggests that they were transported long distances in rivers before burial and fossilization (Figure 6.6). The high proportion of skull cap finds suggest that only the most robust of bones – in this case, the skull caps – survived long journeys. The relative absence of articulated specimens in coastal plain sediments implies that, in life, North American pachycephalosaurs likely lived toward, or even in, mountains, where sediments were more likely to be eroded rather than deposited.

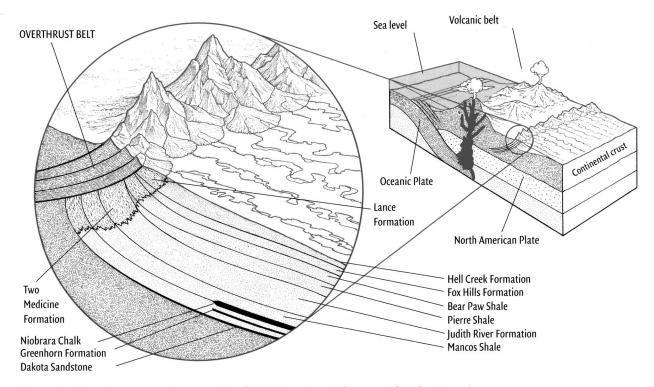

Figure 6.5. Late Cretaceous paleogeographic of North America. As the ancestral Rocky Mountains were uplifted and then drained and eroded by rivers, a thick sedimentary sequence was deposited into geological basins directly to the east of the rising Rockies. Fossil material was carried from the highlands by those rivers, and deposited onto the flat coastal plain to the east.

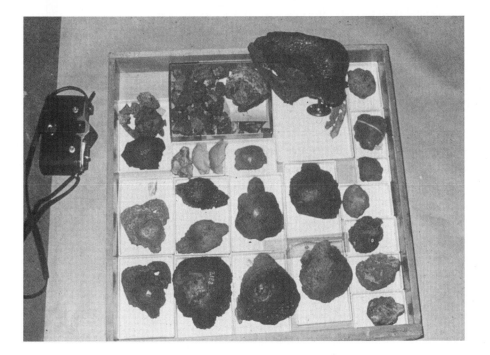

Figure 6.6. A museum tray filled with the isolated skull caps of pachycephalosaurs. Camera is 10 cm.

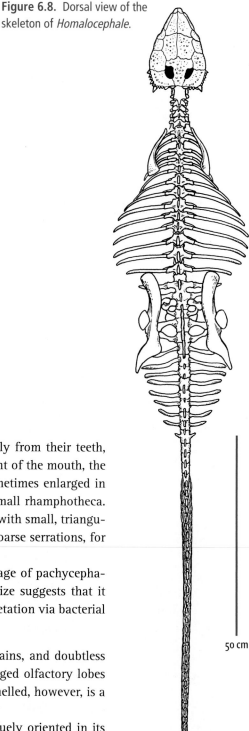

Figure 6.8. Dorsal view of the skeleton of *Homalocephale*.

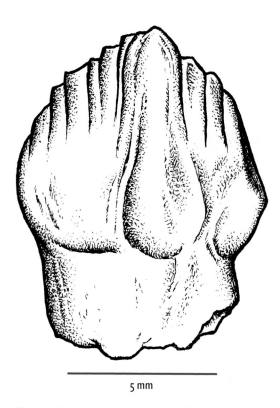

5 mm

Figure 6.7. An upper cheek tooth of *Pachycephalosaurus.*

Feeding The herbivorous habits of pachycephalosaurs are evident not only from their teeth, but also from the impressive volume of their abdominal regions. At the front of the mouth, the jaws contained simple, peg-like gripping teeth, the last of which were sometimes enlarged in a canine-like fashion (see Figure 6.3). These teeth were surrounded by a small rhamphotheca. Further back, the cheek teeth of pachycephalosaurs were uniformly shaped with small, triangular crowns (Figure 6.7). The front and back margins of these crowns bear coarse serrations, for cutting or puncturing plant leaves or fruits.

Turning to the opposite end of the gastrointestinal system, the rib cage of pachycephalosaurs was very broad, extending backward to the base of the tail. Its size suggests that it accommodated a large stomach (or stomachs) which broke down tough vegetation via bacterial fermentation (Figure 6.8).

Pachycephalosaur brains Pachycephalosaurs had typical ornithischian brains, and doubtless aspired to typical ornithischian thoughts. Atypical, though, were the enlarged olfactory lobes of the brain, suggesting a better-than-average sense of smell. What they smelled, however, is a secret that died with them.

For all its unremarkable-ness, the pachycephalosaur brain was uniquely oriented in its skull. The back half of the brain is angled downward, which is reflected in the rotation of the back of the skull (the occiput) to face not only backward, but also slightly downward. It's been shown that the higher the dome, the more downward the orientation of the occiput. And that might be related to pachycephalosaurs using their heads for something other than profound thought.

Using your head … for a battering ram? The morphology of the domes has suggested to many scientists that, incredibly, pachycephalosaurs used their thickened skull roofs as battering

50 cm

rams (Figure 6.9). Internally, the structure of the dome is very dense, with the bone fibers oriented in columns approximately perpendicular to the external surface of the dome. Such an arrangement may be ideal for resisting forces that come from strong and regular thumps to the top of the head and for transmitting such forces around the brain, much as the helmet worn by sparring boxers is supposed to channel forces around the head.

Using special clear plastic cut to resemble a cross-section of the high-domed pachycephalosaur *Stegoceras* (Figure 6.10), paleontologist H.-D. Sues stressed the model in a way that simulated head-butting. The stress lines, seen under ultraviolet light, mimicked the orientation of the columnar bone, reinforcing the suggestion that the fibrous columns evolved to resist stresses induced by head-butting.

Building a better head-butter If head-butting was in fact the preferred means of pachycephalosaur expression, the body had to be set up to allow for it. Recall that the back of the pachycephalosaur skull is rotated forward beneath the skull roof. With the head in a downward position – the only position that makes sense for head-butting – rotation of the back of the skull minimizes the chance of violent rotation or even dislocation of the head on the neck.

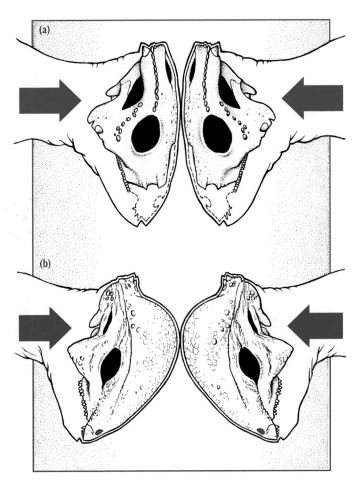

Figure 6.9. Head-on pushing and butting in (a) *Homalocephale* and (b) *Stegoceras*.

We might hope to see some protective measures in the neck as well; unfortunately, we know virtually nothing about the neck of pachycephalosaurs. Still, it is clear from the occiput (Figure 6.11) that the neck musculature was unusually well developed and very strong. We surmise that it was used to position and hold the head correctly for head-butting.

Further down the spinal column, and utterly unique to pachycephalosaurs, the vertebrae are reported to have distinctive tongue-and-groove articulations, which must have provided rigidity to the back. These articulations would have prevented the kinds of violent lateral rotations of the body that would otherwise have been suffered at the time of impact.

Conscientious objectors? But not so fast – or at least not so hard. The thickened skull cap of one North American pachycephalosaur – *"Stygimoloch"* – has been shown to contain abundant microscopic openings for blood vessels. With so much vascularization, the skulls of pachycephalosaurs may not have done well with either front or side impacts; this leaves the domes in this genus, and probably others, principally as display structures rather than WMDs.

Socializing pachycephalosaur style

Either way – display or as a battering ram – social behavior is strongly implied, and some degree of sexual dimorphism might be expected. And so this idea was tested using a large

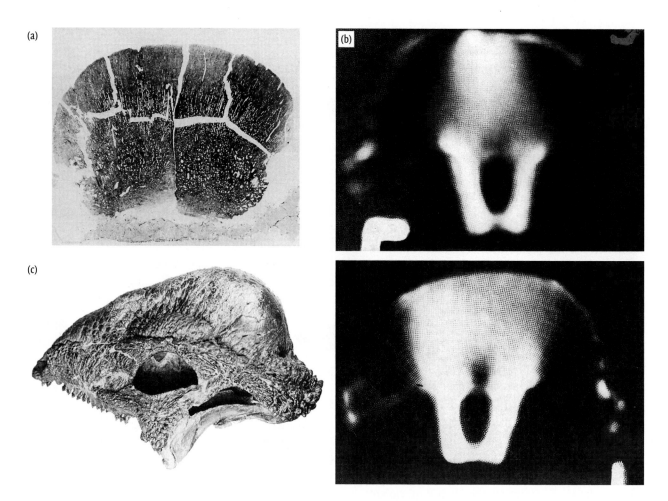

Figure 6.10. (a) Vertical section through the dome of *Stegoceras*. Note the radiating organization of internal bone. (b) Plastic model of the dome of *Stegoceras* in which forces were applied to several points along its outer edge and seen through polarized light. Note the close correspondence of the stress patterns produced in this model and the organization of bone indicated in (c) the left side of the skull of *Stegoceras*.

sample of a single pachycephalosaur species *Stegoceras validum*. It turns out that the domes of *Stegoceras* can be segregated into two groups on the basis of relative size and dome shape (Figure 6.12). One group had larger, thicker domes than the other. Strikingly, the ratio of larger-domed to less-large-domed individuals was one-to-one; exactly what you might expect if one population was male and the other, female. Which was which, however, is anybody's guess.

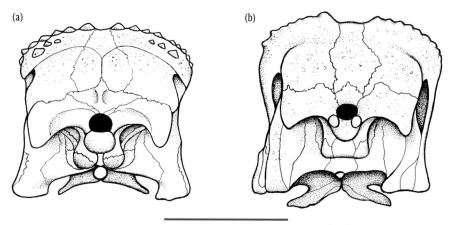

Figure 6.11. Rear (occipital) view of the skull of (a) *Homalocephale* and (b) *Stegoceras*.

10 cm

(a)

(b)

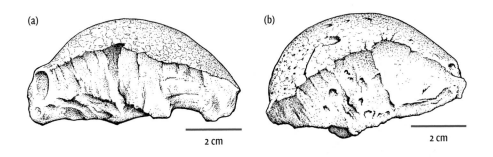

2 cm   2 cm

Figure 6.12. Two forms of the dome of *Stegoceras*. The shallower dome (a) is thought to pertain to a female, while the other dome (b) may pertain to a male.

Sexual selection Pachycephalosaurs all had, along with, and perhaps including the dome, a suite of features related to visual display. Firstly, there are the canine-like teeth. These could have been used in threat display or biting combat between rival individuals, much like pigs and some deer do today. Equally suggestive are the knobby and spiny osteoderms that covered the snout, the side of the face, and most extensively on the back of the marginocephalian shelf (see Figure 6.1, node 2; see also Figure 6.3). These distinctive features were likely all about showing off and establishing dominance.

The establishment of dominance gives one gender – males, if living reptiles are any guide – preferred reproductive access to the other gender, who then select a mate. These same males must also fend off competitors and establish dominance. In general, this practice of establishing dominance hierarchies constitutes sexual selection, selection within one gender (generally, males), rather than among members of a single species.

In pachycephalosaurs, domes, knobs, and spikes all acted in ritual display and, potentially, violent clashes. The winner, likely the male with the best-fashioned cranial hardware, got to perpetuate the family line. But he always had to be vigilant for other males that wanted to literally knock his block off – or at least knock him off the block.

The evolution of Pachycephalosauria

Pachycephalosaurs share a host of derived features, most of them cranial (Figure 6.13). Figure 6.14 maps general trends in pachycephalosaur evolution. The most primitive pachycephalosaurs on the cladogram are Asian, suggesting an Asian origin for the group. Because all but two pachycephalosaurs are from the Late Cretaceous, we infer that they underwent considerable evolution during that time.

There was an early tendency to thicken the skull roof as well as develop nerves for the sense of smell. The increased dome heights in some later forms may reflect the evolution of head-butting from simple pushing to complicated head-to-flank or head-to-head butting, to ritualized display.

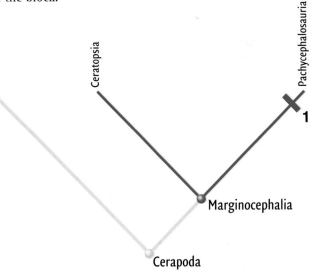

Figure 6.13. Cladogram of Marginocephalia emphasizing the monophyly of Pachycephalosauria. Derived characters include: at **1**, thickened skull roof, frontal excluded from orbital margin, tubercles on caudolateral margin of squamosal, thin, plate-like basal tubera, double ridge-and-groove articulations on dorsal vertebrae, elongate sacral ribs, caudal basket of fusiform ossified tendons, ilium with sigmoidal border, me-dial process on ilium, pubis nearly excluded from acetabulum, tubercles on squamosal, broad expansion of squamosal onto occiput, free ventral margin of the quadratojugal eliminated by contact between jugal and quadrate.

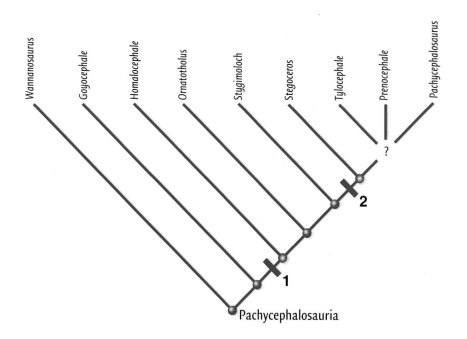

Figure 6.14. Cladogram showing evolutionary changes within selected pachycephalosaurs. Derived characters include: at **1**, broad parietal bone, broad medial process on ilium; at **2**, nasal and postorbitals incorporated into dome.

The broad girth of pachycephalosaurs is clearly derived relative to the narrower more primitive girths seen in most other ornithischians. It suggests a backward migration and enlargement of the digestive tract to occupy a position between the legs and under the tail. As was the case for thyreophorans (see introduction to Part II: Ornithischia; and Chapter 5), simple styles of chewing must have combined with fermentation-based digestion to increase the nutrition available to pachycephalosaurs from the plants they ate.

Marginocephalia: Ceratopsia – horns and all the frills

From the time of their discovery in the second half of the 1800s to the present day, there has hardly been a group of dinosaurs that has evoked more fascination than ceratopsians (Figure 6.15). Some of these quadrupedal, horned, frilled dinosaurs roamed the Great Plains of North America in the Late Cretaceous. They were rhino-like, ranging upward of 6 or 7 tonnes. Equally famous, but for other reasons, is a host of smaller, lighter (25 and 200 kg) non-horned Asian ceratopsians from slightly earlier in the Cretaceous (Figure 6.16).

We know a lot about ceratopsians: the fossil record from Asia and North America is one of the most outstanding of any dinosaur group (Figure 6.17). Primitively small bipeds, these animals evolved into powerful quadrupedals early in their history, developing thick hooves on all toes and reaching sizes to rival that of small tanks.

With or without horns, it is easy to recognize the ceratopsian familial stamp: ceratopsians all had skulls that were narrow, with a hooked beak in front and a skull that flared deeply in the cheek region (Figure 6.18). And at the tip of the snout in the upper jaw was the uniquely evolved rostral bone (Figure 6.19).

As befits their name, many ceratopsians had horns; however, some did not. Those ceratopsians that had them, though, grew some of the most impressive horns ever seen on any vertebrate (Figure 6.20). Like the horns of many mammals, the skulls only preserve the bony

Figure 6.15. *Chasmosaurus,* a ceratopsian from the Late Cretaceous of the Western Interior of North America.

horn cores, covered by keratin sheaths that actually comprised the working end of the horn. The horn visible on the head of the living animal was therefore significantly larger than the horn core alone (Figure 6.21).

Equally memorable is the ceratopsian frill, the marginocephalian shelf run amok. Extending from the back of the skull, frills vary considerably in size, ornamentation, and shape (see Figure 6.20). The largest reach 2 m in length.

Ceratopsian lives and lifestyles

Dressed and ready to chew Ceratopsians chewed. A hooked rhamphotheca, blocks of cheek teeth in both upper and lower jaws, a sturdy coronoid process, and inset jaws, evidence for the existence of fleshy cheeks, all scream CHEWING.

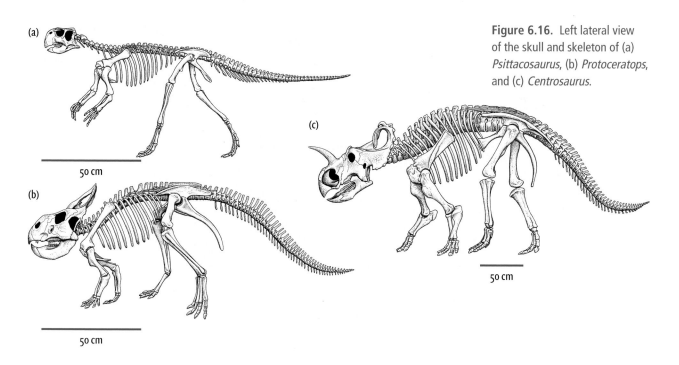

Figure 6.16. Left lateral view of the skull and skeleton of (a) *Psittacosaurus*, (b) *Protoceratops*, and (c) *Centrosaurus*.

The business end of the ceratopsian mouth was the narrow, hooked, beak-tipped snout, suggesting the potential for careful selection of the plants for food. Individually the cheek teeth were relatively small, but they grew stacked and overlapping together into a single functional slicing block in each jaw, the **dental battery** (Figure 6.22a). Worn teeth were constantly replaced, so that the active chewing surface of each of the four dental batteries was continually refurbished. Inexplicably and unique in the animal kingdom, the orientation of the grinding surfaces migrated, becoming more and more vertical, until, in the large, highly derived North American forms, they occurred nearly vertically along the *sides* of the teeth comprising the dental battery (Figure 6.22b).

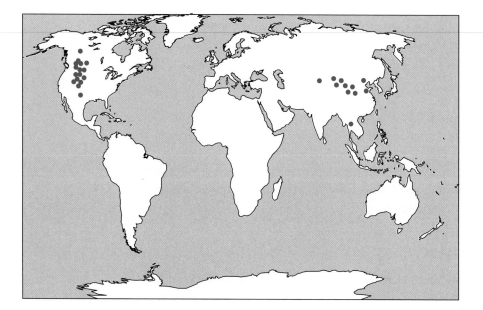

Figure 6.17. Global distribution of Ceratopsia.

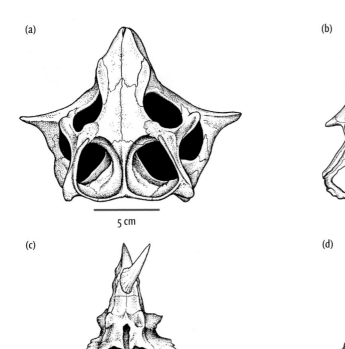

(a)

(b)

5 cm

5 cm

(c)

(d)

50 cm

50 cm

Figure 6.18. Dorsal view of the skull of (a) *Psittacosaurus*, (b) *Protoceratops*, (c) *Styracosaurus*, and (d) *Chasmosaurus*.

Figure 6.19. Snout of ceratopsian; rostral bone highlighted.

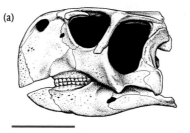

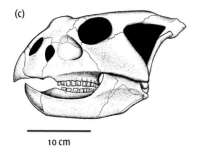

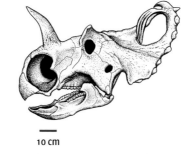

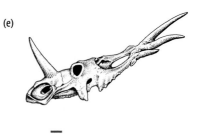

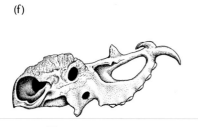

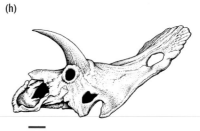

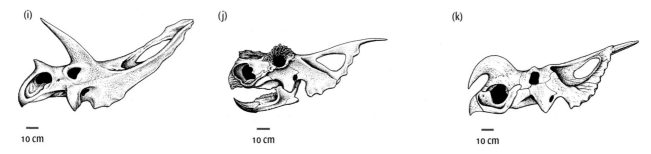

Figure 6.20. Left lateral view of the skull of (a) *Psittacosaurus*, (b) *Leptoceratops*, (c) *Bagaceratops*, (d) *Centrosaurus*, (e) *Styracosaurus*, (f) *Pachyrhinosaurus*, (g) *Pentaceratops*, (h) *Arrhinoceratops*, (i) *Torosaurus*, (j) *Achelousaurus*, and (k) *Einiosaurus*.

The force behind this high-angle mastication derived from a great mass of jaw-closing musculature, which in the frilled forms crept through the upper temporal opening and onto the base of the frill. The other end of this muscle attached to a massive, hulking coronoid process on the mandible (Figure 6.23). All in all, the chewing apparatus in ceratopsians was among the most highly evolved in any vertebrate.[1]

Beyond the mouth Based upon their narrow girth (by comparison with thyreophorans and pachycephalosaurs), the digestive tract does not appear to have been disproportionately large in ceratopsians, and did not likely rely upon wholesale bacterial fermentation for extracting nutrients from plants. Nevertheless, it must have been big enough to accommodate what must have been an endless parade of foliage that formed the diet of these animals.

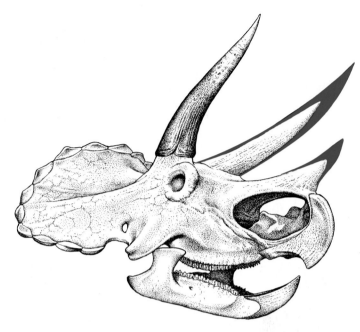

Figure 6.21. The skull of *Triceratops* showing the horn core covered by the keratinized horn as it would have been in life.

Even the largest quadrupedal ceratopsians never browsed particularly high above the ground. The browse height of the largest was probably less than 2 m. Nevertheless, they may have been able to knock over trees of modest size in order to gain access to choice leaves and fruits.

Which plants were preferred by ceratopsians remains a mystery. The principal plants whose statures match browsing heights of ceratopsians were a variety of shrubby angiosperms, ferns, and perhaps small conifers (see Chapter 13).

Locomotion The primitive ceratopsian *Psittacosaurus* appears to have been fully bipedal and must have walked in typical bipedal ornithischian fashion. But the rest were quadrupeds all, and while the back legs were fully erect, the orientations of the front limbs are somewhat controversial. Some have argued that the front limbs were directly under the body, as they are in mammalian quadrupeds. Others have argued, based on the shape of the bones of the front legs, that a more sprawling posture in the front legs is indicated (Figure 6.24).

These considerations naturally affect how we understand the way ceratopsians ran. The sprawling front legs would likely entail slower speeds, and perhaps a more unhurried lifestyle. The mammal-like reconstructions, with fully erect stances for both front and rear legs, suggest faster speeds, reminiscent of a large, Cretaceous rhinoceros. With such uncertainty about stance, speed tentative estimates for walking range somewhere between 2 and 4 km/h (see Box 12.3), while maximum running speeds range from 30 to 35 km/h.

Bringing up baby The first dinosaur egg nests ever found, in 1922, were thought to belong to the small Asian ceratopsian *Protoceratops*, and it was with this dinosaur that, for the next

---

[1] The primitive ceratopsian *Psittacosaurus* lacked the highly refined chewing specializations of the North American ceratopsians, but it is known to have harbored a packet of gastroliths lodged in its gizzard, which would have doubly pulverized its meal. Gastroliths are known from no other ceratopsian.

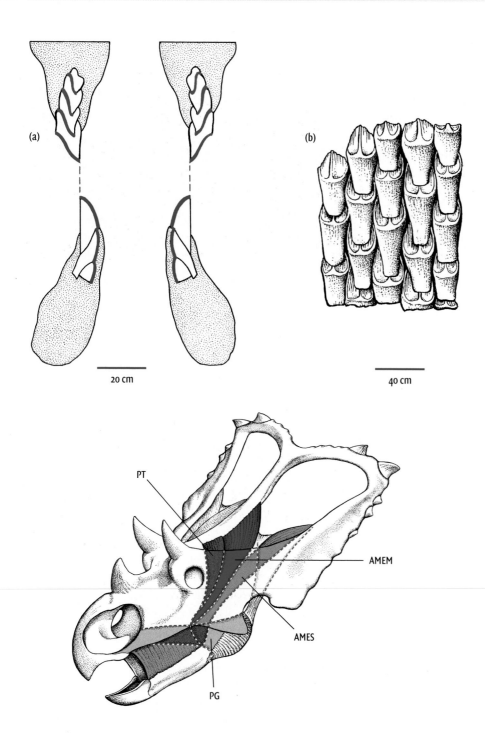

Figure 6.22. Cross-section through the upper and lower jaws of *Triceratops*: (a) high-angle grinding motion of the dental batteries; (b) internal view of the dental battery in the lower jaw of *Triceratops*.

20 cm

40 cm

Figure 6.23. Jaw musculature reconstructed in the skull of a long-frilled ceratopsian. The major jaw closing (adductor) muscles are (1) the adductor mandibularis externus superficialis (AMES); (2) the adductor mandibularis externus medialis (AMEM); (3) the psuedotemporalis (PT); (4) the pterygoideus (PG).

PT

AMEM

AMES

PG

70 years, the eggs were posed in museum displays all over the world. So it was, uh, *informative* to learn – after some 70 years – that the embryos inside those "*Protoceratops* eggs" actually belonged to the theropod *Oviraptor* (see Chapter 9).

Despite the confusion with *Oviraptor*, however, juvenile ceratopsians from Asia are now relatively well known. Hatchlings of *Psittacosaurus*, no more than 23 cm long (with tail!) have been found. And recently, complete skeletons of *Protoceratops* hatchlings – grown somewhat past the newly hatched stage – were found in nests (Figure 6.25). This means that some parental care at the nest after birth is implied for *Protoceratops*. Moreover, all of the growth stages from

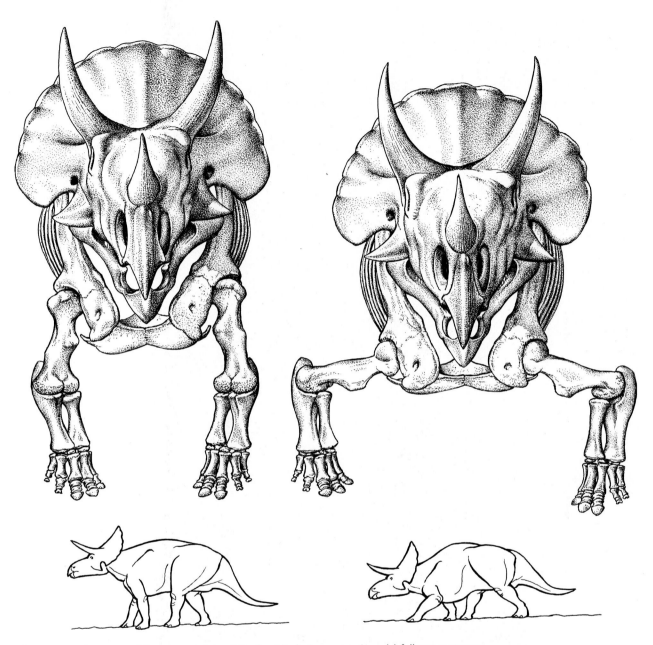

Figure 6.24. Two potential reconstructions of the front limbs in ceratopsians: (a) fully erect stance; (b) semi-erect stance.

hatchling to adult have been documented in *Protoceratops*, making the ontogeny – or growth and development – of this dinosaur perhaps the best understood of all dinosaurs. And the ontogeny of ceratopsian dinosaurs turns out to hold key clues about their behavior.

Horns, frills, and ceratopsian behavior Ceratopsian horns were once thought to have functioned to ward off predators at close quarters. More recent interpretations have not completely ruled this out (see below), but have instead focussed on intraspecific behaviors such as display, ritualized combat, defense of territories, and establishment of social ordering.

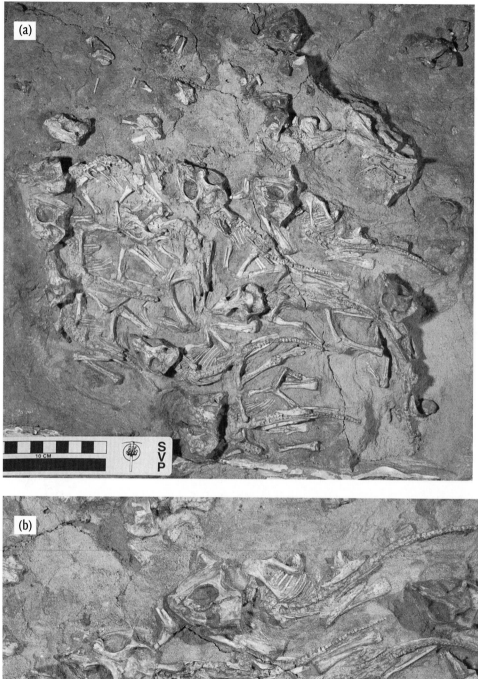

Figure 6.25. A nest of juvenile *Protoceratops* from the Late Cretaceous of Mongolia. (a) The complete nest with 15 juvenile *Protoceratops* clustered inside. (b) Two of the 15 juveniles. An adult *Protoceratops* is about 2 m long; each of these babies is but ~10 cm, suggesting that they were perhaps within their first year of life. Ruler is marked in cm.

The link between dominance, defense, and horns comes from studies of mammals in their natural habitats. In the case of almost all horned mammals, larger males tend to have a reproductive advantage over smaller males. Dominance in these mammals (and in other tetrapods) is accentuated by the development of structures that "advertise" the size of the animal; these obviously include horns and antlers, as well as the bony horn-like knobs (ossicones) of giraffes and the nasal horns of rhinoceroses. In short, the variety of horn and antler shapes in mammals are known to reflect (1) species-recognition mechanisms that aid in preventing interspecific matings (that is, matings between species), and (2) intraspecific differences in displays and ritualized fighting behavior.

Turning to ceratopsians, few have doubted that the horns were used for combat; the question has been "At whom were they aimed?" Using modern horned mammals as analogs, current thought suggests that the large nasal and brow horns of ceratopsians functioned primarily during territorial defense and in establishing dominance. Similarly, the development of elaborate scallops and spikes along the frill margin in many of the more highly derived ceratopsians separates one species from another. Thought of this way, the remarkable variations in the horns and frills in ceratopsians could be used for *interspecific* identification as well as the establishment of *intraspecific* dominance (Figure 6.26).

Striking data that bear upon this have come from *Protoceratops*. Statistical studies of *Protoceratops* show two populations of adult frill and facial morphologies – strong evidence of sexual dimorphism (Figure 6.27). Moreover, the frills don't appear too large in juvenile specimens – they only develop when the animals reach 75% of adult body sizes. This suggests that frill growth is coordinated with sexual maturity and therefore that there is a reproductive connection to frill size and shape. Sound like sexual selection?

Sexual selection is now thought to also occur in other ceratopsians, among them *Centrosaurus* and *Chasmosaurus*. In many of these forms, the development of scallops and spikes on the frill margin would enhance the dimorphic nature of the frill.

All of this implies social interactions, and it thus comes as no surprise to learn that ceratopsians lived in large herds. The evidence for this comes from an ever-increasing catalog of ceratopsian bonebeds, mass accumulations of single species of organisms. Bonebeds are known for at least nine separate species, including several bonebeds exceeding 100 individuals. Such gregariousness makes sense when putting frills and horns into their behavioral context: territoriality, ritualized combat and display, and the establishment of dominance are to be expected in animals that come together in highly social circumstances such as herds.

These ideas suggest that we ought to find puncture wounds inflicted on faces, frills, and bodies of competing ceratopsians. In fact, such wounds are preserved in at least five different ceratopsians. In *Triceratops* and *Centrosaurus*, healed wounds on the cheek region and frill were correlated with the types of horns each animal bears. The horns of *Triceratops* (two large brow horns; a smaller nasal horn) are very different from those of *Centrosaurus* (a single large nasal horn), and if *intraspecific* competition was the main use of the horns, the wounds left on the frills and faces of these animals should look very different from one another. As it turned out, *Triceratops* had a significantly higher number of healed wounds on the frill than did *Centrosaurus*, suggesting that, given the mechanics of fighting, the damage was caused by other *Triceratops* with their massive, highly-placed brow horns. These results provide strong evidence of the blood-letting that comes from head-on engagements between competing members of the same species.

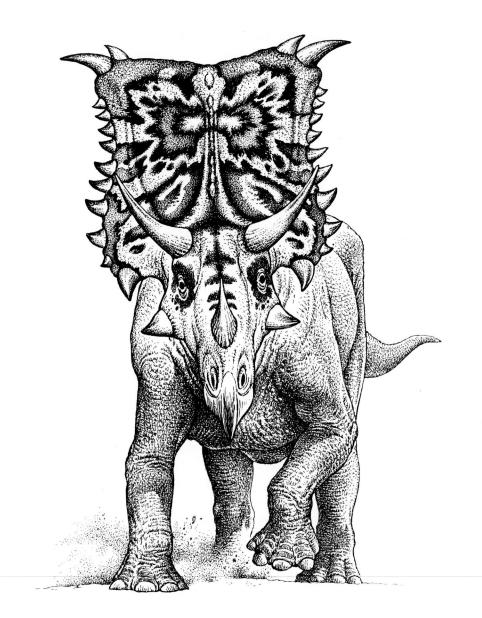

Figure 6.26. "Back off": frill display in *Chasmosaurus.* The very long frill could have provided a very prominent frontal threat display, not only by inclining the head forward but also by nodding or shaking the head from side to side.

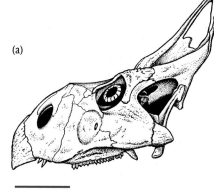

(a)

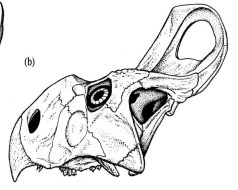

(b)

10 cm

10 cm

Figure 6.27. Sexual dimorphism in *Protoceratops.* Note in (a), a presumed female, the frill is less pronounced and the nasal ridge is less prominent; quite the opposite of (b), a presumed male.

Thoughts of a ceratopsian Given the complex repertoire of inferred ceratopsian behaviors, it comes as a bit of a surprise that their brains were not particularly large (see Box 12.4). Despite being near opposites in terms of body size and display-related anatomy, both *Protoceratops* and *Triceratops* had brains less than the size expected of a similarly sized crocodilian or lizard. Cerebrally, they were above sauropods, ankylosaurs, and stegosaurs, but commanded proportionally less gray matter than either ornithopods or theropods. Regardless, the variety of exotic morphology in and around the head suggests that ceratopsians, large-brained or no, may have had a relatively complicated behavioral repertoire.

The evolution of Ceratopsia

Ceratopsia is a monophyletic taxon, indicated by a rich array of derived features (Figure 6.28). The two ceratopsians that are thought to represent the primitive condition for the group are *Yinlong* (Figure 6.29) and *Psittacosaurus*, both small, Asian bipeds. All more derived ceratopsians – Neoceratopsia – are quadrupeds. This underlines an important evolutionary event that we can read from the cladogram (Figure 6.28): relatively early in their history, ceratopsians, for whatever their reasons, adopted a quadrupedal stance.

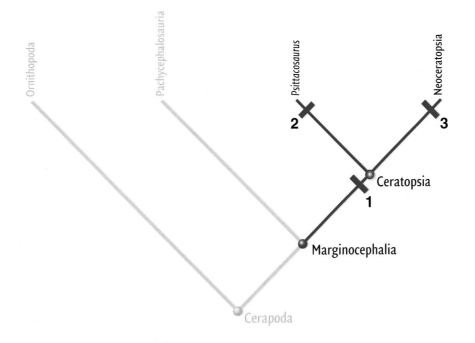

Figure 6.28 *(left)*. Cladogram of Ceratopsia, emphasizing the monophyly of *Psittacosaurus* and Neoceratopsia. Derived characters include: at **1**, rostral bone, a high external naris separated from the ventral border of the premaxilla by a flat area, enlarged premaxilla, well-developed lateral flaring of the jugal; at **2**, short preorbital region of the skull, very elevated naris, loss of antorbital fossa and fenestra, unossified gap in the wall of the lacrimal canal, elongate jugal and squamosal processes of postorbital, dentary crown with bulbous primary ridge, manual digit IV with only one phalanx, manual digit V absent; at **3**, enlarged head, keeled front end of the rostral bone, much reduced quadratojugal, primary ridge on the maxillary teeth, development of humeral head, gently decurved ischium.

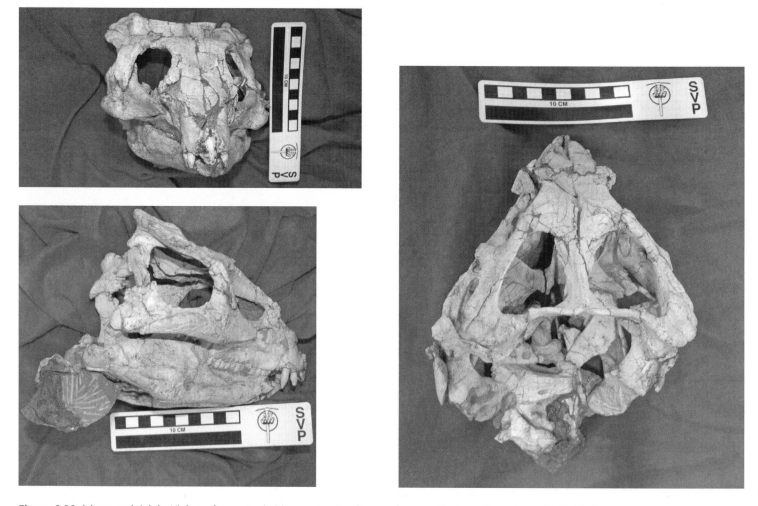

Figure 6.29 *(above and right).*   *Yinlong,* the most primitive ceratopsian dinosaur known.   Photograph courtesy of J. M. Clark.

Those early days also brought with them evidence of a major ceratopsian migration. Neoceratopsia (Figure 6.30) consists of a series of small, relatively primitive forms such as the Asian *Protoceratops* and *Bagaceratops*; the somewhat younger, though still primitive North American *Montanaceratops* and *Leptoceratops*, as well as the more derived, exclusively North American family Ceratopsidae, that group of large, familiar ceratopsians such as *Triceratops* and *Centrosaurus* (Figure 6.31). When we compare the geographical locations of various neoceratopsians, that is, their biogeography, with primitive and advanced ceratopsians on the cladograms shown in Figures 6.28 and 6.29 it becomes clear that, early in neoceratopsian history, a primitive neoceratopsian – looking perhaps a bit like *Protoceratops* – migrated to the New World. The route of choice would likely have been briefly exposed land across the Bering Straits (Figure 6.32).

Once in North America, a few lineages retained the comparatively modest morphology of their more primitive forebears. However, the clade radiated into two spectacular and diverse groups of much larger, flashier ceratopsids: chasmosaurines, after *Chasmosaurus*;

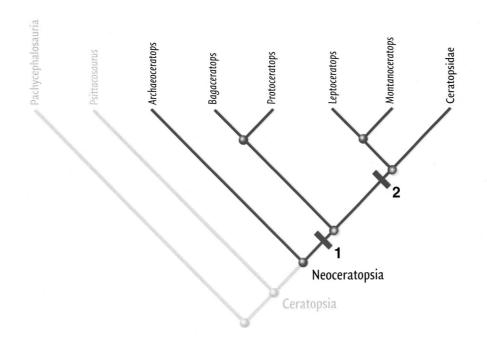

Figure 6.30 *(left).* Cladogram of basal Neoceratopsia, with the more distantly related *Psittacosaurus* and Pachycephalosauria. Derived characters include: at **1**, elongated preorbital region of the skull, an oval antorbital fossa, triangular supratemporal fenestra, development of the syncervical (fusion of cervical vertebrae); at **2**, greatly enlarged external nares, reduced antorbital fenestra, nasal horn core, frontal eliminated from the orbital margin, supraoccipital excluded from foramen magnum, marginal undulations on frill augmented by epoccipitals, more than two replacement teeth, loss of subsidiary ridges on teeth, teeth with two roots, 10 or more sacral vertebrae, laterally everted shelf on dorsal rim of ilium, femur longer than tibia, hoof-like pedal unguals.

Figure 6.31 *(below).* Cladogram of Ceratopsidae. Derived characters for Ceratopsidae. Derived characters (for chasmosaurines) include: at **1**, enlarged rostral, presence of an interpremaxillary fossa, triangular squamosal epoccipitals, rounded ventral sacrum, ischial shaft broadly and continuously decurved. Derived characters (for centrosaurines) include: at **2**, premaxillary oral margin that extends below alveolar margin, postorbital horns less than 15% of skull length, jugal infratemporal flange, squamosal much shorter than parietal, six to eight parietal epoccipitals, predentary biting surface inclined steeply laterally.

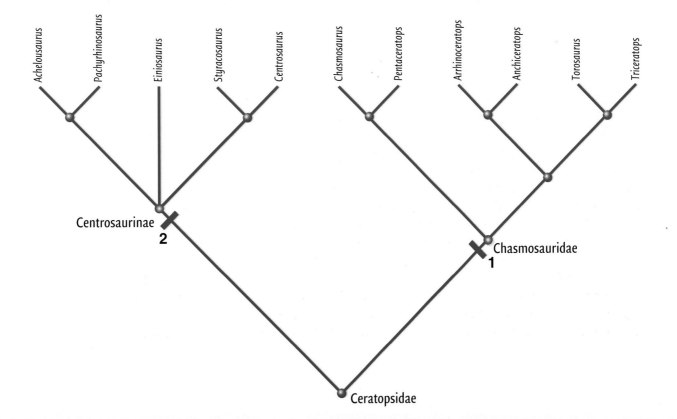

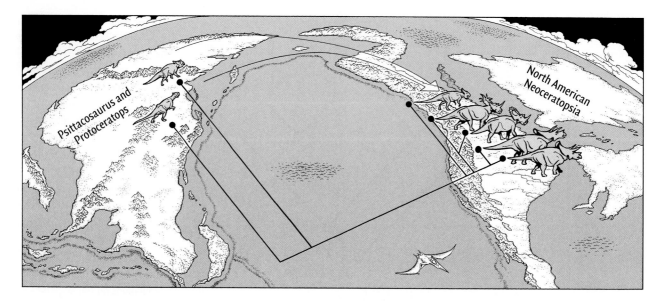

Figure 6.32. Cladogram of ceratopsians superimposed on a map of North America and Asia, showing migration of more derived forms to North America. Arrow is potential route of migration.

and centrosaurines, after *Centrosaurus* (see Figure 6.33). Chasmosaurines are generally called "long-frilled," after a tendency in the group to develop large, open frills, while centrosaurines are sometimes called "short-frilled," after a tendency in the group toward shorter frill lengths.

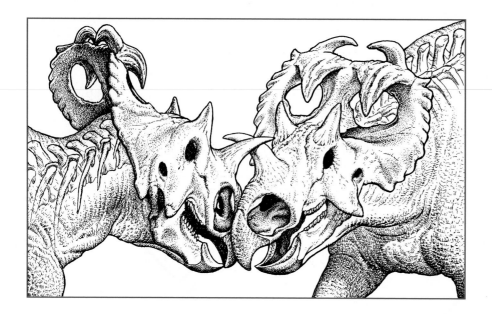

Figure 6.33. "Crossing of the horns": combat between male *Centrosaurus.*

The evolution of behavior If there is some correspondence between morphology and behavior, then the morphological trends identified by all the ceratopsian cladograms should give us insights into the evolution of neoceratopsian behavior. In those ceratopsians with relatively modest frills and horns – forms such as the Asian *Protoceratops*, and the North American *Leptoceratops* and *Montanoceratops* – display perhaps involved swinging the head from side to side. Should this have failed to impress, these animals may have rammed full tilt into the flanks of their opponent.

The more derived ceratopsids share more elaborate frills and either nasal or brow horns. Among the long-frilled ceratopsians (for example, *Chasmosaurus*, *Pentaceratops*, and *Torosaurus*), the display function of the frill may have been emphasized (see Figure 6.26). In contrast, most of the short-frilled ceratopsians (such as *Centrosaurus*, *Avaceratops*, and possibly *Pachyrhinosaurus*) were rather rhinoceros-like in their appearance (Figure 6.33), and likely tried to catch each other on their nasal horns, thus reducing to a degree the amount of damage inflicted on the eyes, ears, and snout.

Horns seem almost to drive the evolution of ceratopsian dinosaurs. In this diverse group, we witness a world where display and competition were all important, where – when push came to shove – it may have been better to vigorously nod than to cross horns.

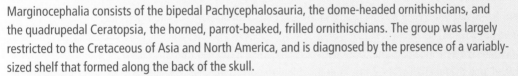

Summary

Marginocephalia consists of the bipedal Pachycephalosauria, the dome-headed ornithishcians, and the quadrupedal Ceratopsia, the horned, parrot-beaked, frilled ornithischians. The group was largely restricted to the Cretaceous of Asia and North America, and is diagnosed by the presence of a variably-sized shelf that formed along the back of the skull.

Marginocephalians were gregarious animals, and sexual selection was likely a driving force in much of their behavior, a fact that is reflected in their morphology. The domes of pachycephalosaurs have been interpreted as structures designed for intraspecific competition: head- and flank-butting have been suggested. The striking morphological variety of horns and frill shapes, and cranial ornamentation in ceratopsians suggests a high level of intraspecific competition. In both groups, sexual dimorphism has been recognized. Ceratopsian gregariousness is also reflected by the presence of large monospecific bonebeds, suggesting that herds of ceratopsians roamed what is now the Great Plains of Canada and the USA.

All marginocephalians are genasaurs, which means that they chewed their food to a greater or lesser extent. While pachycephalosaur teeth don't reveal evidence of remarkable chewing adaptations and pachycephalosaurs likely gut-fermented in their capacious stomachs, ceratopsians developed sophisticated chewing mechanisms including a robust coronoid process, dental batteries, a skull partitioned into cropping, diastem, and grinding sections, and powerful jaw adductor muscles that may have attached high on the frill.

Care of the young is known in ceratopsian dinosaurs. Nests of partially grown ceratopsians have been found, suggesting parental care at the nest.

Ceratopsian evolution was characterized by increasing size, as well as by one or more migrations across the Bering Strait land bridge, from Asia to North America. While intraspecific competition was likely an important behavioral aspect of even the earliest ceratopsians, later forms evolved elaborate frill or horn displays.

Selected readings

Chapman, R. E., Galton, P. M., Sepkoski, J. J. and Wall, W. P. 1981. A morphometric study of the cranium of the pachycephalosaurid dinosaur *Stegoceras. Journal of Paleontology*, **55**, 608–616.

Dodson, P. 1976. Quantitative aspects of relative growth and sexual dimorphism in Protoceratops. *Journal of Paleontology*, **50**, 929–940.

Dodson, P. 1996. *The Horned Dinosaurs*. Princeton University Press, Princeton, NJ, 346pp.

Dodson, P. and Currie, P. J. 1990. Neoceratopsia. In Weishampel, D. B., Dodson, P. and Osmólska, H. (eds.) *The Dinosauria*. University of California Press, Berkeley, pp. 593–618.

Dodson, P., Forster, C. A. and Sampson, S. D. 2004. Ceratopsidae. In Weishampel, D. B., Dodson, P. and Osmólska, H. (eds.) *The Dinosauria*, 2nd edn. University of California Press, Berkeley, pp. 494–513.

Farke, A. A, Wolf, E. D. S. and Tanke, D. H. 2009. Evidence of combat in Triceratops. *PlosOne*, **4**, e4252. doi:10.1371/journal.pone.0004252.

Farlow, J. O. and Dodson, P. 1975. The behavioral significance of frill and horn morphology in ceratopsian dinosaurs. *Evolution*, **29**, 353–361.

Galton, P. M. 1971. A primitive dome-headed dinosaur (Ornithischia: Pachycephalosauridae) from the Lower Cretaceous of England, and the function of the dome in pachycephalosaurids. *Journal of Paleontology*, **45**, 40–47.

Maryańska, T. and Osmólska, H. 1974. Pachycephalosauria, a new suborder of ornithischian dinosaurs. *Paleontologia Polonica*, **30**, 45–102.

Sereno, P. C. 1990. Psittacosauridae. In Weishampel, D. B., Dodson, P. and Osmólska, H. (eds.) *The Dinosauria*. University of California Press, Berkeley, pp. 579–592.

Sues, H.-D. 1978. Functional morphology of the dome in pachycephalosaurid dinosaurs. *Neues Jahrbuch für Geologie und Paläontologie Monatshefte*, pp. 459–472.

You, H.-L. and Dodson, P. 2004. Basal Ceratopsia. In Weishampel, D. B., Dodson, P. and Osmólska, H. (eds.). *The Dinosauria*, 2nd edn. University of California Press, Berkeley, pp. 778–93.

Topic questions

1   Who are marginocephalians?

2   What are the diagnostic characters of Marginocephalia? How are marginocephalians related to other ornithischians?

3   What are the diagnostic characters of Pachycephalosauria?

4   What are the diagnostic characters of Ceratopsia?

5   Describe chewing as practiced by ceratopsian dinosaurs.

6   What do we know about ceratopsian egg-laying and nesting?

7   Give a brief history of ceratopsian biogeography.

8   What is sexual selection? Intraspecific competition?

9   What is in the inferred function of the horns and frill in ceratopsians?

10   What is the inferred function of the dome in pachycephalosaurs?

11   How do marginocephalian features relate to intraspecific competition and sexual selection?

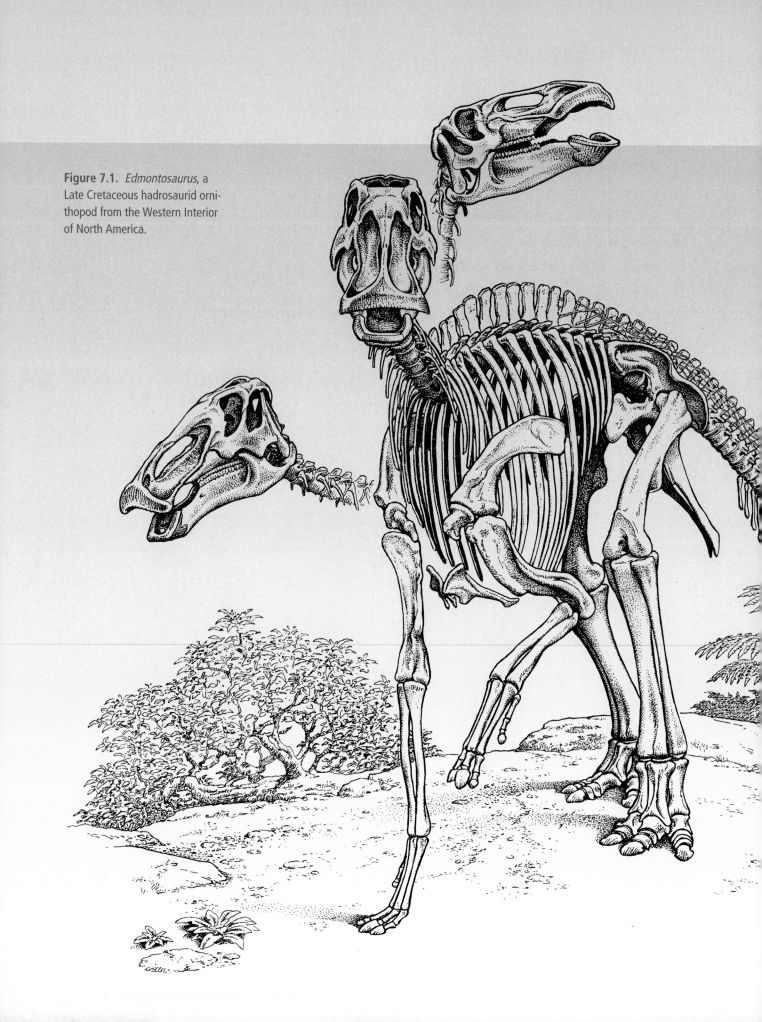

Figure 7.1. *Edmontosaurus*, a Late Cretaceous hadrosaurid ornithopod from the Western Interior of North America.

CHAPTER 7

ORNITHOPODA:
THE TUSKERS, ANTELOPES, AND "MIGHTY DUCKS" OF THE MESOZOIC

CHAPTER OBJECTIVES

- Introduce Ornithopoda

- Develop familiarity with current thinking about lifestyles and behaviors of ornithopods

- Develop an understanding of ornithopod evolution using cladograms, and an understanding of the place of Ornithopoda within Dinosauria

Ornithopoda

Ornithopods (*ornith* – bird; *pod* – foot) were the cows, deer, bison, wild horses, antelope, and sheep of the Mesozoic (Figure 7.1). Magnificent herbivores all, they were one of the most numerous, most diverse, and longest-lived groups in all Dinosauria: from the Jurassic, when they first appeared, until the end of the Cretaceous, when they all went extinct, ornithopods evolved nearly 100 species at present count.

Ornithopods spread all over the globe. They ranged from near the then-equator to such high latitudes as the north slopes of Alaska, the Yukon, and Spitsbergen in the Northern Hemisphere, and Seymour Island, Antarctica, and the southeast coast of Australia in the Southern Hemisphere (Figures 7.2 and 7.3). Local conditions in these regions varied widely, so ornithopods lived in quite diverse habitats and in a wide range of climates.

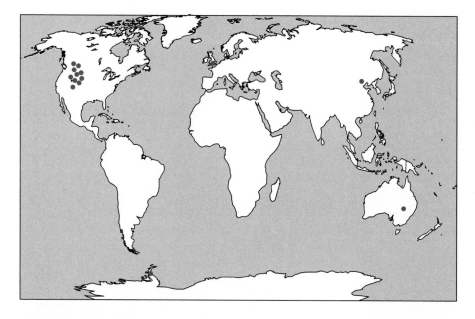

Figure 7.2. Global distribution of Heterodontosauridae and basal Euornithopoda.

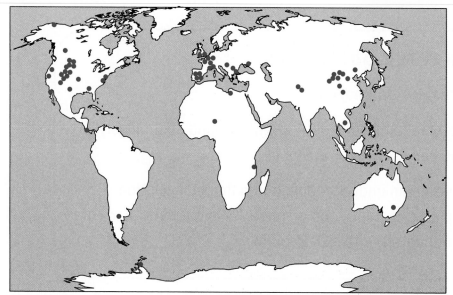

Figure 7.3. Global distribution of Iguanodontia.

(a)

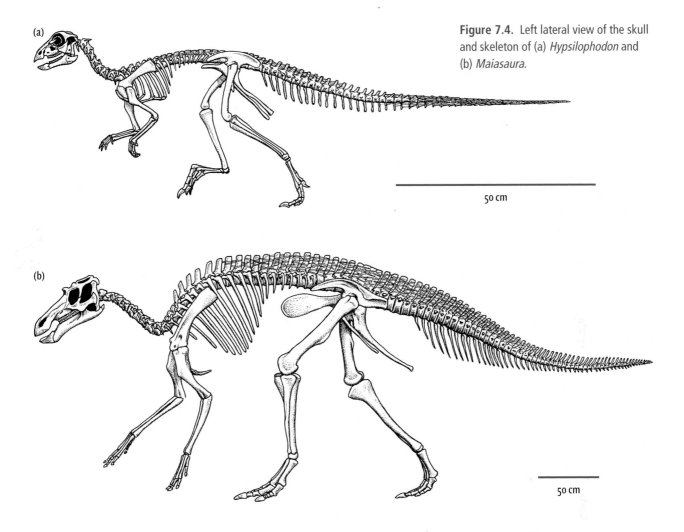

Figure 7.4. Left lateral view of the skull and skeleton of (a) *Hypsilophodon* and (b) *Maiasaura*.

50 cm

(b)

50 cm

They also evolved a range of sizes: early in their history, ornithopods were generally small (ranging from 1 to 2 m in length); however, later some members of the group attained quite large body sizes (upward of 12 m; Figure 7.4).

We know as much about ornithopods as about almost any other group of dinosaurs: *Iguanodon* was a charter member of Sir Richard Owen's original 1842 Dinosauria (see Chapter 14). Hadrosaurids ("duckbills") are known from single bones to huge bonebeds. Their remains include skin impressions – patterns have even been used to distinguish different species – and ossified tendons, as well as delicate skull bones such as sclerotic rings (that support the eyeball), stapes (the thin rod of bone that transmits sound from the eardrum to the brain), and hyoid bones (delicate bones that support the tongue). Paleontologists have also found hadrosaurid eggs and all growth stages represented, from hatchling, to "teenager," to adult. Ornithopod footprints and trackways abound in many parts of the world.

Who were the ornithopods?

As we have seen, ornithopods are genasaurian cerapodans (Figure 7.5). As ornithopod phylogeny is currently understood, there is a basic split between some primitive ornithopods, including *Agilisaurus* and *Hexinlusaurus*, and the remaining ornithopods, Euornithopoda (*eu* – true). Within euornithopods, iguanodontians and hadrosaurids are two important monophyletic groups.

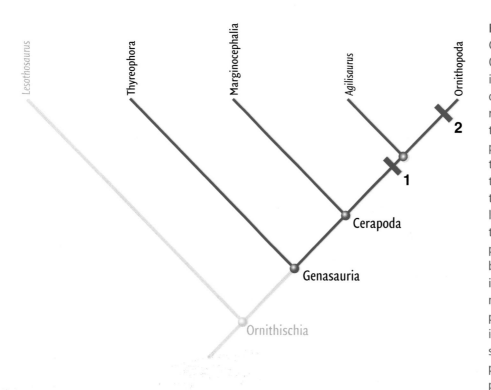

Figure 7.5. Cladogram of Genasauria, monophyly of Ornithopoda. Derived characters include: at **1**, pronounced ventral offset of the premaxillary tooth row relative to the maxillary tooth row, crescentic paroccipital processes, strong depression of the mandibular condyle beneath the level of the upper and lower tooth rows, elongation of the lateral process of the premaxilla to contact the lacrimal and/or prefrontal; at **2**, scarf-like suture between postorbital and jugal, inflated edge on the orbital margin of the postorbital, deep postacetabular blade on the ilium, well-developed brevis shelf, laterally swollen ischial peduncle, elongate and narrow prepubic process.

Ornithopod lives and lifestyles

Gettin' around Ornithopods functioned as bipeds and quadrupeds, with both locomotor modes commonly occurring in the same beast. The smallest were predominantly bipedal, with gracile bones that suggest agility and speed. When eating or standing still, however, they also may have adopted a quadrupedal stance. Some of the larger ornithopods, however, such as *Iguanodon*, may have functioned more as full-time quadrupeds, only going bipedal when in a hurry. A number of larger ornithopods have a sturdy wrist and a hand with thickened hoof-like nails on the central digits, a hand that clearly was capable of considerable weight support. Indeed, some hadrosaurs are now known to have been obligate quadrupeds; they were quadrupedal at all times. For all this, however, reconstructions of their hip musculature suggests more similarities with bipeds than with quadrupeds. Perhaps the strongly bipedal cast to their inferred hip musculature is a function of their ancestry: like ceratopsians, they undoubtedly evolved from bipedal ancestors, but unlike ceratopsians, their hip morphology appears not to have evolved as distinctly quadrupedal. Interestingly, juveniles may have been more bipedal than their fully grown, adult counterparts.

In all cases, the tail was long, muscular, strengthened by ossified tendons, and held at or near horizontal, making an excellent counterbalance for the front of the animal. In general, the powerful hindlimbs tend to be at least as long as, and in some cases more than twice the length of, the forelimbs.

How fast could these dinosaurs have traveled? Larger iguanodontians such as hadrosaurids may have been able to reach 15 to 20 km/h during a sustained run, but upward of 50 km/h on short sprints. Quadrupedal galloping appears unlikely given the rigidity of the vertebral column and the limited movement of the shoulder against the ribcage and sternum. For smaller ornithopods, running speeds were higher. Maximum speeds were probably on the order of 60 km/h (see Box 12.3).

Arms and hands Many euornithopods appear to have had gracile forelimbs, and likely used their hands to grasp at leaves and branches, bringing foliage closer to the mouth so that it could be nipped off by the toothed beak.

Most iguanodontians had very specialized fingers and hands, indicating multiple functions (Figure 7.6). In *Iguanodon*, for example, the first digit (thumb) was conical and sharply pointed. It has been suggested that it was used as a stiletto-like, close-range defensive weapon or for breaking into seeds and fruits (or both, or … ?). The middle digits (II, III, and IV) were tipped with hooves, and were evidently weight-bearing; these would have been key players when the animal was in a quadrupedal stance. Finally, the outer finger (V) was highly flexible, and could bend across the palm, very much as the thumb does in humans: a grasping, opposable pinkie.

Unlike other iguanodontians, in hadrosaurids digit I of the hand was lost, and digit V was relatively small. This left three main fingers, all tipped with hooves, with hardly any way to function other than to support the animal while standing. Hadrosaurids likely spent a lot of time on all fours.

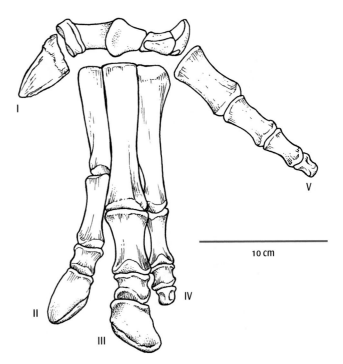

Figure 7.6. The left hand of *Iguanodon*. Note the spiked thumb (digit I).

Dietary fiber Fine dining, ornithopod style, is relatively well understood. For hadrosaurids at least, "mummies," complete with stomach contents, have been found. These spectacular specimens, though not true mummies in the sense that original soft tissue and bone are preserved, apparently first dried to dinosaur jerky before burial. The dried, toughened muscle and flesh tissue didn't decompose, so the whole package – tissue *and* bones – was replaced during burial (see Chapter 1). The startling result is preserved skin impressions, stretched tendons and muscles, and the last supper in the stomach and intestines. Hadrosaurids, we now know, ate twigs, berries, and coarse plant matter.

This selection of food correlates nicely with ornithopod height: they are thought to have been active foragers on ground cover and low-level foliage from conifers and in some cases from deciduous shrubs and trees of the newly evolved angiosperms; that is, the clade of all plants that bear flowers (see Chapter 13). Browsing on such vegetation appears to have been concentrated within the first meter or two above the ground, but the taller animals may have reached vegetation as high as 4 m.

Eating coarse, fibrous food requires some no-nonsense equipment in the jaw to extract enough nutrition for survival, and ornithopods had the necessary goods (Figures 7.7, 7.8, and 7.9). Like all genasaurs, ornithopods had a beak in the front for cropping vegetation, a diastema, a group of cheek teeth for shearing (Figure 7.10), and a large, robust coronoid process for serious mastication. A deeply inset tooth row indicates large fleshy cheeks. But beyond these basics, different ornithopods had different modifications of the jaw, and different kinds of jaw motions are believed to have been used for the processing of food.

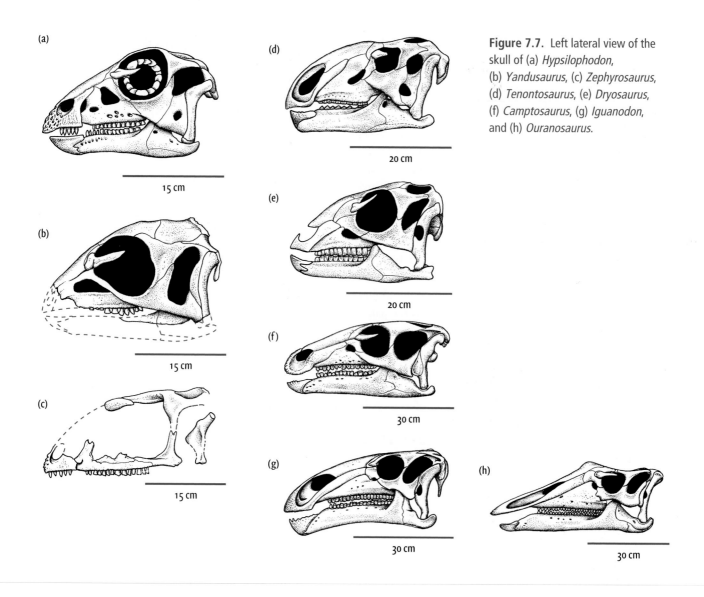

Figure 7.7. Left lateral view of the skull of (a) *Hypsilophodon*, (b) *Yandusaurus*, (c) *Zephyrosaurus*, (d) *Tenontosaurus*, (e) *Dryosaurus*, (f) *Camptosaurus*, (g) *Iguanodon*, and (h) *Ouranosaurus*.

Modern treatments of ornithopod jaw mechanics suggest some differences in ornithopod feeding behavior. In basal ornithopods, the beak was relatively narrow, implying a somewhat selective cropping ability. Euornithopods, by contrast, had broad snouts (Figure 7.11), and in some cases even developed a strongly serrated edge on the rhamphotheca. They were likely not too selective; instead, they hacked and severed leaves and branches without much regard for what they were taking in. Basal ornithopods were likely careful nibblers, while euornithopods were lawn-mowers.

Beyond the diastem, the chewing began. Here it was aided by something that is utterly foreign to humans. Our skulls and lower jaws are akinetic, meaning that, except for the vertical motion of the lower jaws, the bones in our skulls are solidly fused and locked together. Not so with ornithopods. Above and beyond the familiar up and down compression of chewing, slight movements of particular, individual bones within the skull and lower jaws allowed the cheek teeth to grind past each other from *side to side*. A skull in which individual bones move is called kinetic.

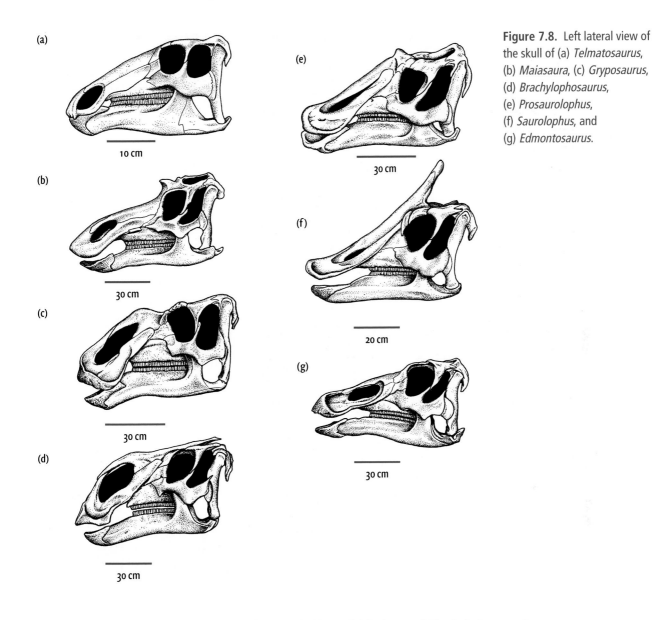

Figure 7.8. Left lateral view of the skull of (a) *Telmatosaurus*, (b) *Maiasaura*, (c) *Gryposaurus*, (d) *Brachylophosaurus*, (e) *Prosaurolophus*, (f) *Saurolophus*, and (g) *Edmontosaurus*.

Euornithopods evolved a unique, kinetic skull, in which they mobilized their upper jaws. This kind of mechanism, called pleurokinesis, involved a slight outward rotation of portions of the upper jaw, especially the maxilla (the bone that contains the upper teeth), with each bite (Figure 7.12). When the upper and lower teeth were brought into contact on both right and left sides, the opposing surfaces of the dental batteries sheared past one another. Pleurokinesis was an important advance for euornithopods, giving them the ability to chew the toughest, most fibrous plants.

Chewing reached its most refined state in hadrosaurids, in which the cheek teeth were fitted tightly together into a *dental battery*, which effectively acted as a single shearing or grinding tool in each jaw (see Figure 7.10d). With constantly replacing teeth, tooth wear was never an issue (as it is in mammals, which only replace teeth once, so their adult teeth have to last their entire lives). The toughest, most fibrous plants undoubtedly succumbed to the hadrosaurid combination of powerful jaw muscles operating on a pleurokinetic skull equipped with constantly replaced grinding surfaces.

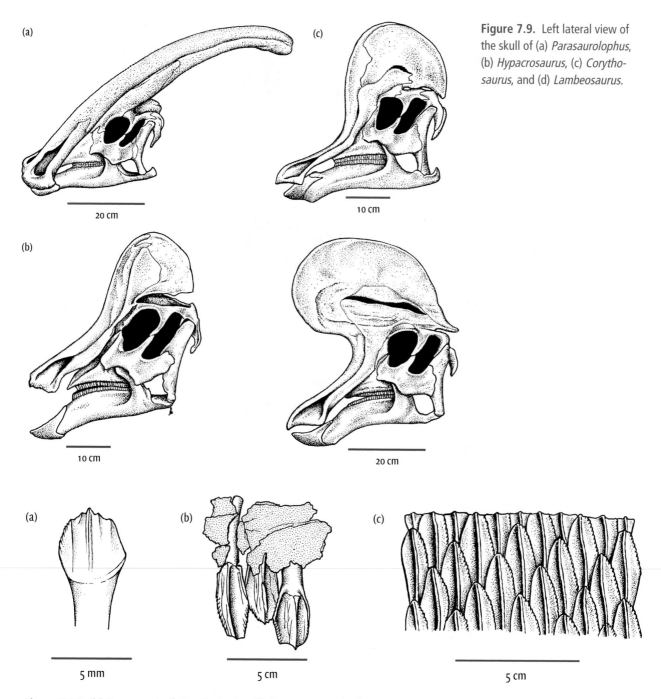

Figure 7.9. Left lateral view of the skull of (a) *Parasaurolophus,* (b) *Hypacrosaurus,* (c) *Corythosaurus,* and (d) *Lambeosaurus.*

Figure 7.10. (a) Upper tooth of *Hypsilophodon,* (b) three upper teeth of *Iguanodon,* and (c) lower dental battery of *Lambeosaurus.* Note that the teeth in these forms are progressively more tightly packed, culminating in the lambeosaur dental battery.

As in all of the other ornithischians that have been discussed, once the food was properly chewed, it was swallowed, and quickly passed through a capacious gut that was present in all ornithopods, and proportionately even larger in the largest iguanodontians (including hadrosaurs). Ornithopods were uniquely equipped to extract the most nutrition out of a low-quality, high-fiber, high-volume diet.

Thoughts of an ornithopod By dinosaur standards, ornithopods were smart – as smart or smarter than might be expected of living archosaurs if they were scaled up to dinosaur size (see Box 12.5). For example, *Leaellynasaura*, a basal euornithopod from Victoria, Australia, was apparently quite brainy and had acute vision, as suggested by prominent optic lobes in the brain.[1] In general, ornithopod smarts may be related to greater reliance on sight, smell, and hearing for protection that, in the absence of other means, may have been their only defense. Moreover, brain size in these dinosaurs may have also been an integral part of a complex behavioral repertoire.

Socializing *à la* Ornithopoda

From the time of their discovery, ornithopods have attracted a good deal of attention, particularly for the extraordinary crests on the heads of hadrosaurids and the lumps on the forehead of *Ouranosaurus*. All of these features hint at sophisticated social behavior.

Song of the saurian Hadrosaurids have attracted the most attention, in large part because of the striking solid or hollow crests – many of them chambered – borne by many genera. The hollow crest morphology was once thought to relate to the aquatic habits of the group or to smell, but studies suggest that the internal chambers of the crests would have made good resonating chambers, producing loud, low-frequency sounds – a kind of Mesozoic *alpenhorn*.

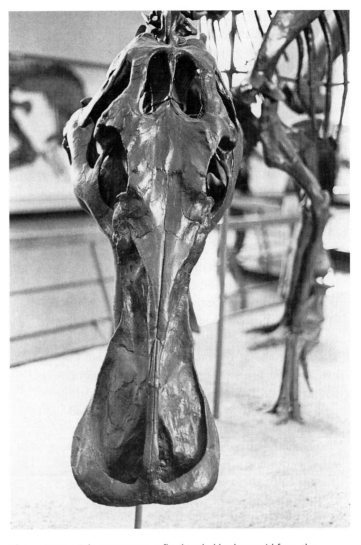

Figure 7.11. *Edmontosaurus,* a flat-headed hadrosaurid from the western USA.

With that insight, much of the discussion now centers on intraspecific competition (see Chapter 5) and sexual selection (see Chapter 6). To convey information about species, sex, and even rank, crests had to have been *visually* and, if part of their function was as a resonating chamber, *vocally* distinctive. Only then can they have promoted successful matings between consenting adults. How strongly is the role of sexual selection implied by hadrosaurid crests?

Paleontologist J. A. Hopson made five predictions that test the hypothesis that hadrosaurid crest morphology was all about sexual selection.

1 If communication and display were important, hadrosaurids must have had both good hearing and good vision.
2 If the crest served the dual role of visual display *and* as a vocal resonator, then its external shape need not necessarily mimic the internal shape of the resonating cavities inside.

---

[1] The animal had an estimated encephalization quotient (EQ; see Box 12.5) of 1.8; J. A. Hopson estimated that the average EQ of other ornithopods is about 1.5.

3   If crests acted as visual signals, then they should be species-specific in size and shape, and they should also be sexually dimorphic.

4   If the crests were a visual cue, they ought to be increasingly distinctive as the number of hadrosaurids living together increases.

5   The crests should become more distinctive through time as a consequence of sexual selection.

How did these hypotheses fare? Hypothesis (1) is relatively well supported, in that hadrosaurids, to judge from their sclerotic rings (see Figure 4.6), had relatively large eyes, implying acute vision. Similarly, preserved middle and inner ear structures suggest that a wide range of frequencies was audible to these animals. Hypothesis (2) is upheld in virtually all cases, in that the profile of the crest is more elaborate or extensive than the walls of the internal plumbing (Figure 7.13). Hypothesis (3) is amply upheld in large part thanks to studies on the growth and development in lambeosaurine hadrosaurids, which show that crests become most prominent when an animal approaches sexual maturity (Figure 7.14). Moreover, adult lambeosaurines are known to be dimorphic, particularly in terms of crest size and shape. Could these "morphs" have been male and female?

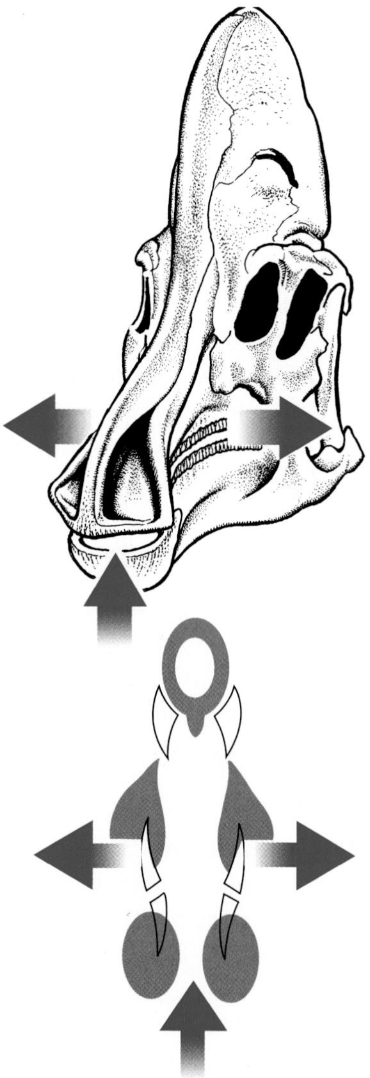

Hypothesis (4) is based on the idea that distinctiveness would be an advantage during the breeding season. It was tested at Dinosaur Provincial Park in Alberta, Canada, where five distinctively equipped species of hollow-crested hadrosaurid and one species of solid-crested hadrosaurid all lived together in multi-species bliss. In support of the hypothesis, elsewhere where hadrosaurid diversity is lower, the distinctiveness of the crests is decreased. Interestingly, however, hypothesis (5) is not well supported, for lambeosaurines' crests, at least, arguably become less distinctive over time.

If the crests were used for species recognition, ritualized display, courtship, parent–offspring communication, and social ranking, the accentuated nasal arch seen in *Gryposaurus*, *Maiasaura*, and *Brachylophosaurus* may have been used for broadside or head pushing during male–male combat (Figure 7.14). Inflatable flaps of skin possibly covered the nostrils and surrounding regions (Figure 7.15); if present, these could have been inflated and used for visual display, as well as noise-making – more a Mesozoic *bagpipe* (Figure 7.15b). In *Prosaurolophus* and *Saurolophus* (see Figure 7.8f), a sac might have extended onto the solid crest that extended above the eyes (Figure 7.15c), while in *Edmontosaurus* (see Figures 7.8g and 7.11), where the nasal arch is not

Figure 7.12. Jaw mechanics in Euornithopoda, showing lateral mobility of the upper jaws (pleurokinesis).

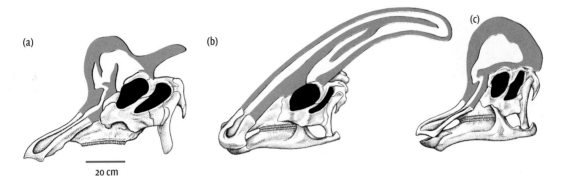

Figure 7.13. Highly modified nasal cavities housed within the hollow crests on the heads of lambeosaurine hadrosaurids. (a) *Lambeosaurus*; (b) *Parasaurolophus*; (c) *Corythosaurus*.

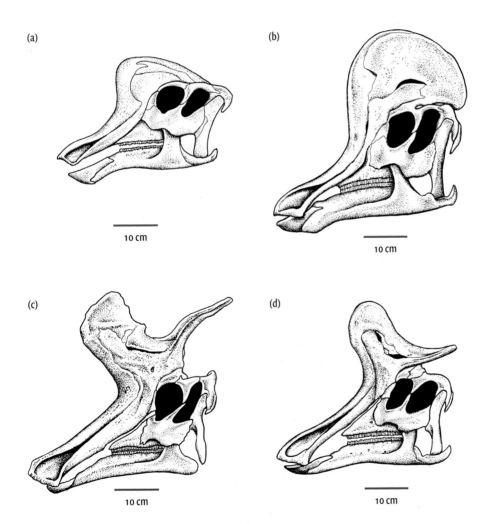

Figure 7.14. Growth and sexual dimorphism in lambeosaurine hadrosaurids. (a) Juvenile and (b) adult *Corythosaurus*. (c) Male (?) and (d) female (?) *Lambeosaurus*.

accentuated nor is there a crest, the complexly excavated nostril region may have housed an inflatable sac. Unfortunately, these soft-tissue-based hypotheses are all speculative.

In lambeosaurine (hollow-crested) hadrosaurids, the crests perched atop the head must have provided for instant recognition (Figures 7.9 and 7.16). This would have been by visual

(a)

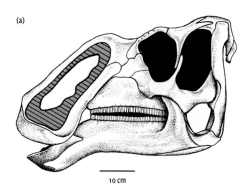

10 cm

(b)

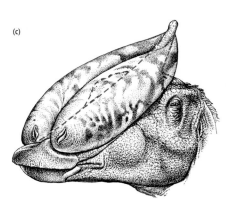

(c)

Figure 7.15. (a) The circumnarial depression in *Gryposaurus* (indicated by cross-hatched region) which may have supported an inflatable flap of skin in hadrosauridines; (b) speculative reconstruction of an inflatable sac in *Gryposaurus*; (c) speculative reconstruction of an inflatable sac in the solid-crested hadrosaurid *Saurolophus*.

cues as well as by low honking tones produced in the large resonating chamber of the crest (see Figure 7.13).

Other ornithopods Other ornithopods show features potentially interpretable in terms of sexual selection and intraspecific competition. Low, broad bumps on top of the head of *Ouranosaurus* and the arched snout of *Muttaburrasaurus* and *Altirhinus* may well have behavioral significance relating to intraspecific competition and sexual selection. *Ouranosaurus* was equipped with extremely tall neural spines, which formed a high, almost sail-like ridge down its back (Figure 7.17). Like *Stegosaurus* (see Chapter 5), it is possible that these long spines were covered with skin and used by the animal to warm up and cool down, and/or they may also have had a display function, providing the animal with a greater side profile than it would otherwise have had.

Display behavior in many ornithopods begins to make even more sense when considered in the context of the discoveries of bonebeds containing just one type of dinosaur. Monotypic bonebeds, that is, bonebeds containing only one type of animal, are known for *Dryosaurus*, *Iguanodon*, *Maiasaura*, and *Hypacrosaurus*, among ornithopods. In the case of hadrosaurids, at least, the evidence suggests that a single herd could have exceeded 10,000 individuals, rivaling the multiple mile-sized bison herds that roamed the Great Plains of North America before the unfortunate pairing of the transcontinental railroad with the .55 caliber Sharps rifle.

Figure 7.16. *Corythosaurus*, a hollow-crested hadrosaurid from the Late Cretaceous of western Canada.

Bringing up baby II The secrets of ornithopod reproductive behavior are just beginning to be told. For the small, basal ornithopod *Orodromeus*, hatchlings had well-developed limb bones, with fully formed joints, indicating that the young could walk, run, jump, and forage for themselves as well as any adult. We thus infer minimal parental care.

As first discovered by paleontologist J. R. Horner, hadrosaurids did not take so *laissez-faire* an attitude toward child-rearing. *Maiasaura*, *Hypacrosaurus*, and probably most others

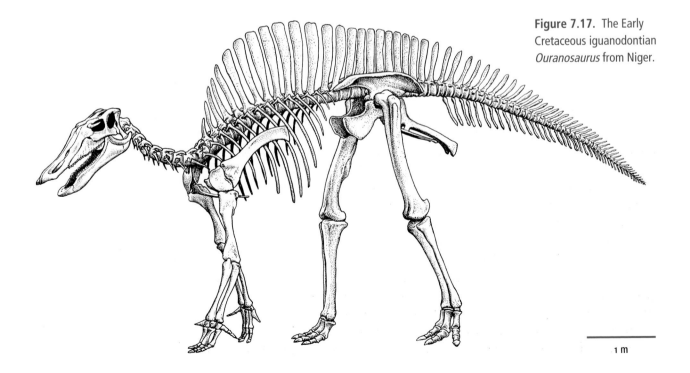

Figure 7.17. The Early Cretaceous iguanodontian *Ouranosaurus* from Niger.

1 m

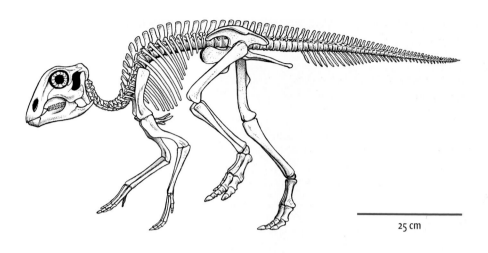

Figure 7.18. Left lateral view of the skull and skeleton of a hatchling hadrosaurid *Maiasaura*.

25 cm

nested in colonies, digging a shallow hole in soft sediments and laying, in the case of *Maia-saura*, up to 17 eggs in each nest. These nests were separated by about a mother's body length, strongly suggesting that they were regularly tended by a parent. Hatchlings (Figure 7.18) are found amid an abundance of eggshell fragments, implying an extended stay at the nest that wreaked havoc on the eggs that once housed them.

With poorly developed joints and limbs, the offspring were helpless during the nest-bound time. They could hardly have foraged far from the nest and must have depended on their parents to provision and protect them. But to go from a 1 m hatchling to a 9 m adult, growth must have come hot and heavy: at approximately 12 cm per month, as fast as or faster than fast-growing mammals and birds (see Chapter 12). This means that hatchlings must have channeled into growth virtually all of the food that their parents brought them.

Recently the development of *Hypacrosaurus*, at least, has been taken to new levels with a detailed analysis of changes in the face as the animals grew up. A complete growth series of skull specimens representing embryos, nestlings, juveniles, subadults (teenagers!), and adults has been reconstructed (*à la Protoceratops*) and from this it is clear that these dinosaurs, like all juveniles, begin life relentlessly cute: short snouts, large eyes, and rudimentary crest development! Just as in all other vertebrates, these features change, becoming less cute, as adulthood is reached.

A new take on ornithopod child-rearing was provided by the hypsilophodont *Orycto-dromeus*, a dinosaur that evidently raised atricial young in a burrow (see Figure 1.8). Only one specimen of the animal is known, but this reveals a burrow with an end chamber, in which were found the remains of an adult and two juveniles. *Oryctodromeus* appears to have some digging specializations in its skull and thoracic region, suggesting a fossorial, or burrowing, lifestyle.

Ornithopod life, therefore, certainly involved much opportunity for interaction: within herds, as breeding pairs, and as families. All of this is part and parcel of the visual and vocal communication we postulated earlier, and affirms complex social behavior, particularly in hadrosaurids.

Ornithopods give us insights into dinosaurian life history strategies; that is, the ways in which particular organisms grow, reproduce, and die. One strategy, called the *r*-strategy, is to produce enormous numbers of eggs that result in thousands of offspring, the vast majority of which do not survive to reproduce during their relatively short lifespans. Think mayflies. Such offspring, born as near-adults, are called precocial. No parental care here – too many children for serious parental investment.

This contrasts with the *K*-strategy, involving fewer offspring, lots of parental care, and longer lifespans. Think whales. Such offspring, born with a longer trek toward adulthood and requiring parental investment to get there, are called altricial.

How do those ornithopods for which we have information conform to either of these two contrasting strategies? *Orodromeus* seems to have been an *r*-strategist, an inference that is based on the precocial nature of the young. In contrast, *Maiasaura*, *Hypacrosaurus*, and perhaps other hadrosaurids had nest-bound, altricial hatchlings, and were likely *K*-strategists.

The evolution of Ornithopoda

The basal split of Ornithopoda from the generalized cerapodan condition likely occurred in the latest Triassic or earliest Jurassic. The clade is diagnosed on the basis of a number of derived features (see Figure 7.5). As we've seen, an early split in Ornithopoda occurred between primitive ornithopods such as *Orodromeus* and *Agilisaurus*, and Euornithopoda (see Figure 7.5), with euornithopods containing much of the future diversity of the clade.

Euornithopoda is a well-diagnosed group (Figure 7.19). It consists of a host of relatively small, agile ornithopods such as *Hypsilophodon* and *Gasparinisaura*, as well as a few somewhat larger, more robust forms (*Parksosaurus* and *Thescelosaurus*), and the diverse clade Iguanodontia (Figure 7.20), residence of such dinosaurian luminaries as *Camptosaurus* (Figure 7.21), *Iguanodon*, and all the hadrosaurids. In general, non-hadrosaur iguanodontians tended to reach their apogee in the Late Jurassic–Early Cretaceous interval.

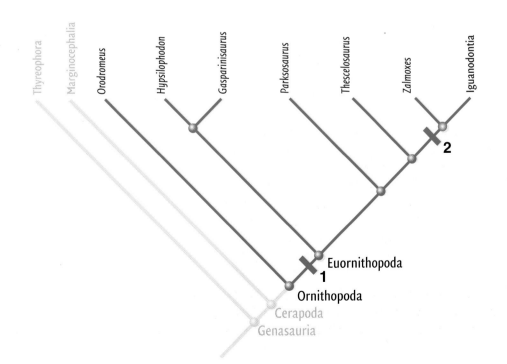

Figure 7.19. Cladogram of basal Ornithopoda. Derived characters include: at **1**, subcircular external antorbital fenestra, distal offset to apex of maxillary crowns, strongly constricted neck to the scapular blade, ossification of sternal ribs, hypaxial ossified tendons in the tail; at **2**, rectangular lower margin of the orbit, widening of the frontals, broadly rounded predentary, dentary with parallel dorsal and ventral margin, absence of premaxillary teeth, 10 or more cervical vertebrae, 6 or more sacral vertebrae, presence of an anterior intercondylar groove, inflation of the medial condyle of the femur.

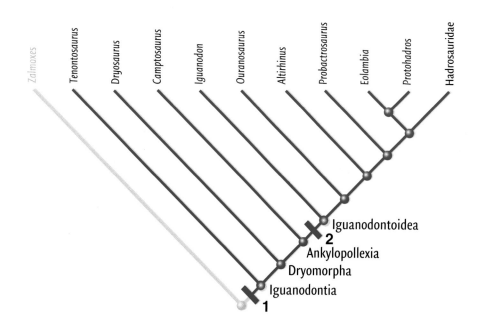

Zalmoxes   Tenontosaurus   Dryosaurus   Camptosaurus   Iguanodon   Ouranosaurus   Altirhinus   Probactrosaurus   Eolambia   Protohadros   Hadrosauridae

Iguanodontoidea
2
Ankylopollexia
Dryomorpha
Iguanodontia
1

Figure 7.20. Cladogram of basal Iguanodontia. Derived characters include: at **1**, premaxilla with a transversely expanded and edentulous margin, reduction of the antorbital fenestra, denticulate margin of the predentary, deep dentary ramus, loss of sternal rib ossification, loss of a phalanx in digit III of the hand, compressed and blade-shaped prepubic process; at **2**, strong offset of premaxilla margin relative to the maxilla, peg-in-socket articulation between maxilla and jugal, development of a pronounced diastema between the beak and mesial dentition, mammillations on marginal denticles of teeth, maxillary crowns narrower and more lanceolate than dentary crowns, closely appressed metacarpals II–IV, deep triangular fourth trochanter, deep extensor groove on femur.

Hadrosaurids are among the best known of all dinosaurs, with a fabulous fossil record that allows, as we have seen, insights into their behavior. Two major clades within hadrosaurids – Lambeosaurinae and Hadrosauridinae – constitute most of Hadrosauridae (Figure 7.22), with a few forms left whose relationships within Hadrosauridae are uncertain.

Several interesting evolutionary trends are present within Ornithopoda. It is perhaps no coincidence that ornithopod diversity seems to parallel gymnosperm and angiosperm diversity (see Figure 13.10), suggesting that these dinosaurs and plants may have been involved in a kind reciprocal *pas de deux*: as gymnosperms developed ways to discourage predation, ornithopods developed more and more efficient ways of extracting nutrients. The reciprocal evolution culminated in the highly efficient pleurokinetic hadrosaurid jaw, with its well-developed, integrated

Figure 7.21. The Late Jurassic iguanodontian *Camptosaurus*, from the Western Interior of the USA.

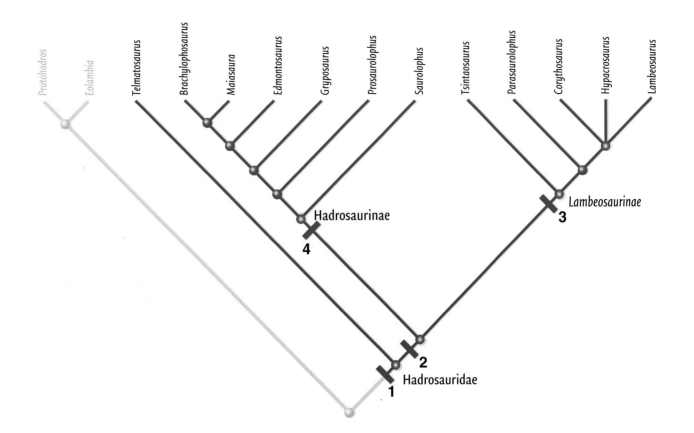

Figure 7.22. Cladogram of Hadrosauridae. Derived characters include: at **1**, three or more replacement teeth per tooth position, posterior extension of the dentary tooth row to behind the apex of the coronoid process, absence of the surangular foramen, absence or fusion of the supraorbital to the orbit rim, long coracoid process, dorsoventrally narrow proximal scapula, very deep, often tunnel-like intercondylar extensor groove; at **2**, absence of the coronoid bone, reduction in surangular contribution to coronoid process, double-layered premaxillary oral margin, triangular occiput, eight or more sacral vertebrae, reduced carpus, fully open pubic obturator foramen, absence of distal tarsals II and III; at **3**, maxilla lacking an anterior process but developing a sloping dorsal shelf, groove on the posterolateral process of the premaxilla, low maxillary apex, a parietal crest less than half the length of the supratemporal fenestrae; at **4**, presence of a caudal margin on the circumnarial fossa.

packages of teeth in continuously replacing dental batteries. Overall patterns within the Late Cretaceous of North America and Asia, at least, suggest that hadrosaurids ecologically replaced large non-hadrosaurid iguanodontians. Hadrosaurids arguably evolved the art of chewing to levels of sophistication unparalleled in the history of life.

Along with the specializations associated with chewing, ornithopod evolution may have also been characterized by a greater and greater investment of parents in their young. We have seen that *Orodromeus* produced relatively precocial offspring; its basal position within Euornithopoda (see Figure 7.19) suggests that precocity may be primitive for at least Euornithopoda. As the diversity of Euornithopoda increased, altricial behavior likely evolved in more derived euornithopods some time prior to the origin of Hadrosauridae, which all are thought to have given birth to altricial young.

Summary

Ornithopods were the most numerous and diverse herbivores of Dinosauria, consisting of iguanodontians, hadrosaurids (duckbills), and a few distinctive, more primitive forms. Although equipped with robust back legs and long tails, and undoubtedly capable of sustained bipedal locomotion, hooves on certain fingers of hadrosaurids and iguanodontians suggest that these animals spent time in a quadrupedal stance as well.

Ornithopods ranged in size from very small (< 2 m) to rather large (>15 m), and evidently colonized virtually every inhabitable region of the globe.

Ornithopods had very advanced chewing capabilities. The skull has an inset tooth row indicating cheeks, as well as the tripartite division of chewing herbivores (including a cropping beak, a diastem, and closely appressed cheek teeth for grinding), and in virtually all cases, a large coronoid process suggests strong jaw adductor muscles. Hadrosaurs took chewing to unprecedented heights with the evolution of a pleurokinetic skull combined with dental batteries. Coprolites and stomach contents suggest hadrosaurs needed all the teeth they had: their diet appears to have been coarse and fibrous.

Ornithopods were very social animals, none more so than hadrosaurs. The discovery of many bonebeds attests to this, as do a remarkable and complex variety of sexually dimorphic head and skull features found in many ornithopods, suggesting that life as an ornithopod involved intensive sexual selection. In hadrosaurs at least, communication may well have been enhanced by a variety of vocalizations.

The genus *Maiasaura* has given us a view of child-rearing, duckbill style. Complete growth series, from hatchlings at nests to adults, are known, and considerable evidence exists that hadrosaurs grew remarkably quickly. Nesting apparently was communal, and parental care was expended on altricial young: duckbills were likely *K*-strategists. Interestingly, other ornithopods, such as the genus *Orodromeus*, may have raised precocial young and favored an *r*-strategy of child-rearing.

Ornithopod evolution was characterized by a increase in the sophistication of chewing specializations. Among the large ornithopods, it could be argued that, in North America and Asia at least, the highly advanced hadrosaurs ecologically replaced non-hadrosaurid iguanodontians.

Selected readings

Butler, R. J., Smith, R. H. M. and **Norman, D. B.** 2007. A primitive ornithischian from the Late Triassic of South Africa and the early evolution and diversification of Ornithischia. *Proceedings of the Royal Society Series B*, **274**, 2041–2046.

Butler, R. J., Upchurch, P. and **Norman, D. B.** 2008. The phylogeny of the ornithischian dinosaurs. *Journal of Systematic Palaeontology*, **6**, 1–40.

Hopson, J. A. 1975. The evolution of cranial display structures in hadrosauridian dinosaurs. *Paleobiology*, **1**, 21–43.

Horner, J. R. 1984. The nesting behavior of dinosaurs. *Scientific American*, **250**, 130–137.

Horner, J. R., Weishampel, D. B. and **Forster, C. A.** 2004. Hadrosauridae. In Weishampel, D. B., Dodson, P. and Osmólska, H. (eds.), *The Dinosauria*, 2nd edn. University of California Press, Berkeley, pp. 438–463.

Norman, D. B. 2004. Basal Iguanodontia. In Weishampel, D. B., Dodson, P. and Osmólska, H. (eds.), *The Dinosauria*, 2nd edn. University of California Press, Berkeley, pp. 413–437.

Norman, D. B., Sues, H.-D., Coria, R. A. and **Witmer, L. M.** 2004. Basal Ornithopoda. In Weishampel, D. B., Dodson, P. and Osmólska, H. (eds.), *The Dinosauria*, 2nd edn. University of California Press, Berkeley, pp. 393–412.

Ostrom, J. H. 1961. Cranial morphology of the hadrosauridian dinosaurs of North America. *Bulletin of the American Museum of Natural History*, **122**, 33–186.

Weishampel, D. B. 1984. The evolution of jaw mechanisms in ornithopod dinosaurs. *Advances in Anatomy, Embryology and Cell Biology*, **87**, 1–110.

Varricchio, D. J., Martin, J. A. and **Katsura, Y.** 2006. First trace and body fossil evidence of a burrowing, denning dinosaur. *Proceedings of the Royal Society Series B*, **274**, 1361–1368.

Topic questions

1   Who are the ornithopods? What are the diagnostic characters of Ornithopoda? How are ornithopods related to other ornithischians?

2   What are the major divisions of Ornithopoda, and what are their diagnostic characters?

3   How could a single ornithopod be both bipedal and quadrupedal?

4   Describe the hands of non-hadrosaurid iguanodontians. How did those differ from the hands of hadrosaurids?

5   Highlight what is known of nests and nesting in ornithopods.

6   What is pleurokinesis? How did it function in the jaws of hadrosaurids?

7   Name another vertebrate with a kinetic skull.

8   Why is it that most paleontologists now think that hadrosaurid head structures were related to intraspecific competition and/or sexual selection?

9   Give a non-dinosaur example of a *K*-strategist, and an *r*-strategist.

10   How does ornithopod diversity parallel gymnosperm and angiosperm diversity?

11   Use a cladogram to make the argument that a *K*-strategy evolved *at least* two times in the history of vertebrates.

PART III

SAURISCHIA:
MEAT, MIGHT, AND MAGNITUDE

Saurischians include the smallest of dinosaurs and the largest animals that ever lived on land; the most agile and ferocious of predatory dinosaurs and the most ponderous of plant-eaters; the brightest and, evidently, the most dim-witted of dinosaurs; the most Earth-bound and the most aerial. And stealth saurischians, birds, remain with us today, very much alive and well!

Saurischians don't have an obvious family resemblance, and you could be forgiven if you were reluctant to suppose that they are all more closely related to each other than they are to anything else. But our best evidence suggests that they are.

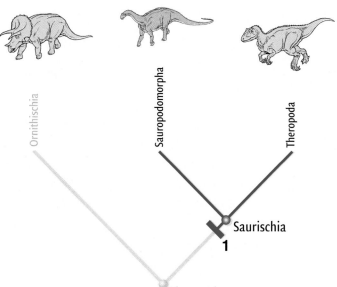

Saurischia: the big picture

What makes a saurischian a saurischian?

For all their differences, Saurischia is monophyletic, and is diagnosed by more than a dozen derived features (Figure III.1), two of which are shown in Figure III.2.

Recall from Chapter 4 that, as early as 1887, H. G. Seeley recognized two great clades within Dinosauria: Ornithischia and Saurischia. Having dwelled on Ornithischia for the past three chapters, we now turn to Saurischia for the next four. Seeley's Saurischia originally consisted of Sauropodomorpha (see Chapter 8) and its sister-taxon Theropoda (*theros* – wild beast; Chapters 9, 10, and 11). A modern view, however, also includes a few dinosaurs that appear to be neither sauropodomorphs nor theropods within Saurischia.

Figure III.1. Cladogram of Dinosauria, emphasizing the monophyly of Saurischia. Derived characters include: at **1**, fossa expanded into the anterior corner of the external naris, the development of a subnarial foramen, a concave facet on the axial intercentrum for the atlas, elongation of the centra of anterior cervical vertebrae, hyposphene–hypantrum articulation on the dorsal vertebrae, expanded transverse processes of sacral vertebrae, loss of distal carpal V, twisting of the first phalanx of manual digit I, well-developed supracetabular crest, restriction of the medioventral lamina of the ischium to the proximal third of the bone.

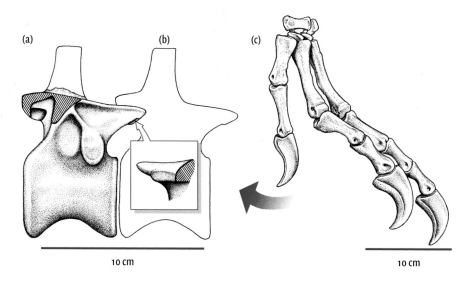

(a)   (b)   (c)

10 cm   10 cm

Figure III.2. (a) Dorsal vertebrae of *Herrerasaurus* indicating the extra hyposphene–hypantrum articulations; (b) hypantrum in medial view; (c) twisted thumb (digit I of the hand).

Selected readings

Gauthier, J. A. 1986. Saurischian monophyly and the origin of birds. *Memoirs of the California Academy of Sciences*, **8**, 1–55.

Langer, M. 2004. Basal Saurischia. In Weishampel, D. B., Dodson, P. and Osmólska, H. (eds.), *The Dinosauria*, 2nd edn. University of California Press, Berkeley, pp. 25–46.

Sereno, P. C. 1999. The evolution of dinosaurs. *Science*, **284**, 2137–2147.

Sereno, P. C., Forster, C. A., Rogers, R. R. and **Monetta, A. M.** 1993. Primitive dinosaur skeleton from Argentina and the early evolution of Dinosauria. *Nature*, **361**, 64–66.

Sereno, P. C. and **Novas, F. E.** 1992. The complete skull and skeleton of an early dinosaur. *Science*, **258**, 1137–1140.

CHAPTER OBJECTIVES

- Introduce Sauropodomorpha

- Develop familiarity with current thinking about lifestyles and behaviors of sauropodomorphs

- Develop an understanding of sauropodomorph evolution using cladograms, and an understanding of the place of Sauropodomorpha within Dinosauria

Figure 8.1. *Diplodocus*, one of the best-known sauropodomorphs, from the Late Jurassic of the Western Interior of the USA.

CHAPTER 8

SAUROPODOMORPHA:
THE BIG, THE BIZARRE, AND THE MAJESTIC

Sauropodomorpha

Life as "large"

We've always heard that brontosaurs were *extremely* large, not too bright, and extinct. Isn't that what dinosaurs are all about?

But what about mighty and majestic? Some of these dinosaurs pushed the extremes of terrestrial body size – to the tune of 75,000 kg and possibly more (Figure 8.1). But some of them were considerably smaller! And in growing gargantuan, they taxed biomechanical and physiological design – weight support, neural circuitry, respiration, digestion, *everything* – to the limit. Viewed from that perspective, sauropods were some of the most sophisticated animals that ever walked the face of the Earth.

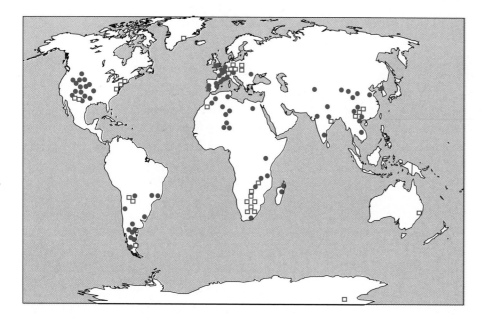

Figure 8.2. Global distribution of Sauropodomorpha. Prosauropoda indicated by solid circles, Sauropoda indicated by open squares.

Sauropodomorphs (*saurus* – lizard; *pod* – foot; *morpho* – form) lived for 160 million years, from the beginning of dinosaur history until its close. Over this long interval, sauropodomorphs managed to walk or be carried to every continent (Figure 8.2), and spawned well over a hundred different species. They must have been doing *something* right!

Who are sauropodomorphs?

Sauropodomorpha is a well-diagnosed group of saurischian dinosaurs (Figure 8.3). The dinosaurs that look like "brontosaurus" – Sauropoda – are but one part Sauropodomorpha; the other consists of a relatively short-lived clade: Prosauropoda (*pro* – before; see Figures 4.5 and 8.4). Sauropodomorphs are split roughly one-third to two-thirds between prosauropods and sauropods.

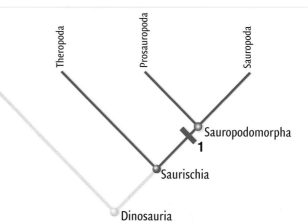

Figure 8.3. Cladogram of Dinosauria emphasizing the monophyly of Sauropodomorpha. Derived characters include: at **1**, relatively small skull (about 5% body length), deflected front end of the lower jaw, elongate lanceolate teeth with coarsely serrated crowns, at least ten neck vertebrae that form a very long neck, dorsal and caudal vertebrae added to the front and hind ends of the sacrum, enormous thumb equipped with an enlarged claw, a very large obturator foramen in the pubis, and an elongate femur.

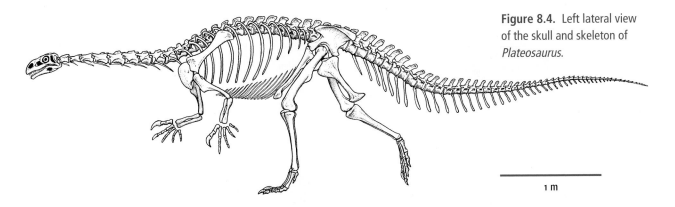

Figure 8.4. Left lateral view of the skull and skeleton of *Plateosaurus.*

1 m

Sauropodomorphs: among the very first dinosaurs

Sauropodomorphs are well represented in among the earliest dinosaurs (see also Chapter 15). They appeared early, and quickly diversified. The changes that they underwent were almost progressive: starting with a bipedal ancestor, heads became smaller, necks became longer, and, as we have seen, hind legs became closer in length to that of the front legs: the animals were becoming quadrupedal. Paralleling this were increasing steps towards unambiguous herbivory.

The link between herbivory and prosauropod evolution is underlined by the fact that the history of prosauropods parallels the rise of gymnosperms – seed-bearing plants (see Figure 13.10). That is, as gymnosperms became an important component of the land plant biota, prosauropods became an increasingly important component of the terrestrial vertebrate fauna. Prosauropods were the first land creatures ever to take advantage of tall-growing plants.

Prosauropoda

Prosauropods were a group of relatively primitive dinosaurs with small heads, long necks, barrel-shaped bodies, and long tails, known from the Late Triassic through Early Jurassic, from all continents except Australia (see Figure 8.2). In general, the front limbs were somewhat shorter than the hindlimbs, and all had five digits. Prosauropod hands were equipped with a large, half-moon-shaped thumb claw (Figure 8.5). Whether for food procurement, defense, and/or some unspecified social activity, the function of this claw remains unknown.

(a)   (b)

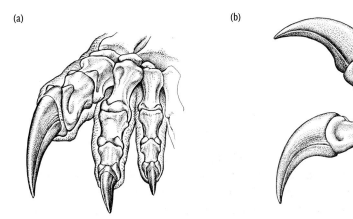

Figure 8.5. Left hand of the prosauropod dinosaur *Plateosaurus,* showing its well-developed thumb claw: (a) reconstructed hand; (b) thumb showing amount of movement permitted by skeleton.

(a)

(b)

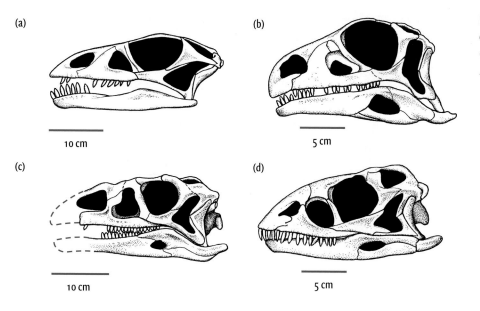

10 cm

5 cm

(c)

(d)

10 cm

5 cm

Figure 8.6. Left lateral view of the skull of (a) *Anchisaurus,* (b) *Coloradisaurus,* (c) *Lufengosaurus,* and (d) *Yunnanosaurus.*

Prosauropod lives and lifestyles

Feeding In the mood for food, sure, but which? The skulls show almost none of the advanced design features associated with chewing (see introduction to Part II: Ornithischia); however, the jaw joint is slightly lower than the tooth row (Figure 8.6). The teeth are generally separated, leaf-shaped (Figure 8.7), and reveal few grinding marks, suggesting puncturing as the dominant tooth function.

Although paleontologists have traditionally considered sauropodomorphs to be herbivores, carnivory has been suggested in some of the earliest and most primitive forms because primitive sauropodomorph skulls, teeth, and general body proportions lack herbivore specializations. Yet, supporting herbivory in prosauropods, the skull is proportionately smaller than that seen in carnivores. Recent treatments of the group split the difference, calling them predominantly herbivores that might have enjoyed an occasional meaty snack.

Once the food was past the mouth, grinding likely took place via gastroliths – which have been found in association with prosauropod skeletons – and by stomach fermention, to judge from their barrel-shaped torsos (see Chapter 5).

Need for speed? In the most primitive of prosauropods, the forelimbs are shorter than the hindlimbs and the trunk region is relatively short, suggesting that these animals walked principally on their hindlimbs

(a)

(b)

(c)

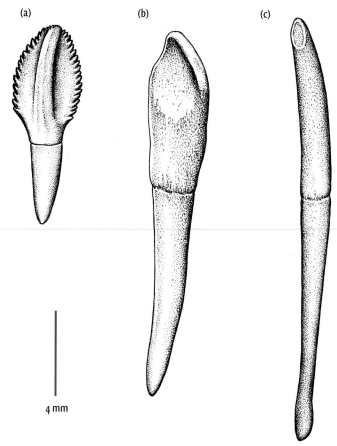

4 mm

Figure 8.7. Teeth in selected sauropodomorphs. (a) Leaf-shaped prosauropod tooth of *Plateosaurus;* (b) spatulate tooth of sauropod *Camarasaurus;* (c) pencil-like tooth of *Diplodocus.* The lower part of each tooth is the root.

rather than on all fours. However, the largest and most derived of prosauropods (among them *Riojasaurus* and *Melanorosaurus*; see Figure 8.19) appear to have become fully quadrupedal. The early history of locomotion in Sauropodomorpha is consistent with the primitive condition for all dinosaurs: bipedality (see Chapters 4 and 15).

For all that, undoubted prosauropod tracks all come from animals walking only quadrupedally. The trackways are broad, with the oval prints of the hindfoot turned outward from the midline. In keeping with a rearward-positioned center of gravity, the imprints of the hands are smaller and somewhat shallower than the feet. Interestingly, the large thumb claw appears to have been held high enough to clear the ground.

Prosauropods appear to have been quite slow. Calculations from trackways (see Box 12.3) suggest speeds of no more than 5 km/h, about the average walking speed of humans. Of course, those are just the tracks that they left one day back in the Late Triassic; in fact, their top speeds were likely much faster.

Socializing Very little is known of prosauropod social habits. The existence of the famous *Plateosaurus* bonebeds in Germany and Switzerland, as well as others elsewhere, however, hints at herds; indeed, herds of prosauropods migrating across the European continent were proposed as early as 1915.

Detailed analyses of *Plateosaurus*, *Thecodontosaurus*, and *Melanorosaurus* (prosauropods for which large numbers of individuals are known) reveal sexual dimorphism in skull dimensions and in thigh bone (femur) size. Sexual dimorphism tends to be pronounced in highly social animals, and thus there may be a connection between the likelihood of herding and sexual dimorphism.

Eggs, nests, and babies Eggs and nests are known for the prosauropods *Mussasaurus* (Argentina) and *Massospondylus* (South Africa). Clutches tended to be small by dinosaur standards (something like ten eggs), and the hatchlings small. Adult prosauropods are roughly 500–1,000 times larger than the hatchlings. How this occurred metabolically is unclear, although rapid growth rates are surely indicated by the disparity in bone size (see Chapter 12)!

Prosauropods were not particularly common beasts and did not hang around on Earth for a very long time, and thus much about them is lost to antiquity. Still, as the first tall-browsing herbivores, they represent the first appearance on Earth of the modern ecosystem that is with us today. Our hope is that, with more attention and finds, the future will bring more insights into this enigmatic, yet fundamental group of dinosaurs.

Sauropoda

Design

Getting *really* big takes some serious evolution, and many sauropods were really big dinosaurs. Yet, the sophisticated sauropod design, once it appeared, remained unique and little changed during their 140 million years on Earth (Figure 8.8). It obviously worked!

The skull itself was distinctive: the tooth row was not inset, as one sees in mammalian and ornithischian herbivores. The teeth, depending upon the the sauropod, had simple crowns and were triangular, spatulate, or slender and pencil-like (see Figure 8.7). There is even a tendency in the clade to limit the teeth to the front of the jaws. In most cases, there is not even a complete mouthful of teeth. Yet, like their ornithischian cousins, advanced sauropods inclined

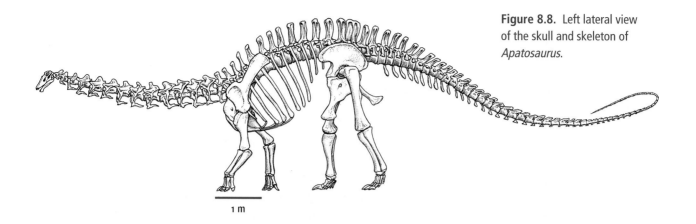

Figure 8.8. Left lateral view of the skull and skeleton of *Apatosaurus*.

1 m

towards chewing. Some likely had cheeks; there is evidence for tooth occlusion in some, and in at least one case (*Nigersaurus*) there is even evidence for the development of a dental battery (see introduction to Part II: Ornithischia, and Chapter 7).

Sauropod skulls tended to be delicately built, with large openings. The skulls appear absurdly tiny – until you realize that only an idiot would design a large, heavy skull at the end of an extremely long neck. The external nares, instead of residing at the tip of the snout, had an as yet unexplained phylogenetic tendency to migrate upward, toward the top of the head (Figures 8.9 and 8.10).

The "extremely long neck" turns out to have been made up of a complex and very sophisticated system of girders and air pockets that maximized lightness and strength. Unlike the vertebrae in mammals, most sauropod veretebrae have an internal system of cavities, called

(a)   (b)

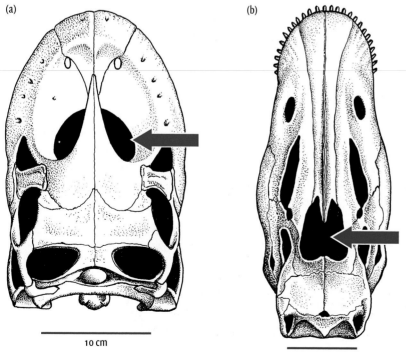

Figure 8.9. Dorsal view of the skull of (a) *Brachiosaurus* and (b) *Diplodocus*. Note the dorsally placed external nares (arrows), especially in *Diplodocus*.

10 cm

10 cm

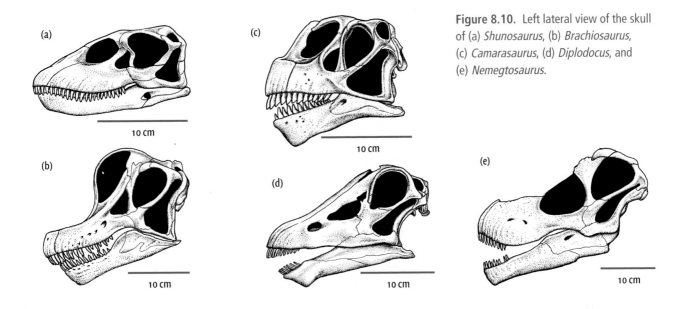

Figure 8.10. Left lateral view of the skull of (a) *Shunosaurus*, (b) *Brachiosaurus*, (c) *Camarasaurus*, (d) *Diplodocus*, and (e) *Nemegtosaurus*.

pleurocoels, as well as small openings called pneumatic foramina (see Chapters 9 and 10). These same features are found in modern birds, which augment the volume of air in their lungs by a complicated series of interconnected auxillary air sacs (see Box 8.1). The auxillary air sacs in modern birds reach (and expand into) the internal cavities via the pneumatic foramina. Such bone is called pneumatic in modern birds, and it would seem that sauropods, too, had truly pneumatic bone. Pneumatic bone is hardly the stuff of primitive vertebrates!

Also distinctive in sauropods were the Y-shaped neural arches on the vertebrae. These held the nuchal ligament, an elastic rope of connective tissue that ran down the back of the animal and supported the head and neck, so that it was not held up exclusively by muscles (Figure 8.11).

Sauropods were quadrupeds, having, as we have seen, secondarily evolved their stance from their bipedal, basal saurischian ancestors. The limbs were pillar-like, and would have done a Greek temple proud. The bones are composed of denser material than that found in the upper parts of the skeleton, an adaptation locating the weight and strength in the skeleton where it was most needed. The hindlimbs articulated with an immense, robust pelvis.

Figure 8.11. Anterior neck vertebrae in *Diplodocus*. The neural spines are bifurcated, and are thought to have held a ligament supporting the neck, the nuchal ligament (shown in solid blue) running from the head, down the neck, and beyond.

The forefeet (the "hands," as it were, on the forelimb) were digitigrade, which means that the animal was standing on its finger tips. The fingers were arranged in a nearly symmetrical horseshoe-shaped semicircle, and the first digit (the thumb) carried a large claw (Figure 8.12a). By contrast, the hindfeet were semi-plantigrade, which means that the animal's weight was supported along the lengths of its toe bones.[1] The foot was asymmetrical, and generally had three large claws (on digits I, II, and III; Figure 8.12b). Sometimes the trackways reveal the

---

[1] For reference, humans are supported along the lengths of their toe and foot bones, and are thus fully plantigrade.

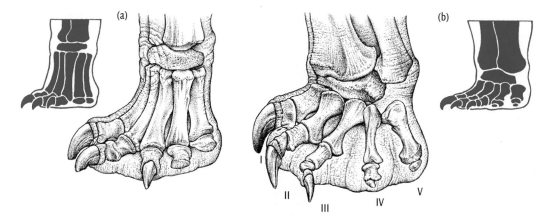

Figure 8.12. Sauropod left (a) forelimb and (b) hindlimb. The "hand" is far more digitigrade than the foot, which is nearly plantigrade (as shown in the limb cross-sections).

impression of a heel pad which nestled behind the claws of both the fore- and hindfeet, and supported the body as well. Sauropod footprints are immense, a single print not uncommonly spanning as much as 1 m!

Despite the fact that the trunk of sauropods was relatively broad (although not as proportionately broad as that seen in ankylosaurs), most sauropod trackways tend to be quite narrow, with the feet aligned toward the midline of the body. Most significantly, relatively few trackways include a tail-drag mark, providing strong evidence that many sauropods carried their immense, whip-like tails – as long as 15 m in the longest cases – entirely off of the ground (Figure 8.13).

Figure 8.13. Five parallel trackways of Late Jurassic age, Morrison Formation, Colorado, USA. Tracks are thought to have been made by diplodocids walking alongside each other. Notice the absence of any mark made by the tail.

Thoughts of a sauropod

This section will necessarily be quite short because, if brain size alone is considered, sauropods did not have obvious pretensions to deep thought. The fact is that sauropods had the smallest brains for their body size (and the lowest EQs; see Box 12.5) of any dinosaur. Yet their long, successful record of survival speaks volumes; as we shall see, their behavioral repertoire may have been more sophisticated than one might expect for an animal with proportionally so small a brain.

Lifestyles of the huge and ancient

A place to roam When we find the remains of these magnificent animals, they come from a myriad different environments, from river floodplains to sandy deserts. Some environments, such as those of the Upper Jurassic Morrison Formation in the American West, required sauropods to cope with long dry seasons during the year. Annual droughts may have been severe enough to have forced sauropods to migrate, a point to which we will return when considering sauropod herding.

Yet, at Tendaguru (see Box 14.7) in southeastern Tanzania, as well as in the USA in northern Texas at the famous Glen Rose trackway sites and in Maryland where the remains of the brachiosaurid *Astrodon* have been uncovered, there is strong evidence that these environments were once close to the sea and quite humid. Perhaps these were some of the conditions that sauropods found most congenial.

Quagmired? For many years, reconstructions commonly showed sauropods as swamp-dwellers, their great bulk buoyed up by water. In this way, so the story went, they could have remained deeply submerged, breathing with only their high nostrils poking out of the water. But what evidence is there for this?

Paleontologists have examined the barometric consequences, that is, the changes in atmospheric pressure, that would occur by submerging a sauropod. Because the thorax (in vertebrates, the part of the body between the neck and stomach) – and hence the lungs – would be under a column of water some six or more meters deep, the thorax would be under nearly double the pressure it would experience on land. This would tend to push whatever air was in the lungs out of the body. How the next breath might be taken is hard to say, since the lungs would have to be expanded against pressures well beyond those experienced in any vertebrate. Unless sauropods had exceedingly powerful chest muscles, they would have been unable to inhale.

In fact, close study of sauropod habitats and anatomy gives no evidence that sauropods whiled away their palmy days buoyed up in Mesozoic swamps. Our best evidence, supported by biomechanical studies of sauropod limbs, is that the dense-boned, massive, and pillar-like limbs were designed for fully terrestrial locomotion.

Beating hearts and necking Sauropod necks have been likened to those of giraffes, inviting the inference that they fed in tall trees. Recent reconstructions, however, indicate that the head in most sauropods was generally held at or near the height of the shoulder, and that a giraffe-like, vertically oriented neck was not physically possible.

For *Brachiosaurus*, things may have been different. Not only was the neck very long, but the front limbs were longer than those in the rear (Figure 8.14). With this "extra boost," its head could potentially have been raised to as high as 13 m, providing the opportunity to feed on foliage to which virtually no one else had access. But animals like *Brachiosaurus* would

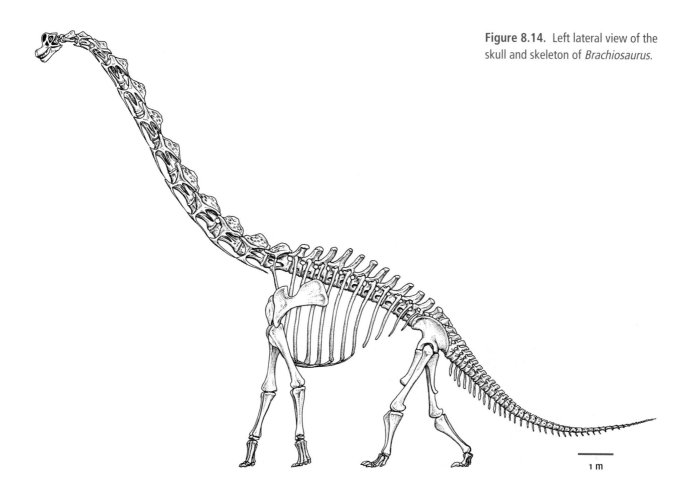

Figure 8.14. Left lateral view of the skull and skeleton of *Brachiosaurus*.

1 m

have had to pay a price for such posture. If the head was truly perched so high, its brain (the relatively smallest among dinosaurs) would have towered about 8 m above its heart.

To push blood through the arteries up its 8.5 m long neck, the heart of a *Brachiosaurus* must have pumped with a pressure exceeding that known in any living animal – indeed, double that of a giraffe. It would indeed take a very muscular heart – some estimate one weighing as much as 400 kg – to do the pumping (Figure 8.15). How fine capillaries in the brain might have withstood such pressures is again a matter for speculation as, unfortunately, is the basic question of how high a *Brachiosaurus* could actually reach in life.

Some paleontologists have suggested *Diplodocus* and other long-necked sauropods may have gained access to foliage at high levels in the trees by adopting a tripodal posture, rearing up on their hindlimbs and using their tails as a "third leg" (Figure 8.16). In tripodal posture, these dinosaurs would have had to pay the same price as *Brachiosaurus* (which itself was probably not able to rear up): elevated blood pressure and a large, powerful heart to produce it. Interestingly, however, a number of modern studies have suggested that the shape of the pelvis in most sauropods precluded rearing back on the hind legs. This would leave dramatic images,[2] such as a mother *Barosaurus* rearing back and protecting its young from a ravenous *Allosaurus* (see Figure 1.12), more in the realm of science fiction than science.

[2] The root of the word "imagination" is, after all, image!

Figure 8.15. Systolic blood pressures compared: (a) a sauropod (approximately 630 mm), (b) a giraffe (320 mm), and (c) a human (150 mm).

(a)   (b)   (c)

None of these considerations addresses yet another challenge posed by long-necked life: the extraordinary amount of unused, wasted, air contained in the neck if sauropods simply breathed in and out, bidirectionally, as do mammals and most tetrapods. If, however, sauropods used a unidirectional, avian style of respiration, in which the lungs are pumped by auxillary air sacs, more oxygen could be extracted from the inhaled air, and the problem of the amount of air contained within the neck would be diminished (Box 8.1).

Good idea... and the presence of pneumatic bone and pleurocoels (see above) is convincing evidence that sauropods must have breathed unidirectionally (Box 8.1). Thus, air sacs and unidirectional breathing, rather than being something exclusively associated with birds, might actually be saurischian characteristics (or earlier; see Box 8.1; Chapter 9).

Considerations of blood pressure, heart size, lung capacity, and breathing style leave us unsure of how sauropods really functioned, but remind us that, in these respects at least, sauropods were a highly evolved, very specialized group of animals.

Feeding

Tooth form and especially tooth wear indicate that sauropods nipped and stripped foliage, unceremoniously delivering a succulent bolus to the gullet, variably chewed, depending upon which sauropod is under consideration. Here, too, the well-developed thumb claw could have played a role, ripping vegetation off plants into bite-sized strips. Swallowing sped the bolus of

Figure 8.16. A sauropod reconstructed in a tripodal posture, using the tail as a "third leg."

food down its long journey through the esophagous (tube leading to the stomach) whereupon it entered the abdomen.

In all sauropodomorphs, the gut must have been capacious, even considering the forward-projecting pubis (in contrast to all ornithischians, which rotated the pubis rearward to accommodate an enlarged gut; see introduction to Part II: Ornithischia). Sauropods likely had an exceptionally large fermentation chamber (or chambers) that would have housed endosymbionts; that is, bacteria that lived within the gut of the dinosaur. The endosymbionts would have chemically broken down the cell walls of the plant food, thereby liberating whatever nutrition was to be had. Considering the size of the abdominal cavity in sauropodomorphs, these animals probably fed on foliage with high fiber content (see Chapter 13); perhaps they also had slow rates of passage of food through the gut in order to ensure a high level of nutrient extraction from such low-quality food. We can only conclude that these huge animals, with their comparatively small mouths, must have been constant feeders to acquire enough nutrition to maintain themselves. The digestive tract of a sauropod had to have been a non-stop, if low-speed, conveyor belt.

Box 8.1 **Every breath you take**

For prosauropods and especially sauropods, the **trachea** (windpipe) would have been exceptionally long, approximately the same length as the arteries carrying blood from the heart to the brain. The trachea brings oxygen into contact with the **alveoli** in the lungs, sites where oxygen is transmitted to the blood and where carbon dioxide is passed back to the air.

In animals that pass air *bidirectionally* into and out of the lungs (that is, during inhalation and exhalation) like a bellows (mammals, lizards, crocodilians, and snakes), the trachea creates physiological dead space: some portion of the inhaled air never reaches the lungs. It is simply brought into the respiratory system and returned without being involved in oxygen–carbon dioxide exchange.

By contrast, birds have *unidirectional* air flow, in which when inhaled air passes into the lungs, nearly all the oxygen is absorbed into the bloodstream, and the now oxygen-depleted air is run through a series of air sacs around the lungs and back into the trachea for exhalation. Obviously, the avian system wrings more oxygen out of the air than the bidirectional, bellows-style lungs found in mammals (Figure B8.1.1).

In animals that have long necks, the problem of physiological dead space can be acute without unidirectional air flow. Long-necked bidirectional breathers, such as giraffes, circumvent the problem of dead space by having an inordinately narrow trachea: dead space is reduced by limiting the internal volume of the trachea. In fact, it is thought by some that giraffe may be the longest-necked animals capable of combining bellows-style lungs and a very long trachea.

The evidence for sauropods now is near overwhelming: with their pleurocoels and pneumatic foramina, sauropodomorphs must have had unidirectional, avian-style lungs which would have eliminated the problems associated with all of that physiological dead space engendered by the very long trachea. Sauropod pleurocoels are best interpreted as early stages in the development of air sacs and reflective of unidirectional breathing, as in some theropod dinosaurs and living birds. It's not so far-fetched: in 2010 biologist Coleen Farmer and colleagues reported uni-

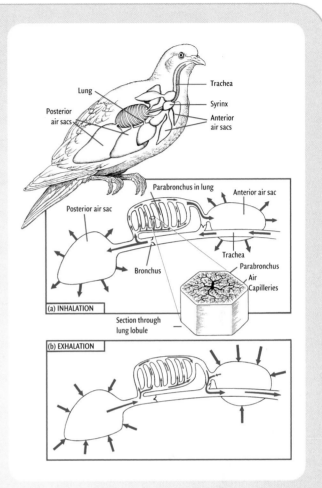

Figure B8.1.1. Unidirectional respiration, shown diagrammatically. As the animal inhales (a), air enters the lungs and posterior air sacs (here represented by a single sac), which expand. Air that goes into the lungs is deoxygenated, and then stored in the anterior air sacs (here represented by a single sac), which expand and fill with deoxygenated air. As the animal exhales (b), the posterior air sac contracts, and its air – still oxygenated – is pumped through the lungs, where it is deoxygenated. The rest of the deoxygenated air, in the lungs and anterior air sac, is expelled via contraction out of the trachea.

directional breathing in alligators! Now, with strong evidence of unidirectional air flow in sauropods, theropods (including birds), and crocodylians, isn't it possible that unidirectional breathing should be described as *archosaurian* and not "avian?!"

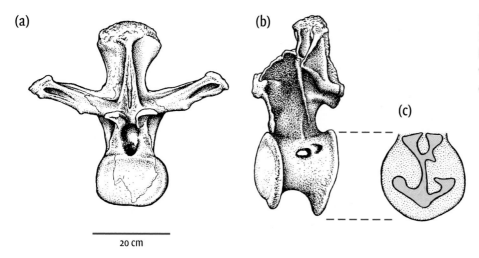

(a)

(b)

(c)

Figure 8.17. Front view (a) and left lateral view (b) of one of the back vertebrae of *Brachiosaurus*, with pleurocoels indicated in cross-section (c).

20 cm

Locomotion

The top speeds of *Brachiosaurus*, *Diplodocus*, and *Apatosaurus* have been calculated to be between 20 and 30 km/h, a reasonable clip for animals the size of a house and weighing in excess of three to ten elephants. They undoubtedly walked a good deal more slowly most of the time, perhaps at rates of 20–40 km/day, as calculated from sauropod trackways (see Box 12.3).

Hanging with the big boys

The many now-famous mass accumulations in the Morrison Formation in the USA (see Figure 8.13), the Tendaguru bonebeds of Tanzania (see Box 14.7), the Lower Jurassic sauropod sites of India, and most recently the Middle Jurassic of Sichuan, China, together with the vast sauropod footprint assemblages, all speak loudly to the existence of gregariousness of sauropods, including *Shunosaurus*, *Diplodocus*, and *Camarasaurus*.

Sauropods living in large groups must have been capable of wreaking severe damage on local vegetation, either by stripping away all the foliage they could reach or by trampling into the ground all of the shrubs, brush, and trees that might have got in the way. So while herds of sauropods likely depleted their food sources and had to move on for more, other sauropods, including *Brachiosaurus* and *Haplocanthosaurus* from the Morrison Formation of the western US, and *Opisthocoelicaudia* from Mongolia, are not so numerous and may thus have lived a more solitary existence, not needing to keep traveling for sustenance. Other than these generalities, we know little about how sauropods communicated within herds, who ran the show, or what might have caught a sauropod's wayward eye.

Defense

In sauropods, size was surely the best deterrent against an attack; since, depending upon the sauropod, they were between 50% and 300% larger than co-existing predators. Living in herds, healthy animals must have been nearly invulnerable. The whip-like tail was likely employed to brush aside would-be predators. In the case of an attack from a pack of predators, both size and the tail may have been brought into play (but see Chapter 9 for how predators might have handled sauropod prey).

Growth and development

Until recently, we knew next to nothing about sauropod nesting, and indeed, as recently as the early 1990s, it was seriously proposed that sauropods gave birth to live young (which, if it were true, would have made them unique among dinosaurs).

In 1997, however, a sauropod nesting ground was discovered in Patagonia. This site, known as Auca Mahuevo ("Auca more eggs"), consists of a massive nesting ground covering more than a square kilometer and littered with tens of thousands of large, unhatched eggs. Upon further investigation, four layers of eggs were uncovered and, in each layer, the eggs were organized into clusters of 15–34 linearly paired eggs, thought to represent individual nests or clutches. Most spectacularly, a high proportion of these eggs contained embryonic skeletons, some with impressions of embryonic skin (Figure 8.18)!

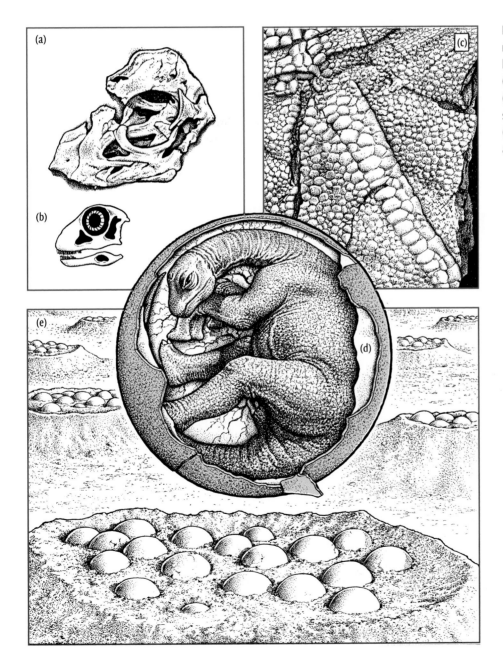

Figure 8.18. Titanosaurian remains from the Auca Mahuevo locality of Patagonia, Argentina. (a) Titanosaur skull (fossil); (b) reconstructed skull; (c) titanosaur skin (fossil) impressions; (d) reconstructed egg/embryo; and (e) schematic field of nests.

The geographical extent of the nesting horizons reaffirmed gregarious behavior in sauropods. Clearly several enormous colonies were preserved, to which mothers are thought to have regularly returned. Because there is no fossil evidence of adults preserved at Auca Mahuevo, the females may have left the site after laying their eggs, although it is possible that they may have communally guarded the whole nesting area from its periphery. If so, the eggs may also have been covered by mounds of vegetation to keep them at optimal temperature and humidity.

Since the find at Auca Mahuevo, eggs have been associated with particular sauropods in the Upper Cretaceous of southern France, Mongolia, and India, where, in 2007, a second, massive sauropod nesting ground was also uncovered. Does any of this suggest that sauropods were *r*-strategists (see Chapter 7)?

Beyond these bonanzas, what do we know about the general aspects of sauropod reproduction, growth, and life histories? Sex in these animals assuredly involved coupling between a tripodal male and a quadrupedal female; however, beyond this most elemental of positions all else remains speculative. For example, was the trenchant thumb claw used in this aspect of sauropod behavior as well?

Once hatched, sauropodomorphs apparently grew at very high rates. New studies of the microscopic structure of sauropod bone indicates rapid and continuous rates in both prosauropods and sauropods. Rather than imagining animals taking about 60 years to reach sexual maturity and having a longevity of perhaps 100–150 years, estimates are that it took about 30 years or less for a sauropod (and probably for a prosauropod as well) to become sexually mature. Similarly, lifespans for these animals were probably on the order of not much more than 100 years.

The evolution of Sauropodomorpha

Sauropodomorpha is a diverse and long-lived clade, containing two major groups of dinosaurs – Prosauropoda and Sauropoda. The group is easily diagnosed by more than a dozen derived features (see Figure 8.3).

Prosauropoda

Prosauropods were once thought to be the early, primitive forebears of sauropods. Now, the group is commonly reckoned to be too specialized to have been directly ancestral to sauropods; rather, they retain many of the characters of the common ancestor of prosauropods and sauropods, an unidentified sauropodomorph that likely must have lived in the early Late Triassic. Prosauropods are monophyletic, and thus united by a suite of diagnostic characters. Recent phylogenetic work indicates that Prosauropoda can be subdivided into two monophyletic groups, with a few genera falling outside these subclades (Figure 8.19).

Sauropoda

Sauropoda is supported by more than a dozen unique features, many of which relate to the attainment of great size and weight on land (Figure 8.20).

Evolution within Sauropoda has only recently been evaluated using cladistic approaches. As currently understood, sauropods consist of several primitive taxa (among them *Blikanasaurus*, *Vulcanodon*, and *Kotasaurus*) on the one hand, and the more derived clade Eusauropoda

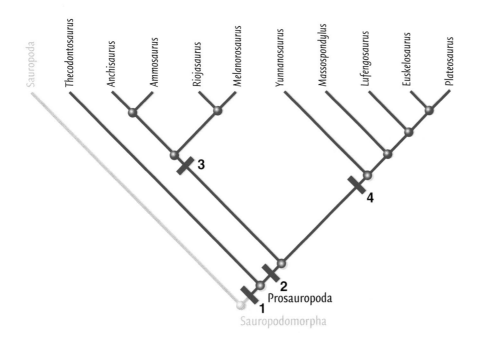

Figure 8.19. Cladogram of a monophyletic Prosauropoda. Derived characters include: at **1**, lateral lamina on the maxilla, strap-like ventral process of the squamosal, ridge on lateral surface of the dentary, elongate posterior dorsal centra, distal carpal I transversely wider than metacarpal I, phalanx I on manual digit I with a proximal heel, a 45° twisting of the large thumb claw; at **2**, separate opening for vena cerebralis media above the trigeminal foramen, axis centrum that is three times longer than high, short and robust metacarpal I, an acetabulum that is completely open medially, subtriangular distal end of ischium, increased robustness of metatarsals II and III; at **3**, prefrontal length approximately that of the frontal, frontal excluded from supratemporal fossa, at least five premaxillary teeth, forelimb length greater than 60% hindlimb length, straight femoral shaft, fourth trochanter is displaced to the caudomedial margin of the shaft, hour-glass-shaped proximal end of metatarsal II; at **4**, long retroarticular process on the lower jaw, longest postaxial cervical centrum at least three times as long as high, a dorsal vertebra added to the sacrum, proximal carpals present, large obturator foramen.

on the other. Eusauropods are diagnosed by many features (Figure 8.20). The most primitive known member of the group, *Shunosaurus*, was a 9 m long sauropod from the Middle Jurassic of China (Figure 8.21). Its skull is relatively long and low, and vaguely reminiscent of the primitive sauropodomorph condition, with nostrils near the front of the snout and a mouth filled with many small and spatulate teeth.

As sauropod evolution proceeded, various aspects of jaw mechanics and body form appear to have been linked. Early in their evolution, sauropods developed a fully quadrupedal stance. At the same time, there occurred a significant increase in body size.

In time, the snout broadened, the lower jaw strengthened, and wear indicating front and rearward movement of the jaws is found on the teeth. Within Neosauropoda, that great clade of sauropods that includes camarasaurs, brachiosaurs, and titanosauroids (Macronaria) on the one hand, and Diplodocoidea on the other (see Figure 8.20), the skull shows additional strengthening (closure of the antorbital fenestra). Macronarians generally show a shortening and elevation of the skull, indicating a more powerful biting force. Among the most distinctive characters uniting macronarians (see Figure 8.20) are the modified hollow spaces, the pleurocoels, in the back region (see Figure 8.17). Within Macronaria are the smaller, stouter camarasauromorphs (in the shape of camarasaurs) including the very familiar *Camarasaurus* (Figure 8.22).

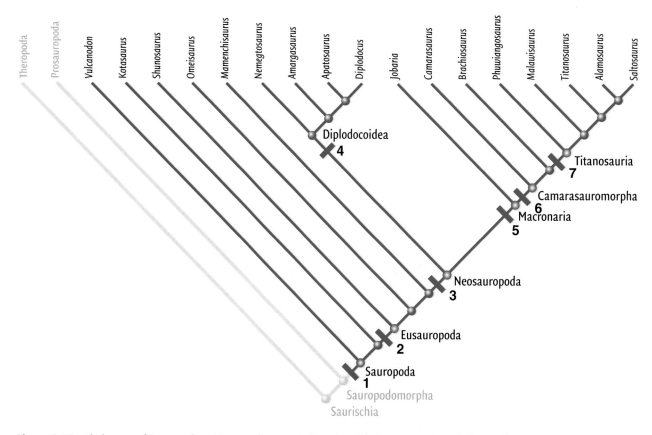

Figure 8.20. Cladogram of Sauropoda, with more distant relationships with Prosauropoda and Theropoda. Derived characters include: at **1**, special laminar system on forward cervical vertebrae, forelimb length greater than 60% hindlimb length, triradiate proximal end of ulna, subrectangular distal end of radius, length of metacarpal V greater than 90% that of metacarpal III, compressed distal end of ischial shaft, reduced anterior trochanter on femur, femoral shaft elliptical in horizontal cross-section, tibia length less than 70% femur length, metatarsal III length less than 40% tibia length, proximal end surfaces of metatarsals I and V are larger than those of metatarsals II, III, and IV, metatarsal III length is >85% metatarsal V length, ratio is 0.85 or higher; at **2**, broadly rounded snout, caudal margin of external nares that extends behind the posterior margin of the antorbital fenestra, lateral plate on premaxillae, maxillae, and dentaries, loss of the anterior process of the prefrontal, frontals wider than length, wrinkled tooth crown enamel, most posterior tooth positioned beneath antorbital fenestra, at least 12 cervical vertebrae, neural spines of the cervical vertebrae that slope strongly forward, dorsal surface of sacral plate at the level of dorsal margin of ilium, block-like carpals, metacarpals arranged in U-shaped colonnade, manual phalanges wider transversely than proximodistally, two or fewer phalanges for manual digits II–IV, strongly convex dorsal margin of ilium, loss of the anterior trochanter of femur, lateral muscle scar at mid length of fibula, distally divergent metatarsals II–IV, three phalanges on pedal digit IV, ungual length greater than 100% metatarsal length for pedal digit I; at **3**, subnarial foramen on premaxilla–maxilla suture, preantorbital fenestra in base of ascending process of maxilla, quadratojugal in contact with maxilla, pedal digit IV with two or fewer phalanges; at **4**, subrectangular snout, fully retracted external nares, elongate subnarial foramen, reduction of angle between midline and premaxilla–maxilla suture to 20° or less, most posterior tooth rostral to antorbital fenestra; at **5**, greatest diameter of external nares greater than that of orbit, subnarial foramen found within the external narial fossa; at **6**, nearly vertical dorsal premaxillary process, splenial extending to mandibular symphysis, acute posterior ends of pleurocoels in anterior dorsal vertebrae, metacarpal I longer than metacarpal IV; at **7**, prominent expansion of rear end of sternal plate, very robust radius and ulna.

Figure 8.21. The eusauropod *Shunosaurus*, from the Middle Jurassic of Sichuan Province, China.

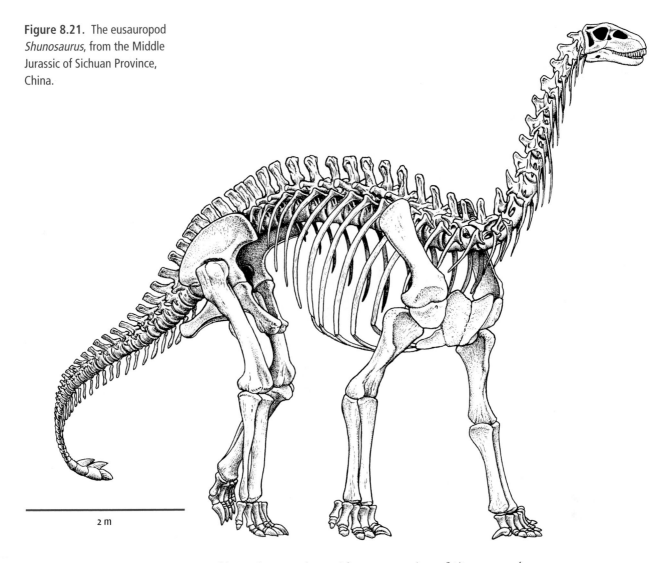

2 m

Our visit to Macronaria would not be complete without a mention of titanosaurs (see Figure 8.20), an extremely diverse group of sauropods, known from all continents, and among the last sauropods to grace the planet Earth. Titanosaurs achieved incredible sizes: the heaviest animals ever known were titanosaurs, and some, such as *Argentinosaurus* (from Argentina) and *Paralititan* (from Egypt) likely exceeded 100 tons. Yet sauropod evolutionary history is not entirely one of getting bigger. In Transylvania is found the titanosaur *Magyarosaurus*, a creature 5–6 m in length, much smaller than contemporary sauropods elsewhere in the world. These smaller forms were living on islands, a common phenomenon in the Mesozoic as now. This "dwarf" sauropod was a mere 6 m long.

Among the most famous of titanosaurs is the Late Cretaceous *Alamosaurus* from the American southwest which, for all its fame, is ironically very poorly known (Figure 8.23). Several titanosaur forms, including *Saltasaurus*, from the Late Cretaceous of Argentina, and *Malawisaurus* (Malawi) are known to have been covered with scattered immense (up to 22 cm in length), hollow osteoderms (Figure 8.24). Why scattered? Palentologist Kristina Curry Rogers proposed that as juveniles, the osteoderms were close together and served, as we saw in ankyosaurs, as a kind of armor pavement. When the animals grew, the osteoderms, she proposed, grew apart, and then their hollow interiors came to serve as reservoirs for minerals and other nutrients during times of environmental stress.

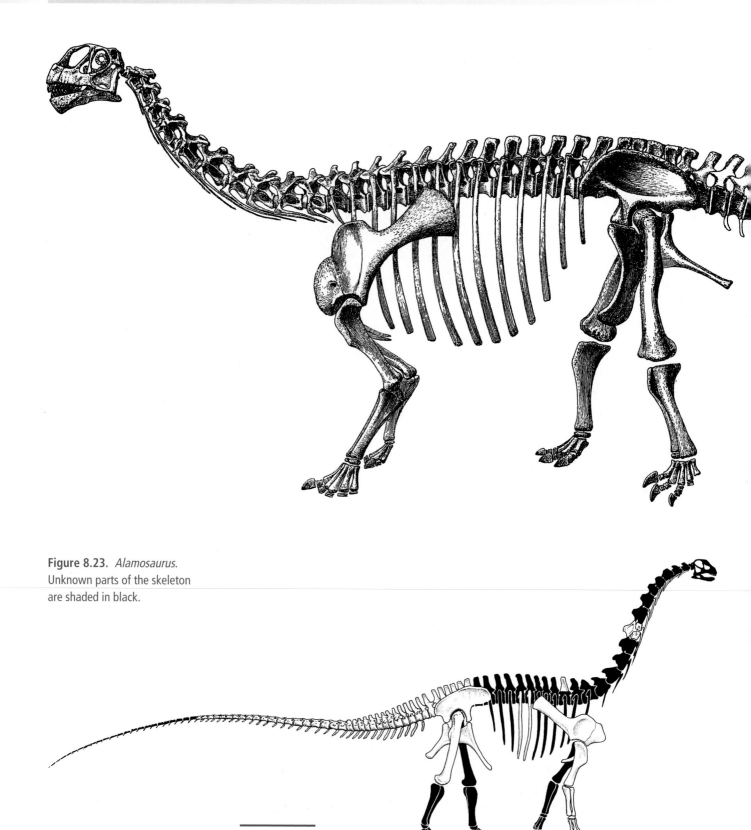

Figure 8.23. *Alamosaurus.*
Unknown parts of the skeleton
are shaded in black.

1 m

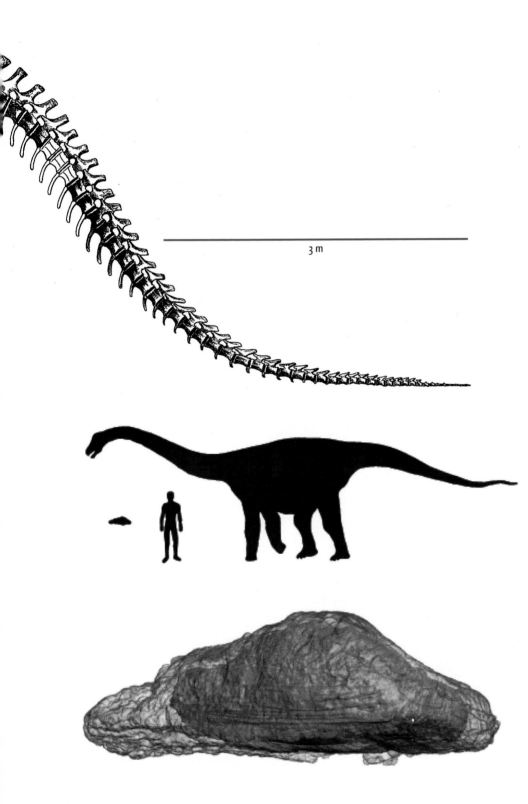

3 m

Figure 8.22. The skeleton of *Camarasaurus*.

Figure 8.24. CT scan of the gigantic, hollow osteoderms of the titanosaur *Malawisaurus*. Note hollowed, central region. Silhouetted dinosaur and human for scale. *Inset*: reconstruction of *Malawisaurus*.

The long-necked, long-tailed, gracile diplodocoids (can an animal whose weight is in double-digit tons *ever* be called "gracile?") developed a distinctive skull: they restricted their dentition to just a series of highly evolved, pencil-like teeth to the front of the jaws. Tooth wear is apparent at the apices of the teeth (instead of along the side of the tooth), and elongation of the snout in diplodocoids suggests that the group abandoned the front–back jaw movement.

Some remarkable dinosaurs are among the diplodocoids, including *Amargosaurus*, with its extraordinary neural arches, from the Early Cretaceous of Argentina (Figure 8.25). The 21 m long *Apatosaurus* (Late Jurassic, Western Interior, USA) is best known by its incorrect name "*Brontosaurus*" (Box 8.2). *Diplodocus*, renowned for its long neck and tail, may also have carried dragon-like osteoderms or scutes along the length of its back (Figure 8.26).

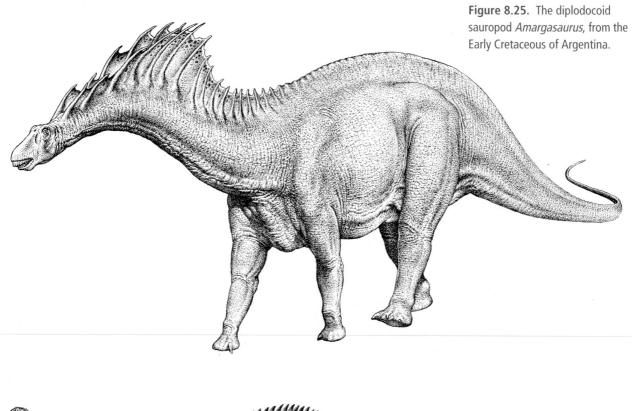

Figure 8.25. The diplodocoid sauropod *Amargasaurus*, from the Early Cretaceous of Argentina.

Figure 8.26. Osteoderms or scutes reconstructed along the back of the Late Jurassic North American diplodocoid, *Diplodocus*.

Box 8.2 The recapitation of *"Brontosaurus"*

With the discoveries of dinosaurs in the Western Interior of the USA during the late nineteenth century, box-car-loads of brand-new, but often incomplete, sauropod skeletons were shipped back east to places such as New Haven and Philadelphia. It was Yale's O. C. Marsh who described one of these new sauropods as *Apatosaurus* in 1877. With further shipments of specimens and more studies, Marsh again named a "new" sauropod in 1879 – *Brontosaurus*.

Years went by and – thanks to the burgeoning popularity of many kinds of dinosaurs – the public came to know the name *Brontosaurus* much better than it did the earlier-discovered *Apatosaurus*. Nevertheless, there was the suspicion by many sauropod researchers that *Apatosaurus* and *Brontosaurus* were the same kind of sauropod. In fact, this case was made in 1903 by E. S. Riggs, of the University of Kansas. Since then, most sauropod workers have regarded *Brontosaurus* as synonymous with *Apatosaurus*. If *Apatosaurus* and *Brontosaurus* are two names for the same sauropod, the older name, *Apatosaurus*, should be applied to this Late Jurassic giant.

But the more interesting story is not in the names, but in the heads. Again we go back to O. C. Marsh. Lamenting in 1883 that his material of *"Brontosaurus"* (now *Apatosaurus*) had no head, he made his best guess as to the kind of skull this animal had: one like *Camarasaurus*. And it was thus that *Apatosaurus* donned the short-snouted profile of its Morrison Formation cohort.

Enter H. F. Osborn, curator of vertebrate paleontology and powerbroker of the American Museum of Natural History in New York, and W. J. Holland, curator of fossil vertebrates at the Carnegie Museum of Natural History in Pittsburgh and equally stalwart in his pursuit of "getting it right" about sauropods. Contemporary dinosaur researchers in the early part of the twentieth century, these two skirmished over the issue of whose head should reside on the neck of *Apatosaurus*. Osborn followed Marsh and had his mount of this majestic sauropod topped with a camarasaur head (Figure B8.2.1a), while Holland was strongly persuaded that *Apatosaurus* had a more *Diplodocus*-like head (based on a somewhat removed yet associated skull found near an otherwise quite complete skeleton at what is now Dinosaur National Monument in Colorado). But Holland gained no adherents and his mount of *Apatosaurus* in the Carnegie Museum remained headless in defiance of Osborn's dogma. After Holland's death, however, the skeleton was fitted with a camarasaur skull, almost as if commanded by Osborn himself.

Whose head belongs to whom was finally resolved in 1978 by Carnegie Museum Curator of Paleontology D. S. Berman and sauropod authority J. S. McIntosh. Through some fascinating detective work on the collection of sauropod specimens at the Carnegie Museum, these two researchers were able to establish that *Apatosaurus* had a rather *Diplodocus*-like skull – long and sleek, not blunt and stout as had previously been suggested. As a consequence, a number of museums that display *Apatosaurus* skeletons celebrated the work of Berman and McIntosh (and Holland) by conducting a painless head transplant – the first ever in dinosaurian history (Figure B8.2.1b).

Figure B8.2.1. **(a)** *"Brontosaurus"* on display at the American Musuem of Natural History in the early 1970s. This dinosaur was a chimera with a camarasaur head on an *Apatosaurus* body. **(b)** *Apatosaurus* being readied for display at the American Museum of Natural History in the mid 1990s.

Sauropodomorphs were the largest terrestrial vertebrate life forms of their time and indeed of all time. We often think of *Brachiosaurus*, from the Late Jurassic of the western USA, as well as from Tanzania (see Figure 8.14), which captured several decades' worth of people's imaginations as the largest land-living animal of all time (measuring 23 m long and weighing in excess of 50,000 to 60,000 kg). Now supplanted by the likes of *Argentinasaurus* and *Seismosaurus*, *Brachiosaurus* nevertheless is still by far the best known of all of these earthly giants.

Sauropods were the fast-growing, yet slow-paced high-browsing giants of the Mesozoic. Today we view them as evolutionary marvels, as they continue to baffle, surprise, and inspire with biomechanical and evolutionary consequences of "living large."

Summary

Sauropodomorpha consists of the great herbivorous saurischian quadrupeds Sauropoda, and an early offshoot, Prosauropoda. Prosauropods were primitive large dinosaurs, appearing in the Late Triassic at the dawn of Dinosauria, and surviving through the Early Jurassic. Initially thought to be the ancestors of Sauropoda, they are now considered to represent an early saurischian radiation, representing the world's first high-browsing herbivores.

Sauropoda were the largest land animals ever to walk the Earth, reaching 40 m from the tip of the snout to the tip of the tail. These obligate herbivorous quadrupeds were highly evolved, with many biomechanical adaptations for large size and weight, including four pillar-like limbs and a massive pelvis, a tendency to lighten bones not immediately involved in support functions, and a complex girder-like neck design, tipped by a small skull, to maximize leverage and lightness. Among the many striking features in the design of sauropod bones are the presence of pleurocoels, hollow spaces suggestive of avian-style air sacs. The likely presence of air sacs, as well as the inefficiency, in an animal of such great size, of mammalian-style bellows breathing, suggest that sauropods may have used avian-style unidirectional breathing.

The skulls of sauropods were relatively small, and the group showed a general tendency toward migrating the nostrils or nares to the top of the skull. Dentition varied, from simple leaf-shaped teeth to pencil-like teeth restricted to the front of the mouth. Sauropods in general lack the chewing adaptations present in ornithischians (particularly Genasauria).

Sauropods appear to have been social animals, particularly as reflected in sauropod bonebeds and in trackways. The trackways clearly indicate that sauropods did not drag their tails. Sauropod gregariousness is also reflected in the recent discoveries of extremely large sauropod nesting grounds. Sauropod lifespans, once thought to be in the hundreds of years, are now thought to have been around 100 years, with extremely rapid growth of juveniles. The large number of eggs and babies associated with sauropod nesting grounds implies that these dinosaurs may have been *r*-strategists.

Sauropod defense was likely accomplished mainly by size, with perhaps an assist from the whip-like tail and the broad, trenchant claw on the forefoot.

The immense size of sauropods makes analogizing them with living terrestrial vertebrates extremely difficult. Although not likely posessed of a "warm-blooded" metabolism – at least as adults (see Chapter 12) – they nonetheless must have consumed copious quantities of food, and must have required virtually incessant food consumption through the comparatively small mouth. In good health and with its full complement of individuals, it is not hard to imagine wholesale defoliation of a region by a herd. This in turn leads to at least the possibility that sauropods were often on the go, searching out new foliage to consume.

Selected readings

Carpenter, K. and Tidwell, V. (eds.). 2005. *Thunder Lizards*. Indiana University Press, Bloomington, IN, 495pp.

Curry Rogers, K., D'Emic, M., Rogers, R., Vicaryous, M. and Cagan, A. 2011, Sauropod dinosaur osteo-derms from the Late Cretaceous of Madagascar. *Nature Communications*, **2: 564**, 1–4. doi: 10.1038/ncomms1578.

Galton, P. M. and Upchurch, P. 2004. Prosauropoda. In Weishampel, D. B., Dodson, P. and Osmólska, H. (eds.), *The Dinosauria*, 2nd edn. University of California Press, Berkeley, pp. 232–258.

Henderson, D. M. 2003. Typsy punters: sauropod dinosaur pneumaticity, buoyancy, and aquatic habits. *Proceedings of the Royal Society of London, Series B (suppl.)*, **271**, S180–S183.

Seymour, R. S. and Lillywhite, H. B. 2000. Hearts, neck posture and metabolic intensity of sauropod dinosaurs. *Proceedings of the Royal Society of London, Series B*, **267**, 1883–1887.

Upchurch, P., Barrett, P. M. and Dodson, P. 2004. Sauropoda. In Weishampel, D. B., Dodson, P. and Osmólska, H. (eds.), *The Dinosauria*, 2nd edn. University of California Press, Berkeley, pp. 259–321.

Wedel, M. J. 2003a. The evolution of vertebral pneumaticity in sauropod dinosaurs. *Journal of Vertebrate Paleontology*, **23**, 344–357. doi:10.1671/0272–4634(2003)023[0344:TEOVPI]2.0.CO;2.

Wedel, M. J. 2003b. Vertebral pneumaticity, air sacs, and the physiology of sauropod dinosaurs. *Paleobiology* **29**, 243–255. doi:10.1666/0094–8373(2003)029,0243: VPASAT.2.0.CO;2.

Topic questions

1   What is Sauropodomorpha? How does it relate to Saurischia?

2   What are the diagnostic characters of Sauropodomorpha? Which two groups comprise Sauropodo-morpha? How do they differ? How are they alike?

3   Outline what is known of sauropod reproductive strategies. Would you classify them as *r*-selected, or *K*-selected? Why?

4   Describe some features that are associated with large size in sauropodomorphs.

5   What is the evidence that sauropods didn't really chew their food? How can sauropods have been herbivores, but not have developed much chewing ability?

6   Why are sexual dimorphism and gregariousness often linked?

7   What kinds of design constraints are associated with having an extremely long neck?

8   What is the evidence that indicates that sauropods did not dwell in swamps? Where, then, and in what kinds of environment, did they live?

9   In Box 8.1, we wrote: "isn't it possible that unidirectional breathing should be described as *saurischian* and not "avian?" Can you think of some other reasons why this statement might be true?

10   Why do we think that the quadrupedal stance of sauropods evolved secondarily?

CHAPTER 9

THEROPODA I:
NATURE RED IN TOOTH AND CLAW

CHAPTER OBJECTIVES

- Introduce Theropoda

- Develop familiarity with current thinking about lifestyles and behaviors of theropods

- Develop an understanding of theropod evolution using cladograms, and an understanding of the place of Theropoda within Dinosauria

Theropoda

Eating meat the theropod way

When dinosaurs got around to carnivory, they did it the theropod way: with steak-knife teeth, sinewy haunches, and grasping hands and feet tipped with scimitar claws (Figure 9.1). The combination was at once formidable and successful, and produced a rainbow palette of different types, among them coelophysoids, neoceratosaurs, carnosaurs, therizinosauroids, ornithomimosaurs, oviraptorosaurs, troodontids, dromaeosaurids, tyrannosauroids … and birds.

Grouped together as Theropoda (*thero* – beast; *pod* – foot), these dinosaurs have had a long evolutionary history extending back from the Late Triassic right up until the end, 65.5 Ma. *Past* that "end," really, since birds are still very much with us. But in this chapter, we'll concentrate on non-avian (that is, non-bird) theropods, holding off on the avian side of the story until Chapter 10.

Non-avian theropods (for simplicity, "theropods") have been found on every continent including Antarctica (Figures 9.2 and 9.3). They have been collected from virtually every kind of depositional environment. In most instances, skeletal remains are found in isolation, the delicate, hollow bones fragmented. However, bonebeds of single theropod species are also known (see "Social behavior" below).

Figure 9.1. The king of the tyrant lizards, *Tyrannosaurus rex* (see previous page).

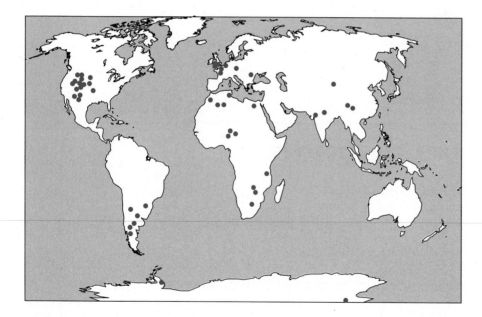

Figure 9.2. Global distribution of non-coelurosaurian Theropoda.

Who are theropods?

Theropoda is a well-diagnosed, monophyletic group with abundant characters (Figure 9.4) within Saurischia (see introduction to Part III: Saurischia): theropods share their closest relationship with Sauropodomorpha and together form a monophyletic Saurischia. Viewed superficially, theropods were all clawed bipeds (Figure 9.5 and still are) and many had sharp, serrated teeth[1] – although some had none at all. All theropods, though, share the distinctive quality of hollow bones, both vertebrae and limbs.

---

[1] The teeth and bipedal stance of theropods appear to have been primitive carry-overs from earlier in ornithodiran history (see Chapters 4 and 15, as well as the introduction to Part III: Saurischia). These characters give the *appearance* that theropods are somehow more primitive organisms than, say, ornithischians; the use of cladograms, however, shows that they're not.

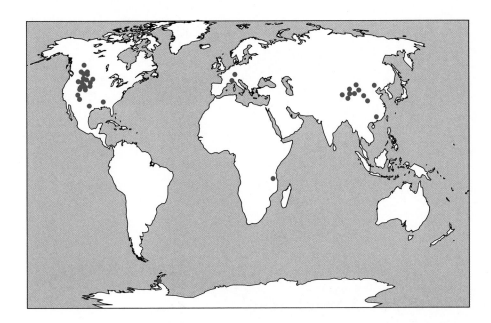

Figure 9.3. Global distribution of non-avian Coelurosauria.

Theropods ranged in size from less than a meter (*Microraptor*) to animals growing to upward of 15 m in length (*Tyrannosaurus*, *Carcharodontosaurus*, *Giganotosaurus*). For all the variety, though, theropod evolution was generally all about tracking, attacking, and feeding.

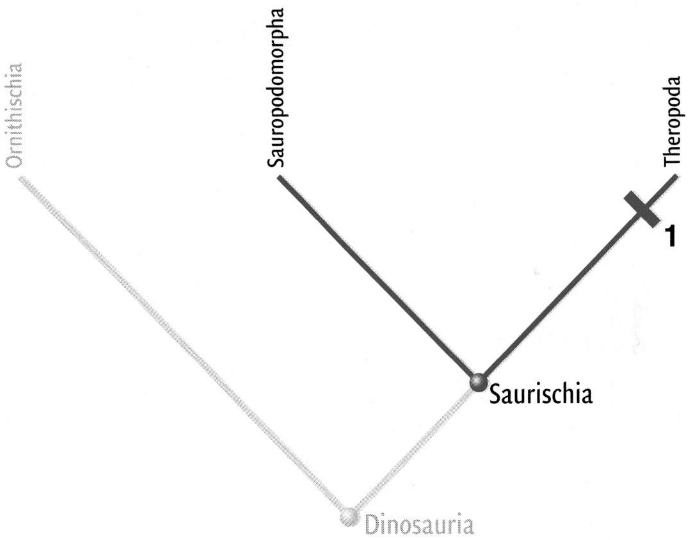

Theropod lives and lifestyles

Theropod imagery is haunted by solitary, drooling brutes like *T. rex* gnawing their way through herds of herbivores. But is that really all there was to it? Hunting in packs and armed with grasping claws and slicing teeth, small, large-eyed, intelligent, and agile theropods were likely the real nightmare terrors of Mesozoic landscapes (Figure 9.6).

Figure 9.4. Cladogram of Dinosauria emphasizing the monophyly of Theropoda. Derived characters include: at **1**, extreme hollowing of vertebrae and long bones, enlarged hand, vestigial fourth and fifth digits, remaining digits capable of extreme extension due to large pits on the upper surfaces of the ends of the metacarpals.

Running for life

All theropods were obligate bipeds, unable to walk or run on anything but their hind legs. The body was balanced directly over the pelvis, with the vertebral column held nearly horizontally (see Figure 9.5). Evidence from the skeleton and trackways indicates that the hind legs were held close to the body, feet so close to the midline that it appears that one foot was placed *ahead* of the other, rather than along its side. The trackways, as well as skeletal material, also indicate that the foot was held in a digitigrade stance (Figure 9.7).

Many small- to medium-sized theropods, especially ornithomimosaurs, must have been fast runners. Their thigh bones were short compared to the length of the rest of the hindlimb; a condition typical of fast-running bipeds (Figure 9.8). Calculations of running speeds on the

(a)

(b)

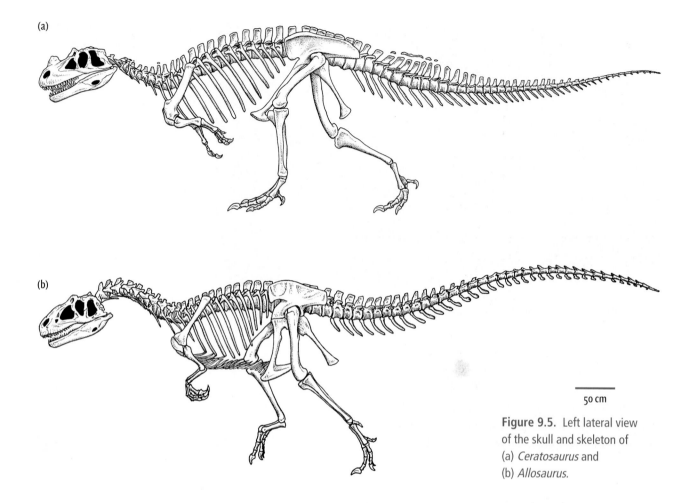

50 cm

Figure 9.5. Left lateral view of the skull and skeleton of (a) *Ceratosaurus* and (b) *Allosaurus.*

basis of hindlimb proportions indicate that the fastest theropods clocked 40–60 km/h. Some footprint evidence bears these numbers out; for example, a trackway in Texas was made by a theropod that thundered away at upward of 45 km/h (see Box 12.3).

The running speeds of large theropods, however, are less clear. Some investigators have calculated, using limb proportions and models of leg motion, that theropods were limited to walking at no more than 4 km/h. Using similar approaches, however, others have deduced much faster running speeds.

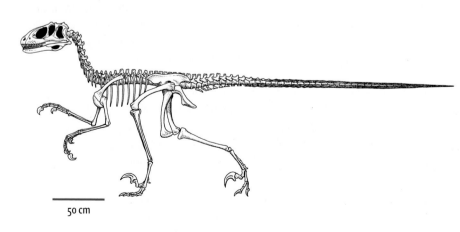

50 cm

Figure 9.6. Left lateral view of the skull and skeleton of *Deinonychus.*

Figure 9.7. A theropod trackway from the Middle Jurassic Entrada Formation, Utah, USA. Note how closely the right and left prints are placed, suggesting a fully erect stance. When spectacular trackways like this are found, it is not hard to imagine the ghostly image of the trackmaker leaving a row of footprints in the soft mud.

Another approach to speed estimations involves reconstructing the muscle mass and volume. In the case of large theropods running at high speeds, so much leg musculature would have been required that the animal would have been grotesquely overmuscled. The balance of dispassionate evidence, therefore, suggests that the largest theropods likely were not the fleetest runners of their time.

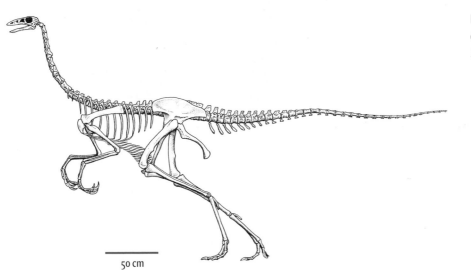

Figure 9.8. Left lateral view of the skull and skeleton of *Struthiosaurus*.

50 cm

For all that running, it is now known that some theropods, at least, could also swim. In 2007 a trackway was discovered in Spain clearly demonstrating the imprints of a medium-sized theropod swimming.

Paws and claws

As in modern birds, the grasping, powerful, clawed feet must have been an important part of the theropod arsenal (Figure 9.9). This character reached unparalleled sophistication in dromaeosaurids and troodontids, in which the claw on the second digit of the foot was especially huge, curved, and sharp, and capable of a very large arc of motion. During normal walking and running, it was held back or up, to protect it from abrasion or breakage. But, when needed, it could be brought forward and, with the powerful kicking motion of the rest of the leg, use to eviscerate the bellies of hapless prey, lethally disemboweling in one rapid stroke (Figure 9.10).

Strong arms and dexterous, three-fingered hands characterized most theropods, particularly small- and medium-sized forms. The digits were long and capable of extreme extension,

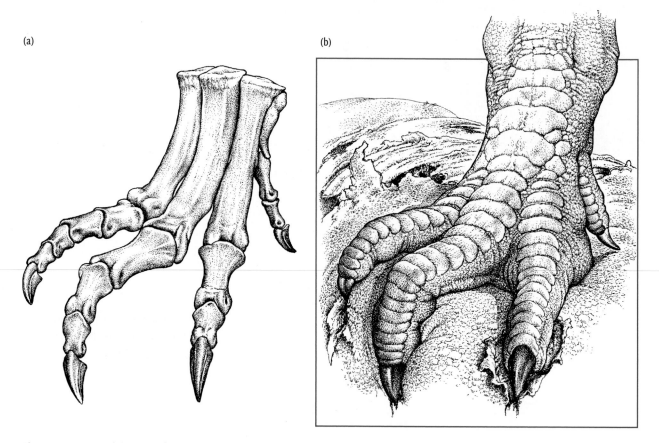

(a)

(b)

Figure 9.9. Typical theropod foot: (a) bones; (b) reconstructed doing what non-avian theropod feet did best.

and tipped with powerful claws. The thumb could fold across the palm in a semi-opposable fashion; that is, somewhat like a human thumb. There is no mistaking the function of this hand: these are all adaptations for grasping.

Even highly specialized large theropods such as *Tyrannosaurus*, *Tarbosaurus*, and *Carnotaurus*, with their notoriously short arms (the hands could not reach the mouth), had stout,

(a)

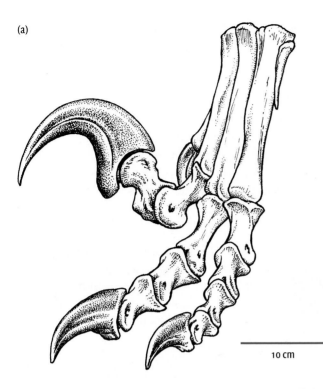

10 cm

(b)

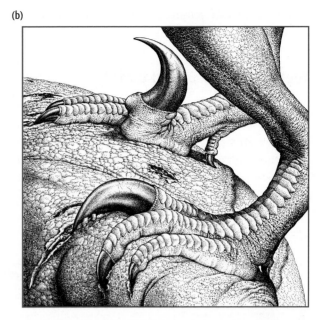

Figure 9.10. (a) Left foot of *Deinonychus* with its disemboweling second-toe claw. (b) Reconstruction of the feet of *Deinonychus* in action.

powerful bones and fingers, suggesting active use (note the hands and arms of *Tyrannosaurus* and *Carnotaurus* in Figure 9.11). But used for what? It has been suggested that perhaps the arms were short in order to balance an overly large head. In bipedal animals that have exceptionally large heads (like *Tyrannosaurus*, but not so much the case in *Carnotaurus*), increasing head size may require downsizing other aspects of the front half of the body to remain balanced with the back half at the hips. It has also been suggested that the arms helped to push the dinosaur up from a resting pose.

Yet the size and design of the arm bones suggests some more compelling function. Indeed, it has been calculated that the arm of *Tyrannosaurus* could have lifted 300 kg. It is clear from the robust bones and large, stout claws at the tips of the strong fingers that the forelimbs

(a)   (b)   (c)

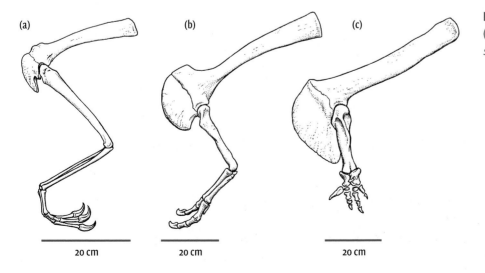

20 cm   20 cm   20 cm

Figure 9.11. Left forelimb of (a) *Struthiomimus*, (b) *Tyrannosaurus*, and (c) *Carnotaurus*.

aspired to greater purposes than weight reduction up front. Tearing flesh? Gripping prey? Lifting? Nobody knows.

Teeth and jaws … and turds?

As with many carnivorous animals, theropod heads tended to be proportionately large. In the case of the biggest, the heads could be upward of 1.75 m in length. In general, theropod skulls are rather primitive, reminiscent of those of many non-dinosaurian ornithodirans. Yet there are differences: tyrannosauroids had robust, deep-jawed skulls, suggesting a powerful bite. Other theropods – even large ones like *Carcharodontosaurus* – had much more lightly built skulls (Figures 9.12 and 9.13).

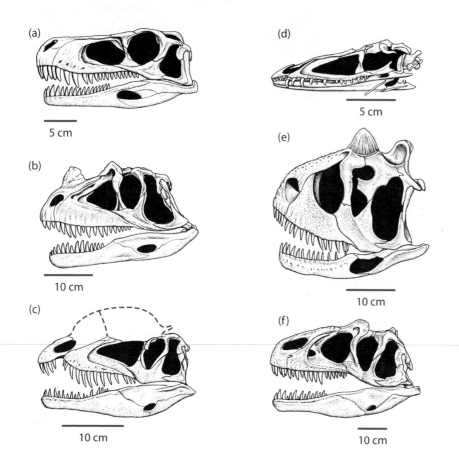

Figure 9.12. Left lateral view of the skull of (a) *Herrerasaurus,* (b) *Ceratosaurus,* (c) *Dilophosaurus,* (d) *Coelophysis,* (e) *Carnotaurus,* and (f) *Allosaurus.*

All theropod teeth – in the case of those that had them – tended to be flattened from side to side, recurved (curved backward), pointed, and serrated. With the jaw joint at the level of the tooth row, the effect was like that of a pair of scissors, slicing from back to front (Figure 9.14). The design clearly lacks the chewing specializations of genasaurs.

It was the sharply pointed, recurved, and serrated teeth in the upper and lower jaws that handled the prey. The recurved shape kept prey from escaping from the mouth. The teeth of smaller theropods such as *Troodon,* with their prominently pointed serrations and narrow cross-sections, sliced like hacksaw blades. Tyrannosauroids, at the other end of the spectrum, with bulbous teeth and rounded serrations, had a weaker cutting ability, but greater strength, suggesting that they could withstand complex, strong, and violent forces, such as might occur

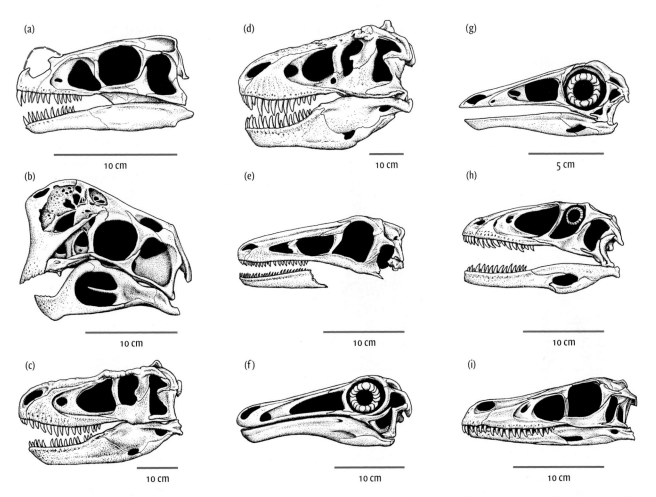

Figure 9.13. Left lateral view of the skull of (a) *Ornitholestes*, (b) *Oviraptor*, (c) *Albertosaurus*, (d) *Tyrannosaurus*, (e) *Saurornithoides*, (f) *Gallimimus*, (g) *Dromiceiomimus*, (h) *Deinonychus*, and (i) *Velociraptor*.

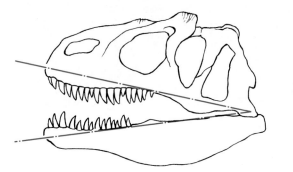

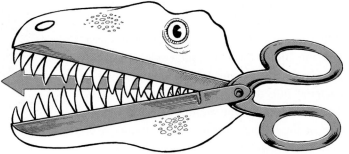

Figure 9.14. In its manner of slicing and in the placement of its hinge, the carnivorous theropod mouth was very similar to the design of a pair of scissors.

with a powerful, actively struggling prey (Figure 9.15). They may have even been able to crush bone (see below).

With the differences in skull and teeth, theropods evidently bit in different ways. Recent studies have paired CT scans and computer-modeled stress analyses to the architecture of

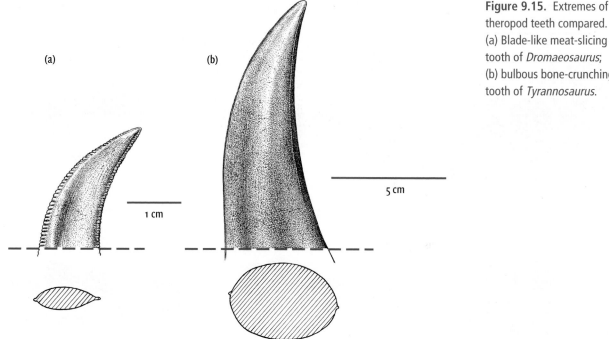

(a)

(b)

1 cm

5 cm

Figure 9.15. Extremes of theropod teeth compared. (a) Blade-like meat-slicing tooth of *Dromaeosaurus*; (b) bulbous bone-crunching (?) tooth of *Tyrannosaurus*.

theropod skulls (Figure 9.16). We now know, for example, that *Allosaurus*, with its relatively lightly built skull, used a "slash-and-tear" attack on its prey, in which powerful neck muscles drove the skull downward rather than delivering a crushing bite with the jaw muscles alone. When the head was retracted, the teeth sliced and tore flesh. Such a wound might not kill prey immediately – but blood loss and possible bacterial infection would work their relentless damage. Tracking and waiting may have been part of the killing technique of dinosaurs with lightly built skulls and thin, blade-like teeth.

This contrasts with tyrannosauroids (see Figures 9.13d and 9.17) or perhaps abelisaurids such as *Carnotaurus* (see Figure 9.12e), whose more bulbous teeth and larger, more heavily built skulls likely delivered a bone-crushing, heart-stopping bite. They may also have suffocated their victim by seizing their snout or neck between their jaws and clamping down. This kind of attack is consistent with the large gape, powerful jaws, and stout teeth of these predators. In

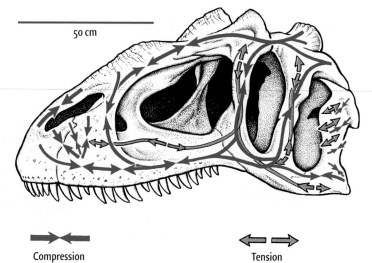

50 cm

Compression

Tension

Figure 9.16. The skull of *Allosaurus* marked with its finite element analysis model, indicating the regions of stress that pass through it as a function of biting.

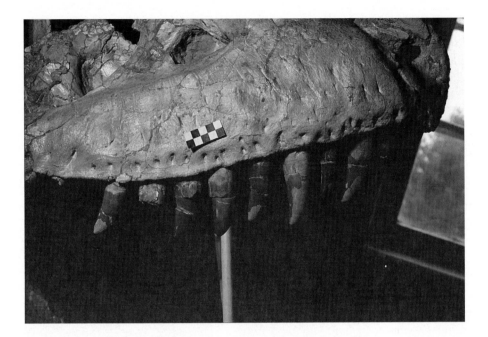

Figure 9.17. The upper teeth of *Tarbosaurus* as seen from the right side of the skull. Scale in centimeters.

all cases, however, the skull had considerable mobility on the neck because of a well-rounded occipital condyle and its articulation with the first part of the cervical (neck) vertebrae.

Is all this who-ate-whom speculation? The answer comes from an unlikely source, a 44 cm long, 13 cm high, and 16 cm wide coprolite. Its age, geographical location, and, err, size point to *Tyrannosaurus rex* as the culprit. The specimen contains between 30% and 50% of bone fragments, thought to be the remains of limb bones or parts of a ceratopsian frill. In combination with other information about theropod diets (see "Eaters and eatees" below), this coprolite provides physical evidence that tyrannosauroids crushed, consumed, and incompletely digested large quantities of bone.

Toothless At least two times in their history,[2] theropods drastically reduced or lost all of their teeth. With the exception of one primitive genus, all ornithomimosaurs, a group of small-skulled, long-legged theropods that look very much like ostriches with long tails (see Figure 9.8), lost all their teeth (see Figure 9.13f,g). Ornithomimosaurs had a beak, and in the case of *Gallimimus*, at least, the beak had a ridged feature that appeared almost sieve-like along its margin, provoking a controversial suggestion that it fed aquatically, much like a modern duck. Later interpretations of the beak edge suggest that it is less a sieve and more of a shearing feature, consistent with grinding fibrous plant matter.

Ornithomimosaurs are also known to have gastroliths, and these indicate the presence of a muscular gizzard for grinding plant matter. These, in combination with its powerfully clawed hands and superb running capability, suggest a terrestrial existence rather than a duck-like lifestyle. Consistent with the evidence from the beak, these animals were likely fast-running creatures that used their gastric mills to grind up fibrous plant matter, as do modern plant-eating birds.

Oviraptorosaurs were also toothless (see Figures 9.13b and 9.18). The skull was very short with apparent pneumaticity (see Chapter 8), and the jaw musculature was very well developed.

---

[2] Three times, when we include birds (see Chapter 10).

Located between their shortened upper jaws is a pair of peg-like projections dead center in the middle of the palate. One analysis of the mechanics of the oviraptorosaur skull suggested that the jaws were designed to feed on hard objects that required crushing, such as clams, oysters, and mussels. Oviraptorosaurs presumably cracked them open by the brute force of their jaw muscles acting on the thick horny bill covering the margins of the mouth and the palate, and especially the stout pegs in the center.

Senses

To locate and track their prey, theropods of all kinds needed a keen awareness of their environment. Thanks to brain endocasts and more recently, CT scan reconstructions of the interiors of braincases (Box 5.2), we've begun to learn a lot more about how these animals operated. Because the brain represents the center for the senses, we look for similarities to the brains of related living creatures, such as birds and crocodiles, whose behavior is known, as well as to regions of enlargement, which suggested increased or heightened function of that part of the brain.

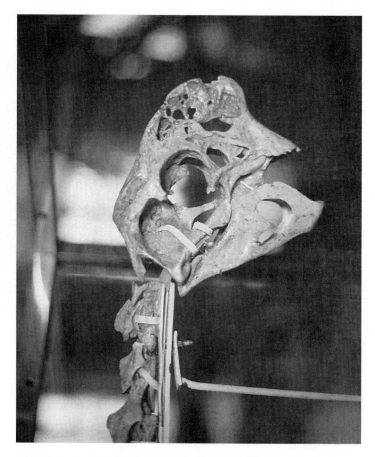

Figure 9.18. The box-like, toothless skull of *Oviraptor*.

Clearly, however, sharp vision was key for theropods, so it is not surprising that their eye size was large. In deinonychids generally, but especially in troodontids, the eyes have migrated to a more forward-looking position, indicating overlapping fields of vision. Overlapping fields of visions almost certainly means that these animals saw stereoscopically – that is, they merged the two separate independent images from each eye into a single image, much as humans and many modern carnivorous birds do today. Recent work suggests that the narrow snouts of even tyrannosauroids allowed a 55° range of binocular vision, not nearly as much as a human or an owl, but far exceeding what one might find in a hadrosaurid (Figure 9.19).

Hearing, too, is important to predatory animals and so it is not surprising that many theropods likely had good sound perception. Indeed, the middle ear cavity of troodontids and ornithomimosaurs was greatly enlarged, suggesting that these theropods were especially able to hear low-frequency sounds. In troodontids, detailed anatomical study of the outer and middle ears suggests that the group was capable of identifying the direction from which sounds came; knowledge that would have been of extreme use to a predator.

Making sense of tyrannosaurs With nothing close to a living analog, the behavior and lifestyle(s) of tyrannosaurs have always been enigmatic, with the mighty *Tyrannosaurus* being perhaps the most enigmatic – yet compelling – of all. Recent CT scan-based work by L. M. Witmer, R. C. Ridgely, and associates at Ohio University has provided some exciting insights about tyrannosaurs, notably, *Tyrannosaurus*, *Gorgosaurus*, and *Nanotyrannus*

(Figure 9.20).[3] With the shape of the brain fully reconstructed, it is clear that despite enlarged cerebral hemispheres, tyrannosaur brains were not the functional – thinking – equivalent of that of a modern bird. No surprise there! Somewhat more surprising is that tyrannosaurs had well-developed senses of smell. By comparison with other theropods, tyrannosaurs had relatively large olfactory bulbs, as well as a particularly large nasal cavity. The likelihood is that these features reflect behaviors that involve the ability to smell.

Like other theropods, tyrannosaurs were particularly adept at hearing low-frequency sounds. In the case of tyrannosaurs, however, the design and construction of the ear bones were such that clearly hearing was an unusually important sense. Witmer and colleagues have reconstructed tyrannosaurs as predatory, with the neural evidence suggesting an ability to track prey rapidly by sight, smell, and sound, using eyes, head, and neck.

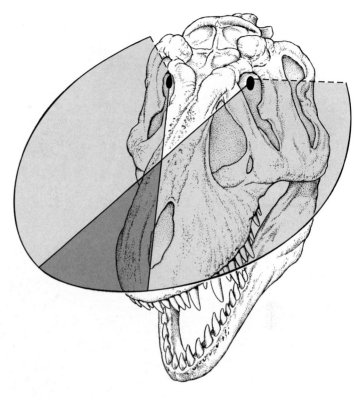

Figure 9.19. Three-quarter view of the skull of *T. rex*, showing a 55° range of binocular vision (see the text).

Balance

The formidable armament of theropods, particularly the small to medium-sized ones, must have been lethally coupled with exceptional balance. The nearly horizontal position of the vertebral column took advantage of the center of gravity being positioned near the hips.

Deinonychosaur balance was aided by a remarkable weapon in their arsenal: a tail stiffened by immensely elongate processes along the neural arches (Figure 9.21). The tail was only flexible at its base, just behind the pelvis. This stiffening allowed the rigid tail to move as a unit in any direction. It thus functioned as a dynamic counterbalancing device against the motions of the long arms, grasping hands, and large skull.

On the basis of their light, yet powerfully built skeletons, dromaeosaurids and troodontids must have had an extraordinary degree of agility. We imagine them flinging themselves at fleeing prey, kicking with great accuracy with one, or even both, of their dangerous feet (Figure 9.22).

Thoughts of a theropod

All theropods for which there is brain-size information have surprising (to humans!) cerebral powers: their brains are every bit as large as one would expect of a crocodilian or lizard "blown up" to the proper body size. Indeed, troodontids had the largest brains for their body size of any non-avian theropods, and were well within the bird range of inferred intelligence

---

[3] *Nanotyrannus* has been a controversial creature since its discovery in the middle of the twentieth century. Originally referred to *Gorgosaurus*, it has since been called *Tyrannosaurus*, then a juvenile of *Tyrannosaurus*. Exactly what it is, taxonomically speaking, remains unclear.

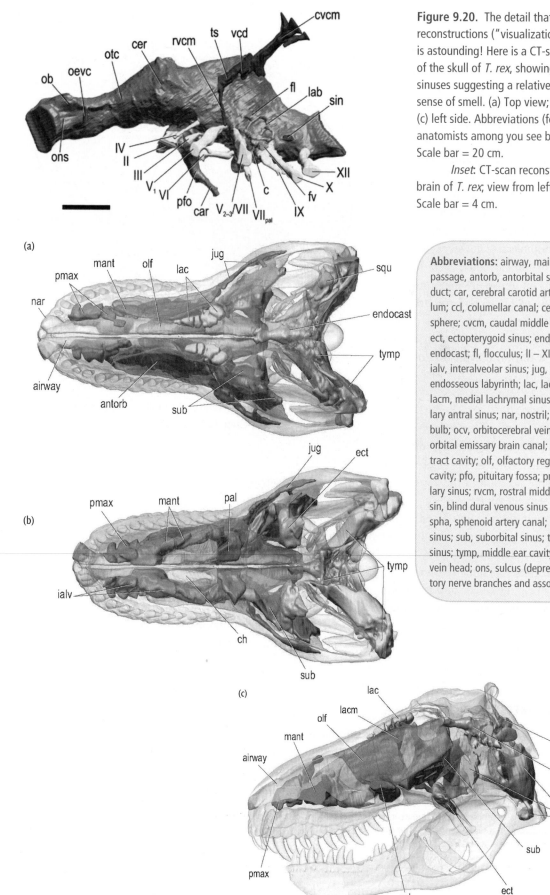

Figure 9.20. The detail that modern CT reconstructions ("visualizations") can achieve is astounding! Here is a CT-scan reconstruction of the skull of *T. rex*, showing elaborated sinuses suggesting a relatively well-developed sense of smell. (a) Top view; (b) bottom view; (c) left side. Abbreviations (for the cranial anatomists among you see box below). Scale bar = 20 cm.

Inset: CT-scan reconstruction of the brain of *T. rex*; view from left side of animal. Scale bar = 4 cm.

Abbreviations: airway, main nasal air passage, antorb, antorbital sinus; c, cochlear duct; car, cerebral carotid artery; cbl, cerebellum; ccl, columellar canal; cer, cerebral hemisphere; cvcm, caudal middle cerebral vein; ect, ectopterygoid sinus; endocast, the brain endocast; fl, flocculus; II – XII, cranial nerves; ialv, interalveolar sinus; jug, jugal sinus; lab, endosseous labyrinth; lac, lachrymal sinus; lacm, medial lachrymal sinus; mant, maxillary antral sinus; nar, nostril; ob, olfactory bulb; ocv, orbitocerebral vein canal; oecv, orbital emissary brain canal; otc, olfactory tract cavity; olf, olfactory region of the nasal cavity; pfo, pituitary fossa; pmax, promaxillary sinus; rvcm, rostral middle cerebral vein; sin, blind dural venous sinus of hindbrain; spha, sphenoid artery canal; squ, squamosal sinus; sub, suborbital sinus; ts, transverse sinus; tymp, middle ear cavity; vcd, dorsal vein head; ons, sulcus (depression) for olfactory nerve branches and associated vessel.

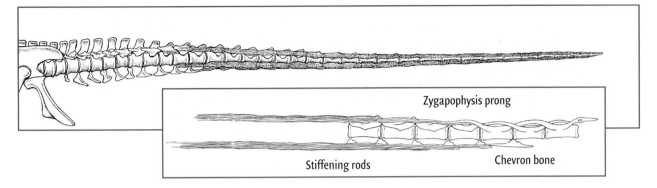

Zygapophysis prong

Stiffening rods

Chevron bone

Figure 9.21. The rigid tail of *Deinonychus*. The elongate, intertwined zygapophyses on each neural arch give the tail its rigidity.

Figure 9.22. *Deinonychus* skeleton mounted in attack mode.

(see Box 12.5). This suggests that these animals probably had more complex perceptual ability and more precise motor-sensory control than some of their smaller-brained brethren. It also likely implies sophisticated inter- and intraspecific behavior. No stegosaurs, they!

The skinny on skin

Until recently, most researchers thought that all (non-avian) theropods were covered with scales of some sort. The only direct information on theropod skin came from the South American neoceratosaur *Carnotaurus*, whose skin was covered with an array of tubercles of modest size surrounded by smaller rounded scales.

When birds and other exceptionally well-preserved fossils began showing up in China in the mid 1990s, paleontologists recognized apparent feathers and feather-like structures on non-avian theropods. Among the most famous of these are *Caudipteryx*, *Protarchaeopteryx*, *Sinornithosaurus*, and *Microraptor* (see Chapter 10). Two other theropods from the same region – *Sinosauropteryx* and *Dilong* – are extensively covered with filaments that have been interpreted as feather precursors. Unsurprisingly, these filaments are now known to have been colored and patterned, suggesting that these animals interacted visually (see "color" below).

We are not so fortunate as regards other theropods. Some kind of insulatory covering has been suggested for highly active ones, such as dromaeosaurids and troodontids, and some paleontologists have speculated that, as juveniles, even large adult theropods may have been covered with a downy feather-like insulation. That "speculation" ceased to be speculation when, as this edition was just being prepared for publication, Chinese paleontologist Xing Xu and colleagues announced the discovery of *Yutyrannus huali*, an 8 m tyrannosaur from China. *Yutyrannus* was evidently covered, at least partially, with 15 cm long filamentous feathers (see Figure 10.10d). With feathers clearly present in *Yutyrannus*, the probability that *T. rex* was also feather-covered becomes very high indeed.

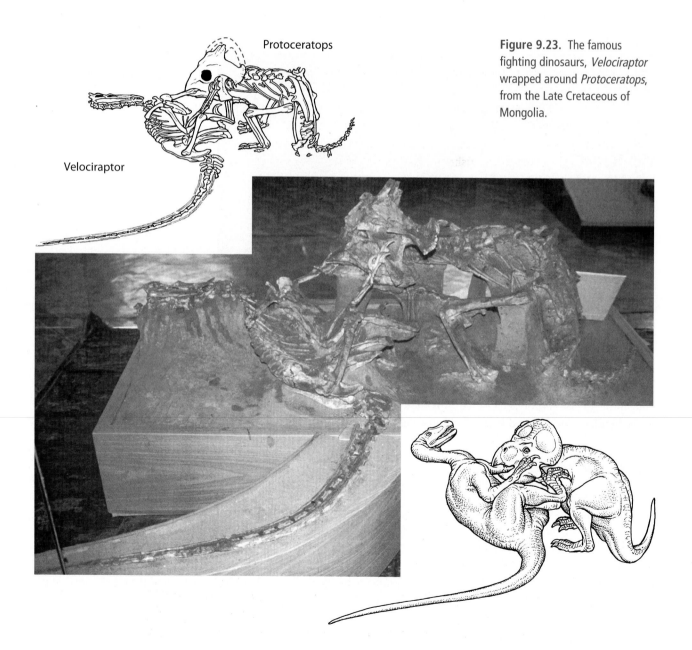

Protoceratops

Velociraptor

Figure 9.23. The famous fighting dinosaurs, *Velociraptor* wrapped around *Protoceratops*, from the Late Cretaceous of Mongolia.

Eaters and eatees

With all the variation in theropods, prey surely varied. The most dynamic and irrefutable evidence about the preferred prey of *Velociraptor* is the so-called "fighting dinosaurs" specimen: *Velociraptor* with its hindfeet half into the belly of a subadult *Protoceratops* and its hands grasping, or being held in, the jaws of the soon-to-be victim (Figure 9.23).

A specimen of the Late Jurassic coelurosaur *Compsognathus* is known that contains most of the skeleton of a fast-running lizard. Not only did *Compsognathus* swallow nearly whole this delectable meal, but it must have captured its victim through its own speed and maneuverability. Other evidence of theropod stomach contents and diet come from *Sinosauropteryx* (lizards and mammals), *Baryonyx* (fish remains), and *Daspletosaurus* (hadrosaurid bones).

Bones of different dinosaurs are often found together, leading to media-fueled speculation about diners and dinners. Teeth of *Deinonychosaurus* have been found in association with the large ornithopod *Tenontosaurus*, suggesting the former ate the latter. But the size difference between the two made this proposal a tall order, unless *Deinonychus* hunted in packs. Thus, the genesis of that idea, distorted almost beyond recognition in the Jurassic Parks!

Along the same lines, the large Cretaceous theropod *Spinosaurus* has been enigmatic; equipped with a row of 2 m high neural spines, a crocodile-like snout, high nostrils, and tall conical teeth (hallmarks of fish-eaters; Figure 9.24), it had generally been assumed that the animal had something to do with fish-eating – but what? *Spinosaurus* teeth have been found in the same deposits as those of the large sawfish *Onchopristis*; this and the fact that identifiable *Onchopristis* fragments were found embedded in the tooth sockets of *Spinosaurus* have led to media[4] speculation that *Spinosaurus* dined on *Onchopristis*. Could the sawfish fragments simply have lodged like pebbles in the empty sockets when the bones were washed together during burial? Unfortunately, yes; so this idea, then, although tantalizing, remains just an idea.

Evenly spaced grooved toothmarks on the bones of the sauropods *Apatosaurus* and *Rapetosaurus* have been attributed to the local large theropods: *Allosaurus* and *Majungatholus* (from the USA and Madagascar, respectively). Toothmarks attributed to *Tyrannosaurus* are known from a pelvis of *Triceratops*, and from a thoroughly crunched tail of the duck-billed dinosaur *Edmontosaurus* (Box 9.1).

Cannibals And then it's clear that some dinosaurs didn't shy away from a bit of cannibalism: it was once argued that the Late Triassic ceratosaur *Coelophysis*, based upon the supposed presence of juveniles within the rib cage, was cannibalistic; however, recent work indicates that the juveniles in the rib cage were actually non-dinosaurian archosaurs. On the other hand, *Majungatholus* apparently didn't avoid the odd conspecific snack: grooved toothmarks matching the spacing of its own teeth have been found on *Majungatholus* specimens (Figure 9.25). With only one carnivore known on Madagascar from that time with that tooth spacing, the evidence is circumstantial, but damning. We still don't know if such cannibalism occurred on the run, or whether it was the scavenging of sick or dead individuals.

Who ate whom? Beyond these few direct observations of dietary preferences, we are left to speculate on who ate whom. Our best guesses are informed by the known theropods and potential prey in a particular place and time – we might pair *Tarbosaurus* with *Saurolophus*, or *Troodon* with small ornithopods or juveniles of much larger co-existing dinosaurs (such as hadrosaurids).

The invention of pack-hunting could have allowed relatively small animals to bring down much larger co-existing prey – hence the pairing of *Deinonychus* (3.5 m) packs with the

---

[4] The British television show *Planet Dinosaur*.

Figure 9.24. Mounted skeletal reconstruction of the Cretaceous tetanuran *Spinosaurus aegypticus*.

large (7 m) ornithopod *Tenontosaurus*, a combination that was first suggested when the shed teeth of *Deinonychus* were found with *Tenontosaurus* specimens. Whether true or not, these are the often the best available data that can be used to address the question of theropod diets (see also Chapter 13).

Of course, successful attack can come from where one least expects it; recent work by J. R. Horner and colleagues suggests that the mighty *T. rex* was under regular, and potentially crippling, attack by a lowly protozoan. These researchers noted that the jaws of many specimens of *Tyrannosaurus* show numerous lesions (Figure 9.26). Once thought to be the work of other tyrannosaurs, these lesions are very similar in morphology to those found in the jaws of modern chickens, turkeys, and falcons. In the modern birds they are the handiwork of a parasitic protozoan, *Trichomonas gallinae*, an organism that infects jaws, eventually eating out

Box 9.1 *Triceratops* spoils or spoiled *Triceratops*?

In 1917, L. M. Lambe of the Geological Survey of Canada suggested that *Gorgosaurus* was not so much an aggressive predator, but instead maintained its sustenance by scavenging. The basis for his remarks was the apparent absence of heavy wear on the teeth of this theropod – these animals therefore must have fed primarily on the softened flesh of putrefying carcasses. This interpretation has appeared on and off again in discussions of theropod diet and hunting behavior, frequently enough to be something like a cottage industry in anecdotal "knowledge" about these animals.

The notion of theropod scavenging rests on the assumption that tooth wear was usually absent and that carcasses were readily available. It is further bolstered by the lack of a convincing account of why the forelimbs of these animals are so small. We take each of these in turn. Firstly, it turns out that tooth wear is present on the teeth of nearly all large theropods. This doesn't "prove" that tyrannosauroids and other large theropods had to have been active predators; both modern scavengers and active predators alike can have a high degree of wear on their teeth.

The commonness of carcasses, putrefied or otherwise, was probably dependent on the season – dry, stressful seasons probably claimed their share of dead hadrosaurids, ceratopsians, sauropods, and the like. Interestingly, this potentially great contribution of carcasses would have been in the form of tough, dry flesh – dinosaur jerky, really – not the kind of soft, predigested carrion postulated by Lambe.

Finally, short forelimbs were likely used to slice and dismember prey, and are fully consistent with a head-first attack.

Recently, the suggestion of tyrannosaurids as carrion eaters has come from the observation that these theropods had surprisingly broad, bulbous teeth. Modern scavengers such as the hyena likewise have broad teeth, which are used to crush the bones of carcasses. This notion has been pooh-poohed by scientists who cannot imagine a dinosaur with the size and obvious carnivorous equipment of *T. rex* being a scavenger.

What *is* clear is that *Tyrannosaurus* teeth clearly exceed the size increase that might be predicted from its enlarged body. For this reason, *T. rex* leaves paleontologists with a mouthful of confusion.

In the end, we think it likely that animals like *Tyrannosaurus* would have been a more than adequate, indeed terrifying, active predator, yet one that wouldn't turn up its nose at a lunch of carrion. Indeed, it is probably better to view all theropods, from the exceptionally large tyrannosauroids down to much smaller troodontids, dromaeosaurids, and coelophysoids, as equal-opportunity consumers, taking large herbivores, smaller flesh eaters, and even the occasional carcass.

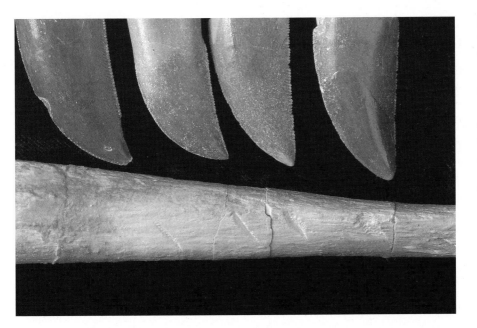

Figure 9.25. Cannibalism in theropods. *Above*: the teeth of the large Madagascan theropod *Majungatholus*. *Below*: toothmarks on *Majungatholus* bone. Note that the spacing of the grooves in the bone matches that of the teeth; note also the scratches next to each groove, reflecting the serrations of the teeth.

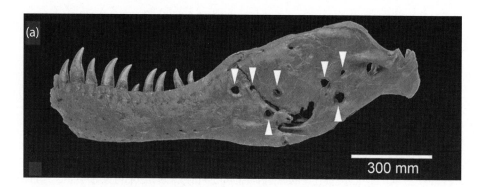

Figure 9.26. (a) Left lower jaw exhibiting multiple lesions (indicated by arrows). (b) Hypothesized reconstruction of the infection in a living tyrannosaur. Lesions within the lower jaw perforate the bone. Reconstruction based on photographs of living birds with similar pathologies.

chunks of the jaw bone, deforming and/or damaging the jaw. Similarities in the morphology of the lesions strongly suggest that this same organism, or a very close relative, was at work on Cretaceous non-avian theropods as well.

Social behavior: sex and the single *rex*

As was the case for various ornithischians, single-species bonebeds remain strong suggestions that theropods functioned in packs. For the most part, these mass graveyards – which include both juveniles and adults – pertain to the coelophysoid group of theropods (see "The evolution of Theropoda" below) and include such luminaries as *Syntarsus* (Zimbabwe, South Africa, and Arizona) and *Coelophysis* (New Mexico).

Other theropods, though, are also known in bonebeds: *Allosaurus* (Utah), *Giganotosaurus* and *Mapusaurus* (Patagonia). Could it have been that some theropods were gregarious, living in large family groups that perished together? Or perhaps each accumulation represents a communal feeding site? Do the bonebeds represent pack hunting? As new finds with one or two individuals get discovered, even that most supposedly solitary of killers – *Tyrannosaurus* – is a potential candidate for pack life. Theropods as gregarious beasts – even large theropods – is becoming a more and more likely prospect.

What can the skeletons of theropods tell us about how these animals related to each other socially? Like crests in hadrosaurids or frills and horns in neoceratopsians, quite a number of predatory dinosaurs – including *Syntarsus, Dilophosaurus, Proceratosaurus*, and possibly *Ornitholestes*, to *Ceratosaurus, Cryolophosaurus, Alioramus*, and *Oviraptor* – sported highly visible cranial crests (see Figures 9.12c and 9.27). Some are made of thin sheets of bone, while others are hollow, presumably part of the cranial air-sinus system. Beyond these, theropods such as *Yangchuanosaurus, Allosaurus, Acrocanthosaurus*, and the tyrannosauroids bore slightly elevated upper margins on the snout and raised and roughened bumps over the eyes. These structures are believed to have been cores for hornlets (small horns) made of keratin, which must have given the face a punk-rock spiky look (Figures 9.12b, e, f, and 9.13c).

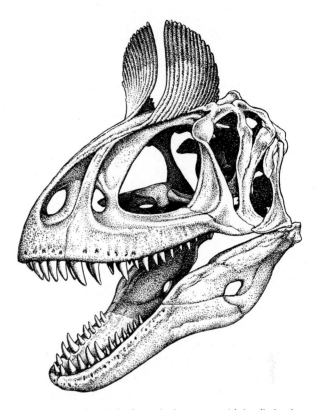

The crests and hornlets assuredly functioned in display and – at least for the latter – may have been used occasionally in head-butting squabbles over territories and mates. If crests and hornlets functioned in visual display, particularly in those theropods that lived in large groups (see above), we might expect them to be species-specific and probably sexually dimorphic so as to signal a given animal's identity and sex. And likewise we might

Figure 9.27. The skull of *Cryolophosaurus*, with its distinctive crest.

expect crests to show their greatest development in reproductively mature individuals; youngsters should have small, poorly developed crests and hornlets.

Are these expectations met in any theropods? To a degree; sexual dimorphism is found in *Syntarsus* and *Coelophysis*. In these two theropods, one of the two morphs had a relatively long skull and neck, thick limbs, and powerfully developed muscles around the elbow and hip, while the other form had a shorter skull and neck, and slender limbs. In *Tyrannosaurus*, the case has been made for sexually dimorphic ornamentation. The larger, more robust morph was hypothesized to be the female. In the cases of other theropods, we just don't yet know; what we would give to see one (or two, or a pack) alive (Box 9.2)!

The wonderful world of color Many diapsids generally, and modern birds in particular, are strongly sexually dimorphic when it comes to color, with the males commonly assuming the brightest hues. These traits are generally related to sexual selection (see Chapter 6). With the evidence of sexual selection so strongly imprinted in the remains of extinct theropods (as well as in the behavior of living theropods), could color have played a role in non-avian theropod sexual selection?

Incredibly, some patterns have actually survived the process of fossilization. For example, a striped pattern in the tail of *Sinosauropteryx* is clearly visible with the naked eye. Beyond rudimentary patterns, however, color in any dinosaur – including those covered with feathers – has been, until recently, a very elusive trait.

In modern birds, feather color is based upon microscopic capsule-shaped organelles called melanosomes. In 2007, using a high-powered electron microscope, melanosomes were identified in the vaned feathers of a primitive bird, *Confuciusornis* (see Chapter 11), as well as in the

Box 9.2 Dinosaur zombies

In western African lore, a *bokor*, or wizard, can revive the dead (the word "zombie" is thought to have came from *nzumbe*, a word in the North Mbundu language of Western Africa). Could scientists ever revive a (very) dead, (very) fossilized dinosaur?

The idea was vetted by Michael Crichton in *Jurassic Park* (1990). In that yarn, dinosaur blood, sucked by mosquitoes, was extracted from the insects when they themselves were trapped in amber. The dinosaur blood contained dinosaur blood cells; the blood cells were cloned, and Sam Neill faced off against *Velociraptor*. At the time, cloning dinosaurs from preserved cells received a bit of play in the serious scientific community, but in the end, DNA was found to degrade within a million or so years, and amber turned out to be semi-permeable, and things could get in and grow and contaminate the samples. And the idea was buried, until the late 1990s when, zombie-like, it was re-exhumed in an entirely different form.

Mary H. Schweitzer, at the time just beginning a Ph.D. at Montana State University, observed red structures, each with a black dot in the center, in a blood vessel channel in a slice of bone from a particularly well-preserved *T. rex*. Her thought was – counter to everything that everybody "knew" about the process of soft-tissue (not bone) preservation – these looked like red blood cells; they were located in the right place; could they actually *be* red blood cells from the dinosaur? As time progressed, she ended up recovering a lot of soft tissue: blood vessels, bone cells, and keratin from the claws of *Rahonavis* (Chapter 11) and from the feathers of *Shuvuuia* (Chapter 10).

Could this stuff be *real*? One way to test it was by checking immune responses. Living bodies, of course, react to foreign objects by creating antibodies. So Schweitzer and her colleagues used mice to generate antibodies to the supposed dinosaur proteins she had extracted. Sure enough, when the antibodies that had been generated were exposed to hemoglobin from rats and turkeys, the antibodies bound to the hemoglobin, suggesting to Schweitzer and her colleagues that the proteins they had extracted from the dinosaur were something very much like hemoglobin. The keratin, too, was tested in the same way: keratin-specific antibodies were generated using the modern claw and feather proteins; these then bound to the ancient keratin. It appeared Schweitzer had actually isolated soft tissue from very extinct, long gone dinosaurs.

It got even crazier when another *T. rex*, astoundingly well preserved over the 65.5+ million years that had elapsed since the animal's death, was found. Schweitzer, who had by then moved to North Carolina State University, found medullary tissue in the fossil bones – tissue that is diagnostic of an animal that is producing eggs. Today medullary tissue is produced only by ovulating female birds; a clear indication that this *T. rex*, at least, must have been female – and ovulating (see Chapter 10). Schweitzer's reaction to her discovery, quoted in the pages of *Scientific American*, was priceless: "'Oh, my gosh, it's a girl – and it's pregnant!' I exclaimed to my assistant, Jennifer Wittmeyer. She looked at me like I had lost my mind."[6] From this *T. rex*, she extracted hollow, transparent, flexible tubes, some filled with some red material, which she interpreted as blood vessels with dinosaur blood.

Schweitzer's work was not accepted uncritically, and there is still some concern that she is actually looking at the remains of modern organisms, rather than dinosaur soft tissue. Underlining the difficulty of this type of work, some authors have suggested that she is looking at modern invading microbes or bacterial biofilms growing on the fossil bone, rather than actual dinosaur tissue. That possibility became less likely very recently, however, when collagen polypeptide sequences were obtained from a hadrosaur (*Brachysaurolophus*) and subjected to the same rigorous suite of tests undergone by the *T. rex* polypeptides. Moreover the work has been duplicated in separate laboratories. Indeed, recent work on fossil collagen has begun to elucidate aspects of the structure of this molecule that led to its unusual resistance to long-term degradation.

It won't end up happening quite as Michael Crichton described it, but the possibility of building backwards from the small fragments of DNA preserved in these fossils has become the kind of thing that some paleontologists dream about. Jack Horner, for one (see Box 14.9), teaming up with molecular biologist colleagues, attempted to grow the tail of a non-avian dinosaur from long-dormant genes in modern birds (a chicken). His idea was that the ancestral dinosaur-tail genes, left over from when theropods regularly grew tails, are still there, waiting to be activated. They never quite got their tail, but their tale, and the dream that motivated it, are described in Horner's book *How to Build a Dinosaur* (2009).

The possibility of seeing a living non-avian dinosaur becomes less and less fantastic, as our knowledge of molecular biology and genetics becomes deeper.

[6] Schweitzer, M.H. 2010. Blood from a stone. *Scientific American*, December, 62–69.

primitive feathers of non-flying theropods *Sinosauropteryx* and *Sinornithosaurus* (see discussion of feathers in theropods in Chapter 10). Because the shape of some melanosomes indicates color, it is, for the first time, possible to learn something about the actual colors present in these long-extinct dinosaurs. The banding in the tail of *Sinosauropteryx* appears to have been black; a crest along its back may have had tan–reddish-brown coloration (Figure 10.11). Even iridescence has been recognized in fossil non-avian theropods! The colors and patterns present in these non-avian theropods reinforce the idea that, like their specialized descendents the birds, non-avian theropods probably used color and pattern for a variety of complex social behaviors.

Mama's (?) little theropod

Regardless of sexual dimorphism in non-avian theropods, what we know about their reproductive biology has been greatly enhanced by the discovery of brooding oviraptorosaurs. Several recently discovered articulated oviraptorosaur skeletons are preserved overlying nests of eggs, laid in a circular pattern of as many as 22 eggs. The embryos are the same species as the adult skeleton overlying them.[5] The oviraptorosaur skeleton (Mom? Dad?) is positioned directly above the center of the nest, with its limbs arranged symmetrically on either side and its arms spread out around the perimeter as if protecting the nest (Figure 9.28). These specimens indicate that incubating eggs on open nests evolved well before the origin of modern birds.

What we know of the post-hatching growth of non-avian theropods mostly comes from bonebeds (for example, Ghost Ranch (New Mexico) and Cleveland-Lloyd (Utah)), as well as from some of the Upper Cretaceous localities of the Gobi Desert in Mongolia. For *Coelophysis* and *Syntarsus*, apparently there was a 10- to 15-fold increase in body size from hatchling to adulthood, and this growth is thought to have been quite rapid. Accompanying this rapid growth were proportional changes in the skull (relatively smaller eye socket, enlargement of the jaws and areas for muscles), relative lengthening of the neck, and relative shortening of the hindlimb. Similar changes – when they can be identified – are thought to occur in *Tarbosaurus* and *Albertosaurus* as well.

The evolution of Theropoda

We've seen that Theropoda is a daunting array of dinosaurs. Despite the amount of evolution such diversity represents, there are many derived features that unite the theropod clade (see Figure 9.4).

Coelophysoidea, Neoceratosauria, and Tetanurae

At its base, Theropoda is the wellspring of the three major groups of descendants: Coelophysoidea (named after *Coelophysis* and including some related, less well-known forms), Neoceratosauria (named after one of its members, the Jurassic *Ceratosaurus* and including some other bad boys, including the formidable Cretaceous-aged *Carnotaurus*), and Tetanurae (*tetanus* – stiff; *uro* – tail).

---

[5] The eggs were first attributed to the ceratopsian *Protoceratops*. Because they were found with theropod skeletons, it was assumed that the theropods were stealing the "*Protoceratops* eggs" and were given the name *Oviraptor* (= egg stealer). *Oviraptor* languished, falsely accused, for 70 years, until the mid-1990s discovery of the unquestionably nesting specimens.

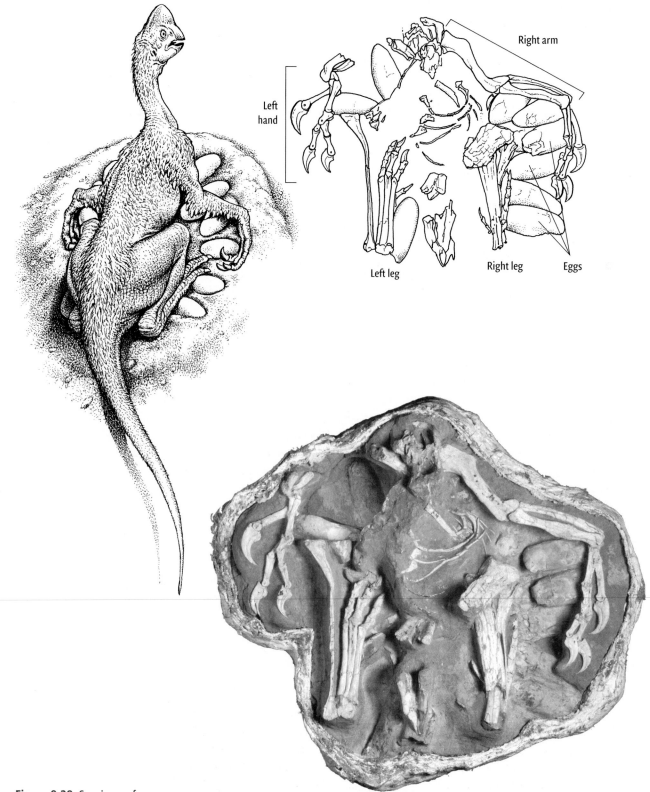

Figure 9.28 Specimen of an
oviraptorosaur adult nesting on
its eggs.

It was in Tetanurae that some of the most remarkable and important theropod evolution took place. For starters, tetanurans evidently took breathing to new places. While saurischians likely had simple air sacs, indicated by pleurocoels, and the rudimentary unidirectional breathing (Box 8.1) to go with it, the development of true pneumaticity, with extensive air sacs, and highly efficient unidirectional breathing such as one sees in modern birds, really began in Tetanurae. A recently discovered basal tetanuran, *Aerosteon*, shows the openings (pneumatic foramina; see Glossary and Chapter 10) and hollowed bones indicative of a complexly branching suite of air sacs like those found in modern birds. These adaptations only increased in complexity through the evolution of Maniraptora and Aves (see Figures 9.31 and 9.36 below) so that highly derived theropods had (and have) some of the most efficient air breathing ever invented. This doubtless enabled them to sustain high levels of activity, and led to a respiratory system capable of supporting the sustained flight of modern birds.

Members of Tetanurae, whose record extends from the Middle Jurassic to the present, share a large number of other derived features (Figure 9.29), but the key feature linking all tetanurans is that the back half of the tail is stiffened by interlocking **zygapophyses**, fore-and-aft projections from the neural arches (Figure 9.30). We have seen that the tail is an important counterbalance in theropod architecture, and it is no surprise that, through tetanuran evolution, there is a marked tendency to *decrease* the flexibility of the tail (except at its base).

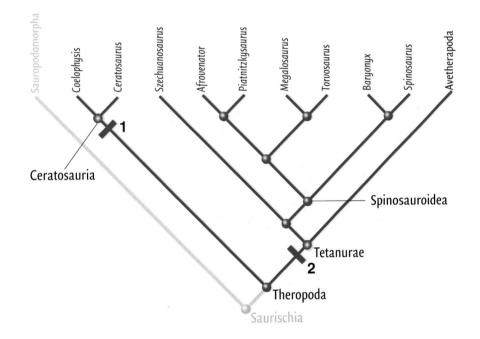

Figure 9.29. Cladogram of Theropoda. Derived characters include: at **1**, modification of the neural spines and transverse processes of the vertebrae, fusion of the sacral ribs with the ilium, ventral and lateral flaring of the crest above the acetabulum on the ilium, modification of the knee joint, and fusion between the upper ankle bones; at **2**, low ridge demarcating the maxillary antorbital fossa, spine table on axis, reduced rod-like axial spinous process, prominent acromion on the scapula, loss of digit IV phalanges, metacarpal II nearly twice the length of metacarpal I, reduced femoral trochanteric shelf, prominent extensor groove on femur, fibular condyle on proximal tibia strongly offset from cnemial crest, broadly triangular metatarsal I attached to distal part of metatarsal II.

Tetanurae includes a host of large theropods – for example, *Spinosaurus, Irritator, Baryonyx, Suchomimus, Torvosaurus, Eustreptospondylus, Megalosaurus,* and *Piatnitzkysaurus* – as well as an important clade of tetanurans known as Avetheropoda.

Avetheropoda

Avetheropoda (*avis* – bird; a reference to bird-like features of many members of this group) is that group of theropods more closely related to birds than are the preceding genera

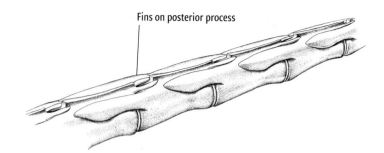

Figure 9.30. Zygapophyses in tetanurans. Note how these processes extend across the adjacent vertebrae both anteriorly and posteriorly, hindering flexibility.

and their near relatives. Avetheropods share many derived features, and consist of two clades, Carnosauria and Coelurosauria (Figure 9.31).

Coelurosauria includes many large forms, most famously the tyrannosauroids. It also includes some small forms, such as *Compsognathus* (Figure 9.31), as well as Maniraptora. Maniraptora includes beasts such as *Ornitholestes* (Figure 9.32), ornithomimosaurs and oviraptorosaurs

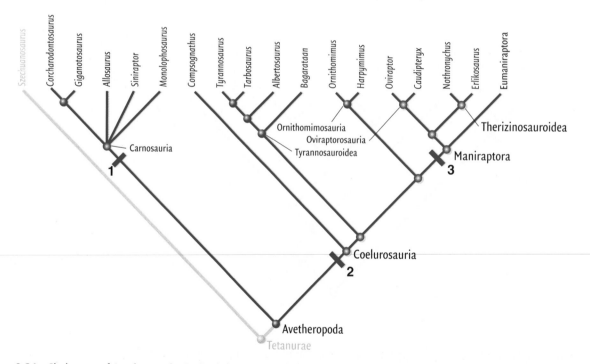

Figure 9.31. Cladogram of Avetheropoda. Derived characters include: at **1**, enlarged external nares, participation of lateral surface of nasal in antorbital cavity, presence of nasal recesses, prefrontal excluded from rostral rim of orbit, supraorbital notch between postorbital and prefrontal, paroccipital processes directed strongly ventrolaterally from occiput to below level of foramen magnum, very short basipterygoid processes, mid-cervical centrum length less than twice the diameter of forward articular surface, front margin of spinous processes of proximal mid-caudal vertebrae with distinct kink, spur along front margin of spinous processes of mid caudals; at **2**, presence of a pterygopalatine fenestra, short mandibular process on pterygoid, presence of sternal ribs (three pairs), reduction of coracoid process, furcula, U-shaped ischial obturator notch, loss of transverse groove on astragalar condyle; at **3**, crenulated ventral margin of premaxilla, parietal at least as long as frontal, U-shaped mandibular symphysis, semi-lunate carpal.

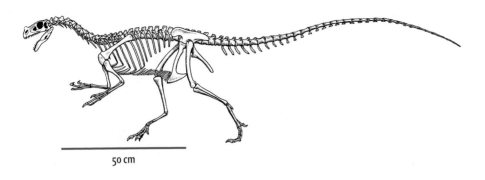

50 cm

(which we've met), therizinosauroids (which we will meet), and the non-avian theropods closest to birds, Eumaniraptora. Maniraptorans adapted the wrist with the development of the semi-lunate carpal, a wrist bone modification that increased manual dexterity and the ability to sever flesh from bone (Figure 9.33).

Therizinosauroids represent a strange theropod venture into herbivory. Large and ponderous, these highly evolved and distinctive maniraptoran theropods were apparently herbivorous and have been compared to giant sloths (Figure 9.34). Their affinities have been a matter of considerable confusion, but the consensus is that they were an aberrant group of theropods.

Convergent evolution in large theropods

Looking within Tetanurae, we see a striking quality of theropod evolution. Superficially, big theropods all resemble each other (they were once all united as "carnosaurs"). Clearly, as theropods evolved to large sizes, lineages *independently* developed some of the same features. Such similar, although independent, evolution is called convergent. In the case of large theropods, features such as proportionally large heads and a tendency toward shorter arms occurred convergently. The same features occur *independently* in neoceratosaurs (for example, *Carnotaurus*), in primitive tetanurans (*Spinosaurus*, *Szechuanosaurus*, *Megalosaurus*, and *Afrovenator*), and in avetheropods (*Giganotosaurus*, *Allosaurus*, and all the tyrannosauroids).

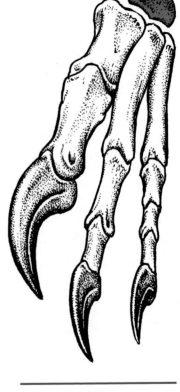

Figure 9.33. Semi-lunate carpal (colored) in the left hand of *Ingenia*, an oviraptorosaurian maniraptoran.

5 cm

As we have seen, however, despite their superficially convergent morphology, these animals behaved very differently. Why, then, did they independently develop similar morphologies? The answer was once thought to reside in the logistics of growing BIG. It was thought that to increase size yet maintain reasonable agility as bipeds, compromises had to be made in the sizes and power of various body parts, and it was assumed that shortened arms and large heads were part of such compromises. All that changed in 2009, however, with the discovery of

Figure 9.34. The therizinosaur *Nothronychus* meets its skeleton.

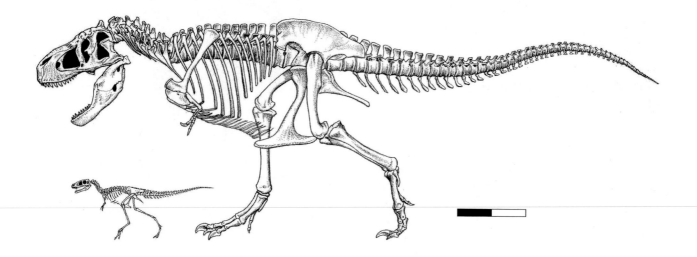

Figure 9.35. The early, primitive tyrannosaur *Raptorex* is dwarfed by the skeleton of *Tyrannosaurus rex*. Scale = 1.5 m.

the Early Cretaceous *Raptorex* from China: a 3 m tyrannosaurid about a fifth the length of the mighty *T. rex* (Figure 9.35)! Here was a small, Early Cretaceous tyrannosaur, indeed, a primitive member of the group, which had short arms and a large head. This suggested that these adaptations, once thought to be part and parcel of life as a large theropod, were, in the case of

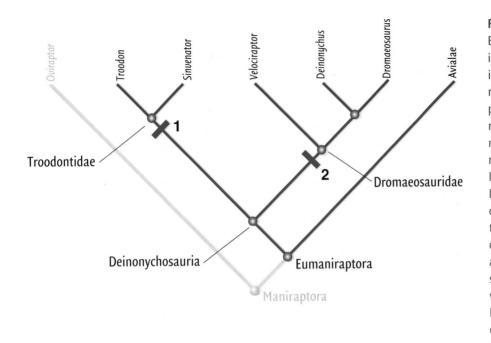

Figure 9.36. Cladogram of Eumaniraptora. Derived characters include: at **1**, pneumatic foramen in quadrate, loss of basisphenoid recess, large number of teeth, close packing of front dentary teeth, reduced basal tubera, asymmetrical metatarsus, slender metatarsal II markedly shorter than metatarsals III and IV, and a robust metatarsal IV; at **2**, short T-shaped frontals, a caudolateral overhanging shelf of the squamosal, lateral process of quadrate that contacts quadratojugal above enlarged quadrate foramen, stalk-like parapophyses on dorsal vertebrae, modified raptorial digit II, chevrons and prezygapophyses of caudal vertebrae elongated and spanning several vertebrae, presence of subglenoid fossa on coracoid.

tyrannosaurs at least, really about life as a tyrannosaur. It was thought that as tyrannosaurs attained large size, the arms got progressively shorter and the head got larger. Yet here, even an early, relatively primitive member of the group still shows the typical *T. rex* morphology of an exceptionally large head and exceptionally short arms. The "why" of this particular morphology remains unanswered.

Eumaniraptora

The remainder of non-avian Theropoda, non-avian eumaniraptorans, were more closely related to birds than to the others (Figure 9.36). The group consists of those highly predaceous and intelligent carnivores, Deinonychosauria and Avialae. Deinonychosaurs rightly ought to evoke more fear and nightmares than *T. rex*, for they include the sickle-clawed troodontids and dromaeosaurids. And with Avialae, we've come to Aves and its very near relatives: subjects rightly deserving their own chapter (Chapter 10).

Summary

Theropods are among the most iconic of dinosaurs, including beasts such as *Tyrannosaurus rex*. Clawed bipeds all, with distinctive hollow bones, the earliest known dinosaurs were theropods, and it is theropods that are still living (as birds). Most of the Mesozoic forms had claws with a semi-opposable thumb on a grasping three-fingered hand; possessed recurved, serrated, laterally compressed teeth; and were carnivorous.

They are a complex group, with a remarkable evolutionary history. The most primitive radiation of theropods is seen in the group Coleophysoidea. A lineage of particular interest (because it includes tyrannosaurs and modern birds) is the tetanurans, a group of theropods whose zygapophysis-stiffened tails were used as dynamic counterbalances to grasping claws; it is among these that some of the most predatory dinosaurs reside. A subset of these, deinonychosaurs, developed eviscerating claws on the legs, grasping, powerful hands, large brains (and inferred high intelligence), and likely stereoscopic vision: pound for pound the most deadly carnivores ever evolved within Dinosauria. And deinonychosaurs are the most closely related non-avian dinosaurs to Aves (living birds).

Non-avian theropods were not strictly carnivorous, and a number of groups developed whose habits are still unknown. There were the long-armed, well-clawed oviraptorosaurids whose toothless mouths may have crushed mollusks or eggs. Then there were ostrich-mimics; toothless, small-skulled forms that may have been among the fastest runners in all Dinosauria. Finally there were therizinosaurs, whose long arms and massive claws are vaguely reminiscent of sloths.

Many theropods appear to be designed for aggressive, active behavior, and it was this aspect of their design that first suggested to researchers that deinonychosaur theropods in particular – and dinosaurs in general – might be endothermic. Deinonychosaurs remain among the best candidates for full-time mammalian-style endothermy.

With modern birds as living (if highly derived) examples, social behavior can be inferred in theropods. A variety of facial features such as hornlets adorned theropods, and some evidence suggests that many of them hunted in packs. Particularly bird-like is theropod maternal behavior: in those forms in which it is known, non-avian theropod mothers (?) incubated clutches of eggs very much like avian theropods.

Selected readings

Barsbold, R. and Osmólska, H. 1990a. Ornithomimosauria. In Weishampel, D. B., Dodson, P. and Osmólska, H. (eds.), *The Dinosauria*, 2nd edn. University of California Press, Berkeley, pp. 225–244.

Barsbold, R. and Osmólska, H. 1990b. Oviraptorosauria. In Weishampel, D. B., Dodson, P. and Osmólska, H. (eds.), *The Dinosauria*, 2nd edn. University of California Press, Berkeley, pp. 249–258.

Clark, J. M., Maryańska, T. and Barsbold, R. 2004, Therizinosauroidea. In Weishampel, D. B., Dodson, P. and Osmólska, H. (eds.), *The Dinosauria*, 2nd edn. University of California Press, Berkeley, pp. 151–164.

Currie, P. J. 1990. Elmisauridae. In Weishampel, D. B., Dodson, P. and Osmólska, H. (eds.), *The Dinosauria*, 2nd edn. University of California Press, Berkeley, pp. 245–248.

Holtz, T. R. 2004. Tyrannosauroidea. In Weishampel, D. B., Dodson, P. and Osmólska, H. (eds.), *The Dinosauria*, 2nd edn. University of California Press, Berkeley, pp. 111–136.

Holtz, T. R., Molnar, R. E. and Currie, P. J. 2004. Basal Tetanurae. In Weishampel, D. B., Dodson, P. and Osmólska, H. (eds.), *The Dinosauria*, 2nd edn. University of California Press, Berkeley, pp. 71–110.

Makovicky, P. J. and Norell, M. A. 2004. Troodontidae. In Weishampel, D. B., Dodson, P. and Osmólska, H. (eds.), *The Dinosauria*, 2nd edn. University of California Press, Berkeley, 184–195.

Makovicky, P. J., Kobayashi, Y. and Currie, P. J. 2004. Ornithomimosauria. In Weishampel, D. B., Dodson, P. and Osmólska, H. (eds.), *The Dinosauria*, 2nd edn. University of California Press, Berkeley, pp. 137–150.

Molnar, R. E. 1990. Problematic Theropoda: "carnosaurs". In Weishampel, D. B., Dodson, P. and Osmólska, H. (eds.), *The Dinosauria*, 2nd edn. University of California Press, Berkeley, pp. 306–319.

Molnar, R. E. and Farlow, J. O. 1990. Carnosaur paleobiology. In Weishampel, D. B., Dodson, P. and Osmólska, H. (eds.), *The Dinosauria*, 2nd edn. University of California Press, Berkeley, pp. 210–224.

Molnar, R. E., Kurzanov, S. M. and Dong, Z. 1990. Carnosauria. In Weishampel, D. B., Dodson, P. and Osmólska, H. (eds.), *The Dinosauria*, 2nd edn. University of California Press, Berkeley, pp. 169–209.

Norell, M. A. and Makovicky, P. J. 2004. Dromaeosauridae. In Weishampel, D. B., Dodson, P. and Osmólska, H. (eds.), *The Dinosauria*, 2nd edn. University of California Press, Berkeley, pp. 196–209.

Norman, D. B. 1990. Problematic Theropoda: "coelurosaurs". In Weishampel, D. B., Dodson, P. and Osmólska, H. (eds.), *The Dinosauria*, 2nd edn. University of California Press, Berkeley, pp. 280–305.

Osmólska, H. and Barsbold, R. 1990. Troodontidae. In Weishampel, D. B., Dodson, P. and Osmólska, H. (eds.), *The Dinosauria*, 2nd edn. University of California Press, Berkeley, pp. 259–268.

Osmólska, H., Currie, P. J. and Barsbold, R. 2004. Oviraptorosauria. In Weishampel, D. B., Dodson, P. and Osmólska, H. (eds.), *The Dinosauria*, 2nd edn. University of California Press, Berkeley, pp. 165–183.

Ostrom, J. H. 1990. Dromaeosauridae. In Weishampel, D. B., Dodson, P. and Osmólska, H. (eds.), *The Dinosauria*, 2nd edn. University of California Press, Berkeley, pp. 269–279.

Rowe, T. and Gauthier, J. A. 1990. Ceratosauria. In Weishampel, D. B., Dodson, P. and Osmólska, H. (eds.), *The Dinosauria*, 2nd edn. University of California Press, Berkeley, pp. 151–168.

Sereno, P. C., Martinez, R. N., Wilson, J. A., Varricchio, D. J., Alcobar, O. A. and Larsson, H. C. E. 2008. Evidence for avian intrathoracic air sacs in a new predatory dinosaur from Argentina, *PLoS ONE*, **3**, e3303. doi:10.1371/journal.pone.0003303.

Sues, H. D. 1990. *Staurikosaurus* and Herrerasauridae. In Weishampel, D. B., Dodson, P. and Osmólska, H. (eds.), The Dinosauria, 2nd edn. University of California Press, Berkeley, pp. 143–147.

Tykoski, R. S. and Rowe, T. 2004. Ceratosauria. In Weishampel, D. B., Dodson, P. and Osmólska, H. (eds.), *The Dinosauria*, 2nd edn. University of California Press, Berkeley, pp. 47–70.

Witmer, L. M. and Ridgely, R. C., 2009. New insights into the brain, braincase, and ear region of tyrannosaurs (Dinosauria, Theropoda), with implications for sensory organization and behavior. *The Anatomical Record*, **292**, 1266–1296.

Topic questions

1 Describe the general features of theropods.

2 Describe how we know that running – cursoriality – was a key feature of theropod behavior.

3 What features of the theropod hand strongly suggest dexterity and grasping ability?

4 What kinds of evidence exist for what theropods ate?

5 Can a toothless animal be carnivorous?

6 Describe the range of skull and tooth design in theropods. How do those relate to our understanding of how theropods bit and killed?

7 What are the key clues that suggest that, as a group, theropods were carnivorous?

8 Highlight the evidence for pack-hunting by theropods.

9 Consider the design of ornithomimosaurs. Why is it difficult to explain the lifestyle of these animals?

10 We said that the most unambiguous features of a carnivorous lifestyle are seen in small- to medium-sized theropods. What features caused us to say this? Speculate on why this might be the case.

11 Theropod evolution was complex, so here is an exercise to help you better understand their relationships. Construct a single cladogram with the groups Theropoda, Coelophysoidea, Neoceratosauria, Tetanurae, Avetheropoda, Coelurosauria, Carnosauria, Maniraptora, Eumaniraptora, and Aves on it.

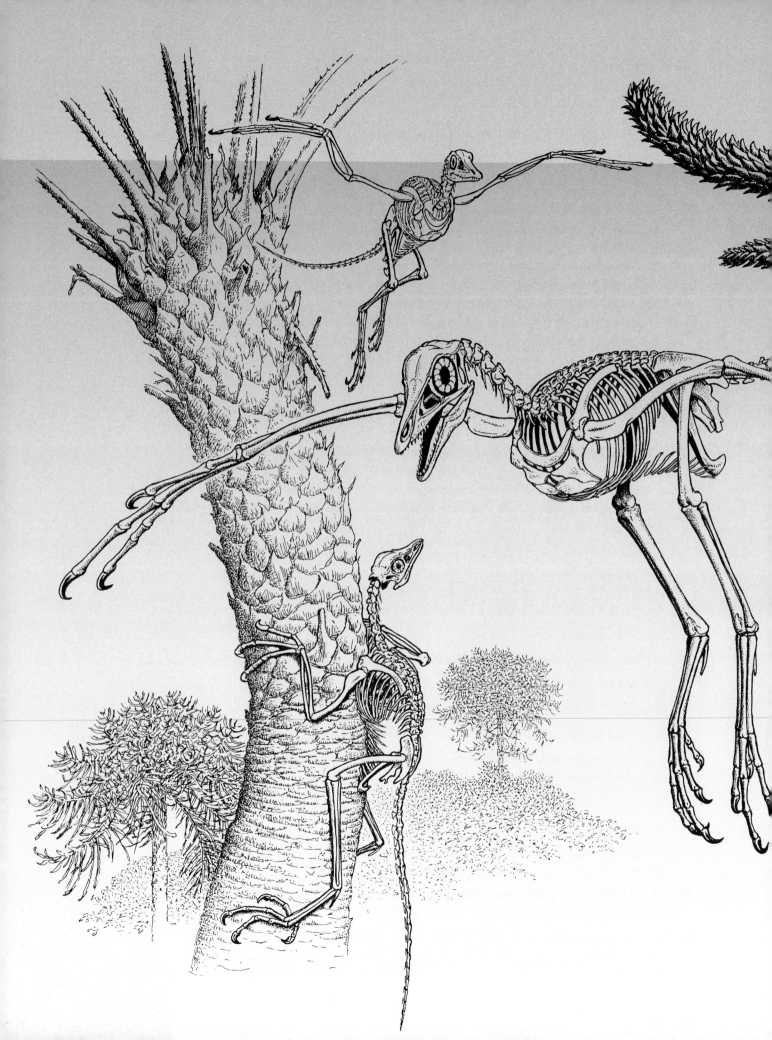

CHAPTER 10

THEROPODA II:
THE ORIGIN OF BIRDS

Q: Which came first, the chicken or the egg?

A: The egg. But it wasn't from a chicken.

CHAPTER OBJECTIVES

- Understand birds as theropod dinosaurs

- Learn about the origin of flight

- Refine our cladograms as a tool for investigating evolution

Birds

Birds are dinosaurs

Birds *are* dinosaurs. We don't mean that they are *related* to dinosaurs – although, if they are dinosaurs, they must be related to them. We don't mean that they *come from* dinosaurs – although they obviously evolved from something that was itself a dinosaur. We mean that birds *are* dinosaurs, a statement that, as this chapter unfolds, will be no more radical than saying that humans are mammals.

So how do we figure out who birds are related to? The same way that we explored in Chapter 3: using diagnostic characters. Here we choose those features that might be easily observed in a fossil.

Diagnostic features of living birds

Among living vertebrates, birds possess a remarkable and largely unique suite of diagnostic features (Figure 10.1 and Table 10.1). There are many; most of those selected here are likely to be preserved in a fossil.

Feathers All living birds have feathers – complex, distinctive structures that consist of a hollow, central shaft that decreases in diameter toward the tip. Radiating from the shaft are barbs, feather material that, when linked together along the length of the shaft by small hooks called barbules, form the sheet of feather material called the vane (Figure 10.1a). Feathers with well-developed, asymmetrical vanes are usually used for flight and are therefore called flight feathers. Feathers in which the barbules are not well developed tend to be puffy, with poorly developed vanes, and are called down as we know from sleeping bags, comforters, and ski parkas, they are superb insulation.

Loss of teeth No living bird has teeth. The jaws of birds are covered with a rhamphotheca.

Large brains Living birds have well-developed brains protected by a large braincase.

Carpometacarpus The wrist and hand bones in the hand of modern birds are fused into a unique structure called the carpometacarpus[1] (Figure 10.1b). The carpometacarpus is composed of three fused fingers, now generally thought to be digits I (the thumb), II, and III.

Legs and feet Birds are fully bipedal, and have an erect stance (see Chapter 4). The twin shin bones (tibia and fibula; together, the "drumstick" on the dinner table) are unequal: the tibia is large, but the fibula thins to a sliver close to the ankle.

[1] Spiced and served with beer, we call them "buffalo wings"!

Table 10.1. Diagnostic features of living birds
Modern birds
Teeth (–)
Swollen braincase
Pygostyle (+)
Carpometacarpus (+); fused digits I, II, III
Legs: 1 Bipedal 2 Tarsometatarsus
Foot: 1 3 toes in front; 1 in back 2 Digit V (–) 3 Claws
Pneumatic bones
Furcula (wishbone)
Rigidified trunk 1 Carinate sternum 2 Synsacrum 3 Some vertebrae (–) 4 Flying adaptations
Feathers (+)
The plus sign (+) indicates character present; the minus sign (–) indicates character absent.

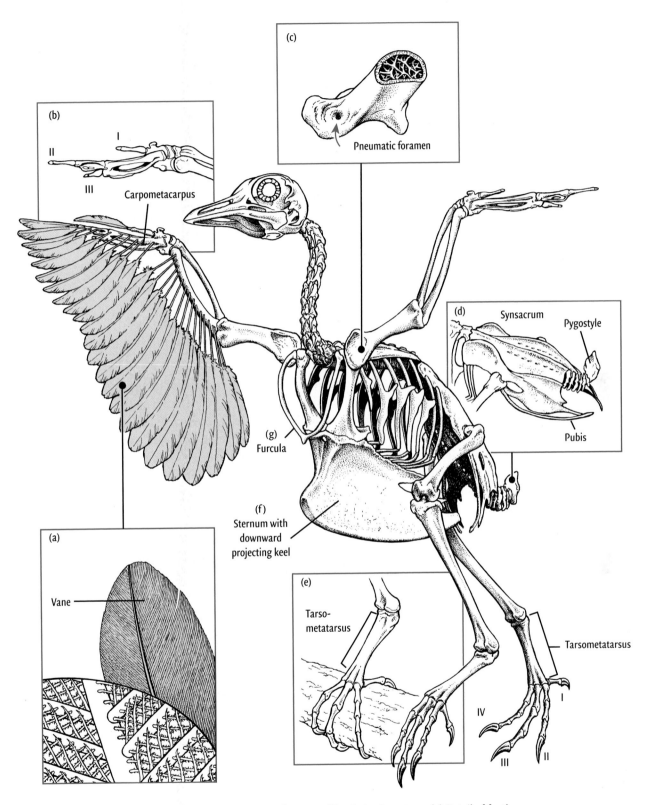

Figure 10.1. The skeleton of a pigeon, showing major features of its skeletal anatomy. (a) Detail of feather structure; (b) carpometacarpus with digits labeled; (c) hollow bone with pneumatic foramina; (d) synsacrum (fused pelvic bones) with pygostyle; (e) tarsometatarsus; (f) sternum with large downward-projecting keel; and (g) furcula.

The feet of all living birds are clawed and have three toes in front (digits II, III, and IV), and a smaller toe (digit I) at the back. The three central metatarsals (foot bones, to which the toes attach; in this case II, III, and IV) are fused together and with some of the ankle bones, to form a unique structure called a tarsometatarsus (Figure 10.1e).

Pygostyle No living bird has a long tail skeleton. Instead, in most cases, the bones are fused into a compact, vestigial structure called a pygostyle (*pygo* – rump; *stylus* – stake; Figure 10.1d).

Pneumatic bones Living birds breathe unidirectionally with a complex system of air sacs (see Box 8.1). Their bones are thus are pneumatic and have pneumatic foramina (Figure 10.1c).

Rigid skeleton Bird skeletons have undergone a series of bone reductions and fusions to produce a light, rigid platform to which the wings and the muscles that power them attach. Fused vertebrae in the back are connected with a well-developed breastbone, or sternum, by ribs with upper and lower segments. The sternum is large and, in flapping flyers, has a broad, deep keel, or downward-protruding bony sheet, for the attachment of flight muscles (Figure 10.1f). The pelvic region is fused together into a synsacrum, a single structure consisting of many sacral vertebrae fused together (Figure 10.1d). The pubis is very slender and points posteriorly.

In the shoulder, pillar-like coracoid bones buttress against the front of the sternum, the shoulder blade (scapula), and against paired, fused collar bones[2] (furcula; Figure 10.1g). No living organism except birds has a furcula.

Flight musculature In modern flying birds, the downward stroke of the wing is obtained by the pectoralis muscle, which attaches to the front of the coracoid and sternum, and to the furcula and humerus. The recovery stroke is carried out by the supracoracoideus muscle. The supracoracoideus attaches at the keel of the sternum, runs up along the side of the coracoid bone, and attaches via a tendon at the top of the upper arm bone through a hole (the trioseal foramen) formed by the coracoid, furcula, and scapula (Figure 10.2). This is an adaptation unique to living birds.

Lessons from history

It turns out that when we look at the features of birds, although unique among *living* vertebrates, many are old friends from our excursion through Theropoda (see Chapter 9). These features include:

- Hollow and/or pneumatic bones (diagnostic of theropods).
- Pleurocoels and pneumatic foramina both present in birds; also present in other saurischians (see Chapters 8 and 9).
- Bipedality (found in all theropods).
- Distinctive foot (found in all theropods).
- Three-fingered hand (found in most theropods).
- Furcula (a diagnostic character of Coelurosauria).
- Large tibia; small fibula thinning toward the ankle (diagnostic character of eumaniraptoran theropods).
- Large braincase and stereoscopic vision (found in eumaniraptoran theropods).
- Feathers (found in non-avian avialan theropods and perhaps going back as far as Ornithischia; see below).

---

[2] At the dinner table we call them the "wishbone."

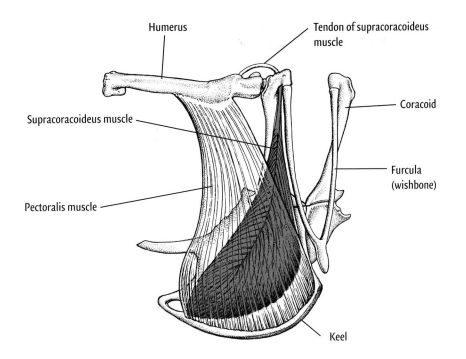

Humerus

Tendon of supracoracoideus muscle

Supracoracoideus muscle

Coracoid

Pectoralis muscle

Furcula (wishbone)

Keel

Figure 10.2. The two major muscles for flight: the *pectoralis* and the *supracoracoideus*. The pectoralis is the muscle used in the downward (power) stroke, while the supracoracoideus is used in the recovery stroke.

The list of shared, diagnostic characters between birds and theropods leads us to the inevitable conclusion that birds are theropod dinosaurs.

Table 10.2 summarizes some of these diagnostic characters, but it also highlights a different problem: while there are many features that birds and theropods share, it's still quite a jump from one to the other. How to bridge that gap?

Feathered theropods and the ancestry of living birds

It has become obvious in the past 20 years that the "gap" between birds and dinosaurs is only a gap when viewed from the perspective of the modern world. Birds certainly look distinctive today, among living veterbrates. But viewed from an evolutionary perspective, and considered in the context of Theropoda, the difference between the first bird and its closest ancestor is a fine line, indeed. With more than a dozen feathered theropods known from the fossil record, what is and what is not a bird becomes, literally, a matter of definition.

Even with so many feathered dinosaurs known, the route to the origin of birds has always run though *Archaeopteryx*. First discovered as a feather in 1860 (Figure 10.3) and as bones shortly thereafter, the half-meter long feathered dinosaur seemed chimeric, because it had "bird" feathers co-existing with "reptilian" features, such as a tail and hands with claws (Figure 10.4).[3]

We'll take a detailed look at *Archaeopteryx*, because this creature beautifully blurs the distinction between dinosaurs and modern birds (Table 10.3).

Anatomy of *Archaeopteryx*

Skull The skull of *Archaeopteryx* (Figure 10.5a) is tyically archosaurian, with nasal, antorbital, and eye openings. Some specimens preserve a sclerotic ring, a series of plates that supported

---

[3] Darwin had just published *On the Origin of Species* in 1859, proposing that species evolved into other species. Here, a mere two years later, was discovered an apparent "missing link" that mixed "reptilian" and avian features.

the eyeball. The temporal region is poorly known but hints of lower and upper temporal fenestrae are preserved. *Archaeopteryx* has blade-like, unserrated, recurved teeth.

Arms and hands The arms are quite long (about 70% of the length of the legs). The hands are about as large as the feet, and each hand bears three, fully moveable, separate fingers. Each finger is tipped with a well-developed, recurved claw. The wrist of *Archaeopteryx* bears a semilunate carpal (Figure 10.5e; see Chapter 9).

Legs and feet The foot of *Archaeopteryx* has three toes in front, and a fourth toe lies to the side (or behind; the specimens are flattened). The three in front are more or less symmetrical around digit III, and all the toes all have well-developed claws (Figure 10.5d).

The ankle of *Archaeopteryx* is a modified mesotarsal joint (see Chapter 4). It preserves a small splint of bone rising up from the center of the astragalus, one of two bones in the ankle (see Figure 4.5), to form a tall ascending process. The three foot bones are unfused. The thighs are considerably shorter than the shins, and the fibula is sliver-like as it approaches the ankle.

Table 10.2. **Characters shared by maniraptorans and living birds**	
Maniraptoran theropods	**Modern birds**
Teeth (+)	Teeth (−)
Braincase slightly enlarged	Swollen braincase
Tail long, well developed	Pygostyle (+)
Hand three-fingered; I, II, & III	Carpometacarpus (+); fused digits I, II, III
Legs: 1 Bipedal 2 Unfused foot	Legs: 1 Bipedal 2 Tarsometatarsus
Foot: 1 3 toes in front; 1 in back 2 Digit V (−) 3 Claws	Foot: 1 3 toes in front; 1 in back 2 Digit V (−) 3 Claws
Hollow bones; some pneumatic	Pneumatic bones
Furcula (wishbone)	Furcula (wishbone)
Trunk not rigid: 1 Sternum small; flat 2 Pelvis unfused 3 All vertebrae (+) 4 Flying adaptions (−)	Rigidified trunk: 1 Carinate sternum 2 Synsacrum 3 Some vertebrae (−) 4 Flying adaptations
Feathers (+)	Feathers (+)

The plus sign (+) indicates character present; the minus sign (−) indicates character absent.

Long bones *Archaeopteryx* has thin-walled long bones with large hollow spaces.

Trunk and tail The axial skeleton of *Archaeopteryx* lacks many of the highly evolved features that characterize modern birds. The body is relatively long and shows none of the foreshortening or fusion that one sees in the vertebrae of birds. The sternum is relatively small, with a small keel. A large, strong furcula is present (Figure 10.5f). Also present are gastralia, or belly ribs, which primitively line the belly in many archosaurs (Figure 10.5c).

Archaeopteryx lacks a synsacrum and instead has a primitive, unfused archosaurian pelvis. The pubis is directed downward. The distal end of the pubis (the footplate) is well developed, although the front part is absent.

Archaeopteryx has a long, straight, well-developed tail. Projections from the neural arches (zygapophyses) are elongate, meaning that the tail has little flexibility and has little potential for movement along its length.

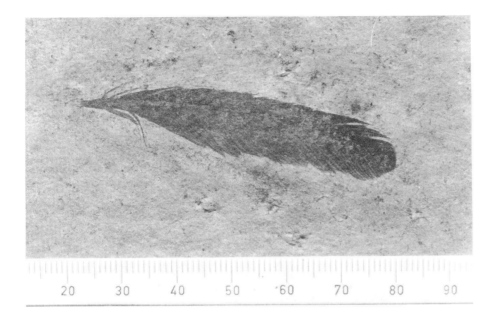

Figure 10.3. The first evidence for Jurassic-aged birds: the feather of *Archaeopteryx lithographica*, described in 1861, from the Solnhofen quarry in Bavaria (scale in centimeters).

Feathers *Archaeopteryx* has well-preserved, unambiguous feather impressions. The best-preserved feathers are clearly flight feathers (Figures 10.5b) and are indistinguishable from those of modern birds. Unlike in living birds, however, there are feathers also lining a long, bony tail. These radiate out from the vertebrae, and form an impressive tail plume.

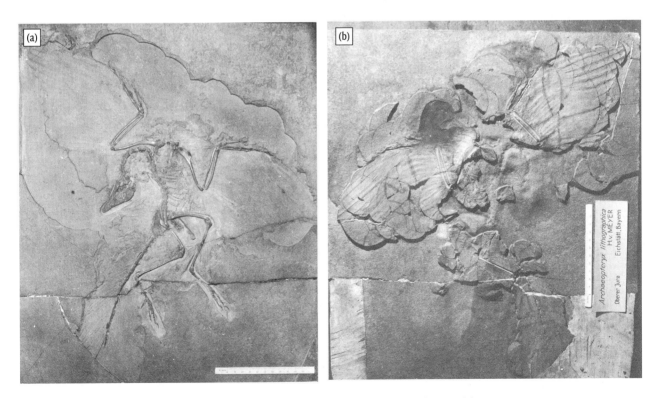

Figure 10.4. The beautifully preserved, complete Berlin specimen of *Archaeopteryx*. (a) Main slab preserving most of specimen; (b) counterslab, preserving opposite side of specimen, primarily impressions. Note the exquisite feather impressions radiating out from the wings and tail.

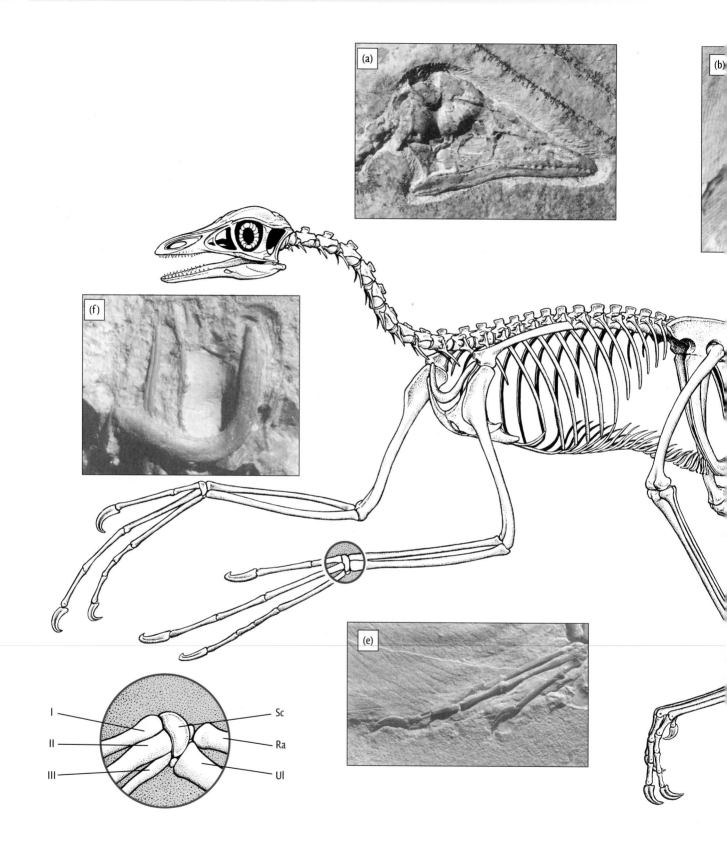

(a)

(b)

(f)

(e)

I   Sc

II   Ra

III   Ul

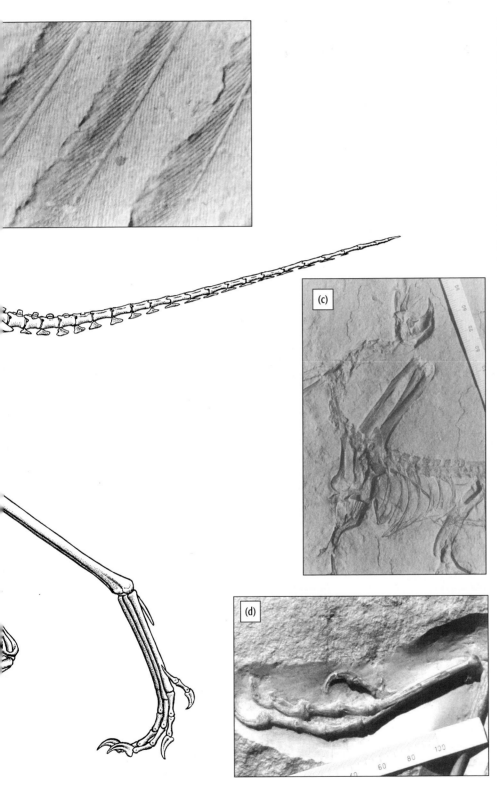

Figure 10.5. A reconstruction of *Archaeopteryx*, surrounded by photographs taken from the actual specimens. (a) Skull, seen from right side, note teeth; (b) feather impressions showing vanes and shaft superbly preserved; (c) trunk region seen from left side, note gastralia; (d) foot (four-toed and clawed, with symmetry around digit III; digit I opposite digits II, III, and IV); (e) right hand and wrist with clawed digits (in ascending order, I, II, and III). *Inset*: drawing of left wrist, showing semi-lunate carpal (Ra, radius; Ul, ulna; Sc, semi-lunate carpal); (f) robust theropod furcula.

Table 10.3. **Distribution of characters among maniraptoran theropods, *Archaeopteryx*, and modern birds**

Maniraptoran theropods	*Archaeopteryx*	Modern birds
Teeth (+)	Teeth (+)	Teeth (–)
Braincase slightly enlarged	Braincase slightly enlarged	Swollen braincase
Tail long, well developed	Tail long, well developed	Pygostyle (+)
Hand three-fingered; I, II, & III	Hand three-fingered; I, II, & III	Carpometacarpus (+); fused digits I, II, III
Legs: 1 Bipedal 2 Unfused foot	Legs: 1 Bipedal 2 Unfused foot	Legs: 1 Bipedal 2 Tarsometatarsus
Foot: 1 3 toes in front; 1 in back 2 Digit V (–) 3 Claws	Foot: 1 3 toes in front; 1 in back 2 Digit V (–) 3 Claws	Foot: 1 3 toes in front; 1 in back 2 Digit V (–) 3 Claws
Hollow bones; some pneumatic	Pneumatic bones	Pneumatic bones
Furcula (wishbone)	Furcula (wishbone)	Furcula (wishbone)
Trunk not rigid 1 Sternum small; flat 2 Pelvis unfused 3 All vertebrae (+) 4 Flight adaptions (–)	Trunk not rigid 1 Sternum small; flat 2 Pelvis unfused 3 All vertebrae (+) 4 Partial flight adaptations	Rigidified trunk 1 Carinate sternum 2 Synsacrum 3 Some vertebrae (–) 4 Flight adaptations
Feathers (+)	Feathers (+)	Feathers (+)

The plus sign (+) indicates character present; the minus sign (–) indicates character absent.

Archaeopteryx as a bird

Archaeopteryx was immediately recognized as a fossil of the most primitive bird known. The feathers identified it as a bird, and its wings suggested flight.[4] Because many other features, particularly the stance, legs, and feet, were remarkably bird-like, together with living birds, *Archaeopteryx* forms a monophyletic Avialae (Figure 10.6). But where did *Archaeopteryx* come from?

---

[4] By all (theoretical!) accounts, *Archaeopteryx* did not fly particularly well, as compared with living birds. *Archaeopteryx* lacks many of the skeletal specializations of modern birds. Recent work suggests that *Archaeopteryx* could flap its wings, attaining moderately high speeds, but could not perform the kind of slow flight that a running take-off might require. For this reason, some suggest that *Archaeopteryx* had to have been primarily a tree-dweller, capable of some powered flight, but not of the kind available to living birds.

Archaeopteryx as a dinosaur

Higher relationships of *Archaeopteryx* *Archaeopteryx* has an antorbital opening; therefore *Archaeopteryx* (and thus modern birds) bears one of the diagnostic features of an archosaur. In the hindfoot of *Archaeopteryx* and living birds, three toes point forward (digits II, III, and IV), and the fourth (digit I) is reduced; the toes are symmetrical around digit III, and all toes are clawed (Figure 10.5d). This condition is diagnostic of ornithodirans (see Figures 4.11 and 10.7). All living birds (as well as *Archaeopteryx*) have a fully erect stance, in which the shaft of the femur is at 90° to the head, and the ankle of *Archaeopteryx* (and all birds) is a modified mesotarsal joint; these characters diagnose Dinosauria.

Archaeopteryx, because it bears characters diagnostic of Dinosauria, is a dinosaur. That being the case, living birds must be a subset of Dinosauria, and both of them should be subsumed within an expanded Reptilia.

Archaeopteryx as a theropod *Archaeopteryx* (and all modern birds, for that matter) have hollow bones: a character diagnostic of Theropoda. Moreover, *Archaeopteryx* bears an enlarged three-fingered hand with the deep pits at the end

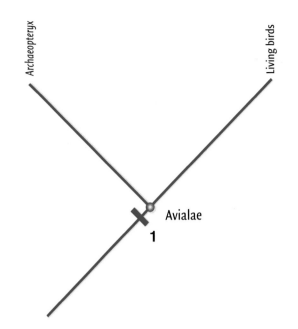

Figure 10.6. Cladogram of Avialae (all living birds + *Archaeopteryx*). Derived characters include: at **1**, very long arms, narrowing of the face and reduction of the size and number of the teeth, enlargement of the braincase, reduction of the fibula toward the ankle, and, significantly, the presence of feathers. As we shall see, although this last character had been valid since the discovery of *Archaeopteryx*, it has become clear in the last seven years that it diagnoses a group significantly more inclusive than Avialae.

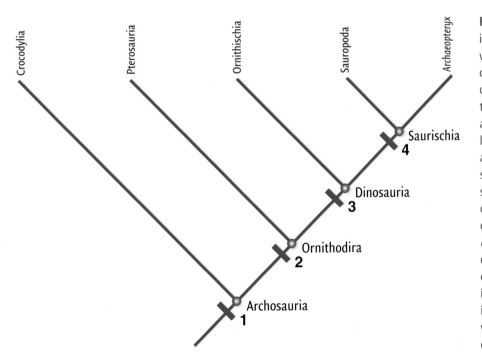

Figure 10.7. Cladogram depicting the position of *Archaeopteryx* within Archosauria. Derived characters include: at **1**, antorbital opening (Archosauria); at **2**, four-toed, clawed foot, with symmetry around digit III, digit I reduced, lying closely appressed to and alongside digit II (Ornithodira); at **3**, semi-perforate acetabulum (Dinosauria); at **4**, ascending process on the astragalus (Saurischia). The cladogram shows that *Archaeopteryx*, and therefore birds, are dinosaurs. Within Dinosauria, the character at **4** among others (see introduction to Part III: Saurischia) indicates that *Archaeopteryx*, while a bird, is also a saurischian dinosaur.

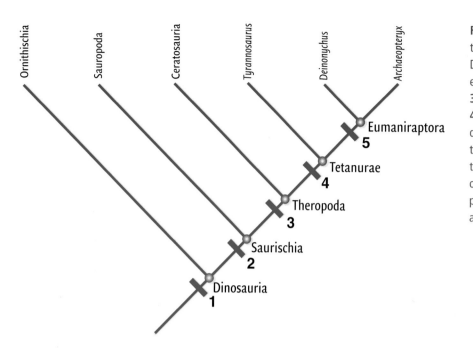

Figure 10.8. Cladogram depicting the position of *Archaeopteryx* within Dinosauria. Derived characters at each node include: **1** is the same as **3** in Figure 10.7; **2** is the same as **4** in Figure 10.7. At **3**, one obvious diagnostic character (of many) is the hollow bones possessed by all theropods. At **4** is shared elongation of the hemal and neural arches, and possession of a furcula; at **5**, opisthopubic pubis.

of the metacarpals so diagnostic of Theropoda (Figure 10.8, and see Chapter 9). Birds must therefore be theropods.

Archaeopteryx as a tetanuran *Archaeopteryx* has a furcula, a character diagnostic of tetanurans. *Archaeopteryx* also has elongate zygapophyses (leading to a stiffened tail), a shortened tooth row, and an astragalar groove. With a high ascending process on its astragalus, *Archaeopteryx* is clearly not at the base of Tetanurae.

Archaeopteryx as a coelurosaur *Archaeopteryx* possesses a shortened ischium (far shorter than the pubis) and large, circular orbits. Coelurosaurs have a furcula, as does *Archaeopteryx*. *Archaeopteryx* is a coelurosaur (Figure 10.8).

Archaeopteryx as a maniraptoran As befits their name, all maniraptoran coelurosaurs have a grasping, three-fingered hand, which is a modification of the ancestral theropod condition. It also has a semi-lunate carpal bone in the wrist, which allows the hand to be folded under the forearm when the elbow is flexed and exended outward when the elbow is extended. Another maniraptoran character on *Archaeopteryx* is a shortened opisthopubic pubis; that is, one pointing directly down, in which the anterior face of the footplate is missing. The distinctive maniraptoran addition to that hand is an elongation of the middle digit (II). *Archaeopteryx* has this feature. Other maniraptoran features found in *Archaeopteryx* include a highly flexed neck, elongate forelimbs, and a distinctive bowed ulna. *Archaeopteryx* is a member of Maniraptora.

Archaeoptyeryx as a eumaniraptoran *Archaeopteryx* has a highly reduced fibula, a eumaniraptoran character, as well as forelimbs that are equal to or greater than the length of the hindlimbs (Figure 10.8).

Archaeopteryx as an avialan In *Archaeopteryx*, as in all avialans, the tail vertebrae all show an extensive elongation of the hemal and neural arches. Likewise, the teeth of avialans lose their serrations – as in *Archaeopteryx*.

What can we conclude from all this? That *Archaeopteryx* is an avialan theropod. *Because* Archaeopteryx *is also a bird, we conclude that birds are avialan theropods as well* (Box 10.1).

Box 10.1 *Plus ça change...*

The relationship of birds to dinosaurs as outlined here is not new. The famous early Darwinian advocate T. H. Huxley, as well as a variety of European natural scientists from the middle and late 1800s, recognized the connection between the two groups. Indeed, one did not have to be a Darwinian to recognize the important shared similarities, and Huxley's opinions were widely accepted at the time. As noted in 1986 by Yale's J. A. Gauthier (Figure 14.12), Huxley outlined 35 characters that he considered "evidence of the affinity between dinosaurian reptiles and birds," of which 17 are still considered valid today.

So what happened? Why is it news that birds are dinosaurs? During the very early part of the twentieth century, Huxley's ideas fell into some disfavor, as it was proposed that many of the features shared between birds and dinosaurs were due to convergent evolution.

What evidence was there to argue for convergence in the case of dinosaurs and birds? Really, not terribly much. But in light of the limited knowledge of dinosaurs at the time, the group just seemed too specialized to have given rise to birds. Moreover, clavicles were not known from theropods (then, as now, the leading contender as the most likely dinosaurian ancestor of birds). Thus, fused clavicles (furcula) in birds had to have originated outside Dinosauria. What was needed was a more primitive group of archosaurs that did not seem to be as specialized as the dinosaurs.

In the early part of the twentieth century, such a group of archosaurs, the ill-defined "Thecodontia" (see footnote 6, Chapter 4), was established by Danish anatomist G. Heilmann as the group from which all other archosaurs evolved. Since this was by definition the group that gave rise to all archosaurs, and since birds are clearly archosaurs, it was concluded that birds must have come from "thecodonts." Heilmann had in mind an ancestor such as *Ornithosuchus* (note the name: *ornitho* – bird; *suchus* – crocodile), a 1.5 m long carnivorous bipedal archosaur that, among living archosaurs, looks a bit like a long-legged crocodile. For over 50 years, Heilmann's detailed and well-argued analysis held sway over ideas about the origin of birds.

Several events caused the thecodont ancestry hypothesis to fall into general disfavor. The first was that clavicles were found in coelurosaurians among theropods. Moreover, it came to be recognized that "Thecodontia" is not monophyletic; that is, it is defined by no unique, diagnostic characters pertaining to all its members and no others. How could one derive birds (or anything else) from a group that had no diagnostic characters?

The renaissance of the dinosaur–bird connection must be credited to J. H. Ostrom of Yale University. In the early 1970s, through a series of painstakingly researched studies, he spectacularly documented the relationship between *Archaeopteryx* and dinosaurs, in particular coelurosaurian theropods. His ideas inspired R. T. Bakker and P. Galton, who in 1974 published a paper suggesting that birds should be included within a new vertebrate class: Dinosauria. The idea didn't catch on, in part because it involved controversial assumptions about dinosaur physiology and because the anatomical arguments on which it was constructed were not completely convincing. In 1986, however, Gauthier applied cladistic methods to the origin of birds, and with well over a hundred characters demonstrated that *Archaeopteryx* (and hence, birds) is indeed a coelurosaurian dinosaur.

Counting a bird's fingers – which is which?

Recall that the carpometacarpus of living birds is a fused structure composed of three fingers. Paleontology clearly tells us which three fingers these are: if Avialae is real, then the fingers must in fact be I, II, and III, since the fingers in avialan theropods, including *Archaeopteryx*, are I, II, and III.

But because the fingers fuse into the carpometacarpus as the hand forms during a living bird's development, embryologists since the 1870s have studied the hand of modern birds as the carpometacarpus develops. They have repeatedly concluded that the fingers of the bird hand appear to be II, III, and IV. How could paleontology unambiguously identify the fingers as I, II, and III, when embryology identifies them as II, III, and IV?

Moreover, at least one primitive theropod, a ceratosaur called *Limusaurus* (*Limus* – mud, or mire), may have reduced digits from both sides of the hand, in a *bilateral* reduction in digit size: both the "thumb" and "pinky" sides (digits I and V) become smaller. The remaining fingers of this animal thus seemed to point toward the embryological II, III, IV digit identification. More advanced theropods, of course, aren't built that way; the thumb is distinctive, with offset heads suggesting partial opposability, and the reduction and loss of digits is *unilateral*: from just the "pinky" side (digits IV and V), meaning that the remaining digits must be I, II, and III. How could the identification of the digits be both II, III, and IV, and yet be I, II, and III?

A possible solution to this conundrum was proposed by paleontologist J. A. Gauthier (see Figure 14.12) and developmental biologist G. P. Wagner. Embryologists identify the sequence of "*condensations*"; that is, early developed buds of material that later become fingers. As early growth occurs, embryologists thought they saw condensation I become digit I, condensation II become digit II, and so on. Gauthier and Wagner thought they saw something else, however, something that they called a "frameshift in the developmental identities" of the fingers. According to them, the bud considered to be embryological condensation II actually becomes adult digit I, condensation III actually becomes digit II, and condensation IV actually becomes digit III. More recent work now suggests that genes coding for the development of the original digit I cause its development in position II; and genes coding for the development of the original digit II cause its development in position III. The "frameshift" appears to be from the ancestral position to a new one.

And what about *Limusaurus*? The most probable explanation for its morphology is that its hand, rather than representing the ancestral condition for all theropods, was something of an evolutionary dead end. The future of theropod evolution – from Tetanurae onwards – appears to have been through a common ancestor that underwent a remarkable frameshift in the development of its hand.

With these new observations, the fingers identified in living birds and those in extinct theropods are the same. Wagner and Gauthier's work resolves the apparent discrepancy between embryology and paleontology, and reaffirms the avialan origin of birds.

Old wives' tales (and feathers)

Pneumatic bones and feathers are commonly singled out as marvelous adaptations to maintain lightness and permit flight. Well, they surely maintain avian lightness, and feathers work well for flight. But did pneumatic bones and feathers evolve for lightness and flight, respectively?

In both cases, now that we have a sense of bird ancestry, the answer is "No." Hollow bones are a theropod character (recall that even the name *Coelurosauria* contains a reference to the hollow bones in these dinosaurs), and pneumaticity is likely related to efficient breathing (we saw this in large sauropods, which we can safely assume didn't fly, as well as in some maniraptoran theropods). Avian pneumaticity likely developed within avialans, likely long before there were things we would call birds. Modern birds inherited pneumaticity along with feathers and a host of other features.

And feathers ... well, the evidence is becoming pretty overwhelming that these didn't evolve for flight either!

Feathers without flight

Feathers in modern birds do lots of things: they also insulate (think down sleeping bags and vests), they are used for display (think peacock), and, most famously, they permit flight (think airfoil). One anatomical structure; multiple uses.

Insulation

The insulatory properties of down feathers are well known. Could feathers have evolved *first* for insulation, and then *only later* gotten co-opted for flight? This idea is less far fetched when we remember that all living birds are warm-blooded; and most warm-blooded creatures, especially small ones, require insulation. Then, it's simply a matter of how far back in the history of theropods warm-bloodedness goes (see Figure 12.12). If it precedes birds and flight, then it likely evolved for something else besides flight.

Our record of modern-looking feathers goes back to the Jurassic; of course for *Archaeopteryx*, but also for other feathered theropods, such as the Chinese theropod *Epidexipteryx*. It makes no sense for a cold-blooded animal, which needs an exterior source of heat to warm up, to be insulated. So if warm-bloodedness goes as far back as the Jurassic, could it go back even further and be shared among all avialans, or perhaps even among eumaniraptorans? Who out there has feathers?

If feathers were actually an adaptation for insulating a warm-blooded creature, then fossils of non-flying theropods ought to be found with feathers, since, as we have seen, eumaniraptoran theropods were likely very active animals. When evidence finally came that showed that feathers originated for insulation (and not for flight), it came from both embryology and paleontology, and it was spectacular.

Stylin'

Modern birds are brightly colored creatures, and it is no secret that along with flight and insulation, birds use feathers for display. While shape is important (think about a lyrebird or a peacock), considering just the feathers themselves (e.g., excluding behavior and calls), color is probably the key criterion for reproductive success. So feathers today have an important display function, we know that sexual selection undoubtedly occurred in non-avian theropods long before there were birds, so, did feathers serve a display role in non-bird theropods? Recently, paleontologists have begin to gain insights into the actual colors and patterns of ancient feathers. Using scanning electron microscopy, chromatophores – pigment cells – have been identified – and are known by their shape and size to represent blacks and browns (Figure 10.11; see also Chapter 9, "The wonderful world of color."). So we now know that in non-bird theropods, as well as in birds feathers bore patterns and colors that were unquestionably important components of sexual selection in non-bird theropods.

Embryology

Feathers were long thought to be an outgrowth of "reptilian" scales. Somehow the scales grew longer and divided into barbs and barbules. Work in the past 10 years, however, suggests that the development of feathers occurs by the interaction of specialized follicles and a series of specialized genes that control the onset and termination of growth. Four sequential stages of feather evolution have been identified, each stage a developmental modification of the previous stage, and each found in living birds (Figure 10.9). These stages are:

1   Formation of a keratin-based, hollow cylinder or filament (the shaft).
2   Loosely associated, unconnected, unhooked barbs (downy feathers).
3   Hooked barbs on a symmetrical vane (contour feathers, such as wrap around the body).
4   Hooked barbs on an asymmetrical vane (flight feathers).

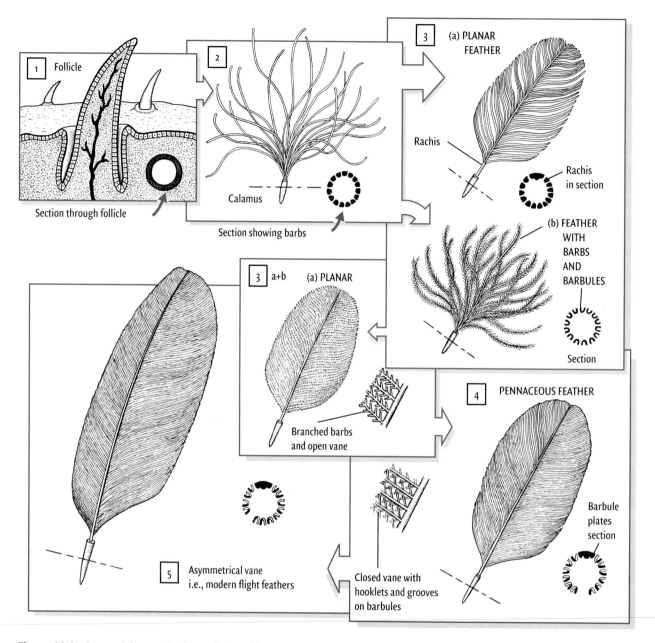

Figure 10.9. Sequential stages in the evolution of feathers. Type 1: simple, hollow, cylindrical filaments. Type 2: tufts of elongate, multiple filaments, attached at one end. Type 3: filament tufts align in a single plane (Type 3a) while also developing barbs and barbules (Type 3b). Eventually, a new planar (vaned) barbed form evolves (Type 3a+b). Type 4: vane becomes "closed"; that is, tiny hooks on the barbule attach to grooves on adjacent barbules, producing an integrated semi-rigid vane that does not allow much air to pass through. Type 5: vane becomes asymmetrical (for example, a flight feather).

Fossils

Archaeopteryx has flight feathers that are indistinguishable from those of living birds, and so considerable evolution had to have taken place prior to the Late Jurassic. Since flight feathers represent the most advanced embryological stage (above), it long seemed as though the fossil record *ought* to produce a feathered, non-flying theropod, particularly if our evolutionary scenario regarding the origin of feathers had any validity.[5]

In the early 1990s, feathered theropod dinosaurs from 124 Ma in Liaoning Province, China, began to be recovered. The fossil-rich rocks of Liaoning look superficially like those of Solnhofen in Bavaria. Preservation is spectacular; the specimens are generally complete *and* completely articulated, and the fine mudstones in which they are preserved show not only the impressions of the animals' coverings but also some darkened staining, possibly representing the original organic matter (Figure 10.10). With such superb preservation, there is little room for doubt about the nature of these dinosaurs or their feathers.

First, there was the 1997 discovery of *Sinosauropteryx*, a small coelurosaur whose design was such that it obviously didn't fly. Yet it was covered with barb-like filaments, a very primitive downy coat insulating a clearly non-flying theropod. Strikingly, patterns are still preserved in this creature's plummage (Figure 10.11 and below). Similar coverings were found on *Pedopenna* and *Dilong*; the former is a maniraptoran theropod of Jurassic vintage; the latter lived somewhat later, and is distantly related to *Tyrannosaurus*. Next came the somewhat larger, toothless *Caudipteryx* and *Protarchaeopteryx*, once thought to be flightless birds but then clearly revealed to be oviraptorosaurs. *Caudipteryx* bears feathers with well-developed barbs, barbules, and symmetrical vanes. Even more startling was *Beipiaosaurus*, a very large (ostrich-sized) therizinosauroid (see Chapter 9), also with no obvious ability to fly. *Beipiaosaurus* has relatively primitive feathers with only barbules. Keratin feather shaft remnants were also discovered on the Mongolian therizinosaurid *Shuvuuia* (*shuvuu* is a transliteration of the Mongolian for "bird," although *Shuvuuia* is far from being a bird). A non-flying deinonychosaur was also found: *Sinornithosaurus*. This organism bears feathers that are comparable in every way to those of living birds. And then, curiously, a flying (?) deinonychosaur: *Microraptor*, covered with flight feathers on its arms *and* legs (Figure 10.12). Also recently discovered is *Anchiornis huxleyi*, a Jurassic-aged troodontid (certainly not flying!), older than *Archaeopteryx*, with well-developed vaned plumage. These were all relatively small animals; in 2012, however, three nearly complete specimens of a full-sized (c. 9 m), feathered tyrannosaur, *Yutyrannus*, were reported from China (see Figures 10.10 through 10.13).

This, in turn, provides real insights into what the origin of feathers was all about. The prediction that feathered non-flying dinosaurs would eventually be discovered was correct: feathers likely first provided the insulation that is a prerequisite for warm-bloodedness, allowing theropods to maintain high levels of activity for the extended periods of time that were eventually necessary for flight (see Chapter 12).

So far our discussion of feathers has been restricted primarily to avalians, with several non-avalian maniraptorans like *Dilong, Caudipteryx, Sinornithosaurus, and Microraptor* suggesting that feathers – or their precursors – go deeper in Theropoda than just Avialae. Just *how* much deeper? The stupefying results came in 2009, with the discovery of *Tianyulong* (see Part II: Ornithischia). This animal is clearly a heterodontosaur: a primtive *ornithischian* dinosaur! *Tianyulong* is thought to have had a row of long, filamentous structures – Stage 1 proto-feathers? – running down its neck, back, and tail. An ornithischian with some kind of filamentous pre-feather coating seemed absurd, until the unpublished discovery of a specimen of the basal ceratopsian *Psittacosaurus* (see Figure 6.16) with similar structures lining its tail (Figure 10.14). Could feathers – or their precursors – go all the way back to basal Dinosauria? The short answer is "yes," a subject that we revisit in Chapter 15.

---

[5] Many paleontologists, including us in the first edition of our book (1996), predicted that non-flying theropods that used feathers for insulation would be found.

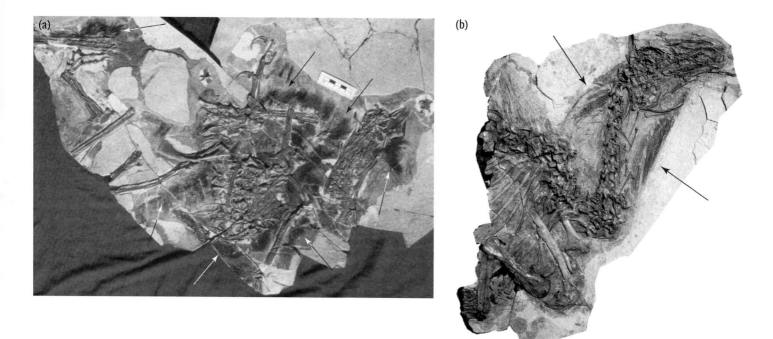

Figure 10.10. Feathered non-flying dinosaurs from Liaoning Province, China. (a) *Sinornithosaurus* - complete skeleton; (b) *Beipiaosaurus* - skull, neck, and anterior portion of rib cage (including scapula and humerus); (c) *Sinosauropteryx* - complete skeleton; (d) Section of tail of *Yutyrannus*. Arrows indicate filamentous features interpreted to be insulating (non-flight) feathers.

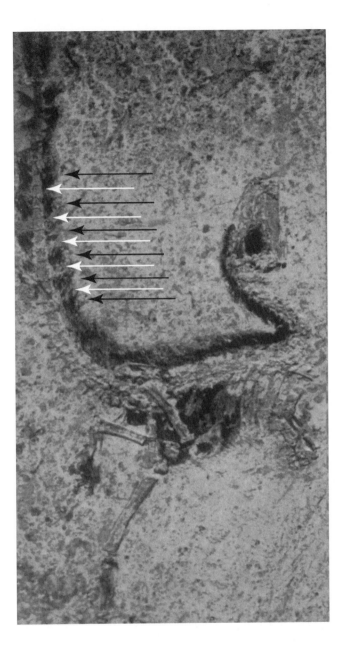

Figure 10.11. Patterns of color on the feathered, non-avian dinosaur *Sinosauropteryx*. Filamentous features – interpreted to be early non-flying feathers – are preserved down the back of the animal, extending the length of the tail. Arrows indicate color banding, particularly well preserved on the tail.

Living tissue

For some scientists, all this talk of ancient fossils seemed like a lot of old fossils telling us about other old fossils. Then, in 2006, a team of scientists extracted pliable, organic soft tissue from the femur of *Tyrannosaurus* (Box 9.2). If that weren't enough, a year later they determined that the molecular composition of one of the proteins, collagen, was more similar to that of living birds (a chicken, in this case) than to that of any other living animal. For some – particularly non-paleontologists – that was the most compelling evidence of the relationship between birds and non-avian theropod dinosaurs.

What, if anything, is a bird?

Clearly, the old equation (feathers = bird) won't fly; there are now many examples of feathered, non-flying dinosaurs *way* below Avialae on the cladogram. Likewise, the equation

Figure 10.12. *Microraptor gui*, also from Liaoning Province, China, with broad apparent flight feathers on all four limbs.

(warm-blooded = bird) also doesn't work; these feathered dinosaurs were surely warm-blooded. After all, why insulate yourself from your heat source if you are cold-blooded? Should Aves – traditionally birds – be restricted to all those organisms bearing the distinctive suite of characters of living birds? That would, of course, exclude *Archaeopteryx*, which certainly has a plausible claim on the designation "bird." Should the equation be bird = flight?

Origin of avian flight

Somewhere in eumaniraptorans, flight evolved. But how? Two opposing endpoints exist as regards the origin of bird flight (Figure 10.16). The first is the so-called arboreal (or "trees down") hypothesis: that bird flight originated by birds gliding down from trees (Figure 10.16a). In this hypothesis, gliding is a precursor to flapping (powered) flight; as birds became more and more skillful gliders, they extended their range and capability by developing powered flight. Perhaps flapping developed as a modification of the motions used in controlling flight paths.

Antithetical to the arboreal hypothesis is the cursorial (or "ground up") hypothesis for the origin of flight (Figure 10.16b). The cursorial hypothesis states that bird flight originated by an ancestral bird running along the ground. In this scenario, perhaps as obstacles were avoided, the animal became briefly airborne. Flapping (powered) flight appeared early on, as the animal strove to overcome more fully the force of gravity. This idea obviously requires a highly cursorial ancestor, in which feathers were already present. In this hypothesis, the legs, feet, and hands of avialans like *Archaeopteryx* are viewed as an inheritance from a cursorial maniraptoran ancestory.

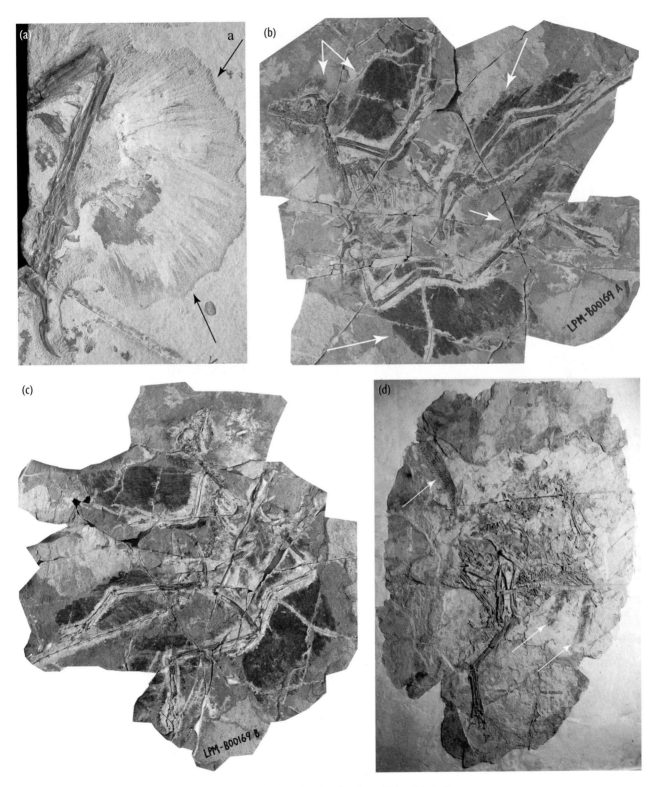

Figure 10.13. More feathered, non-flying dinosaurs from Liaoning Province, China. (a) *Pedopenna* – complete skeleton; (b) *Anchiornis* slab and (c) counter slab (unmarked) – complete specimen; and (d) *Caudipteryx* – complete specimen. Note apparent feather tufts on skull of *Anchiornis*; in *Caudipteryx*, original coloration visible at upper left-hand arrow. Arrows indicate filamentous features interpreted to be insulating (non-flight) feathers.

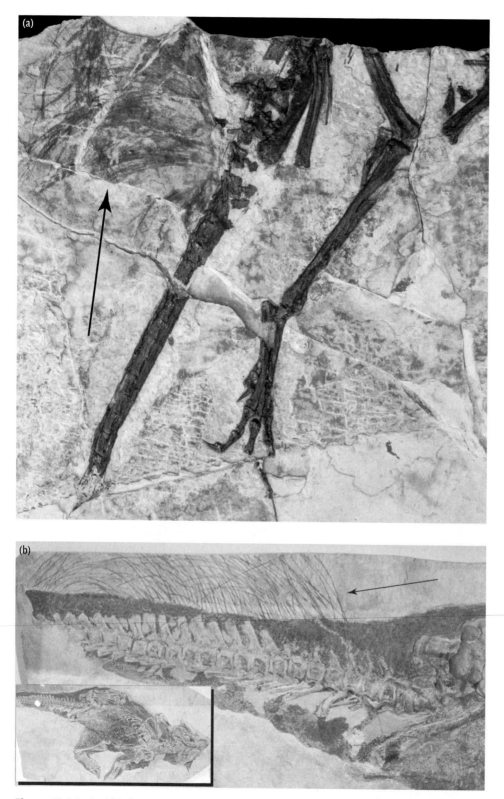

Figure 10.14. Ornithischians with apparent filamentous – feather? – coverings. (a) tail [to left] and left leg of *Tianyulong,* a basal ornithischian; and (b) tail region of *Psittacosaurus,* a primitive ceratopsian. *Inset:* complete specimen. Photographs of *Psittacosaurus* taken from cast of specimen. Arrows indicate filamentous features interpreted to be insulating (non-flight) feathers.

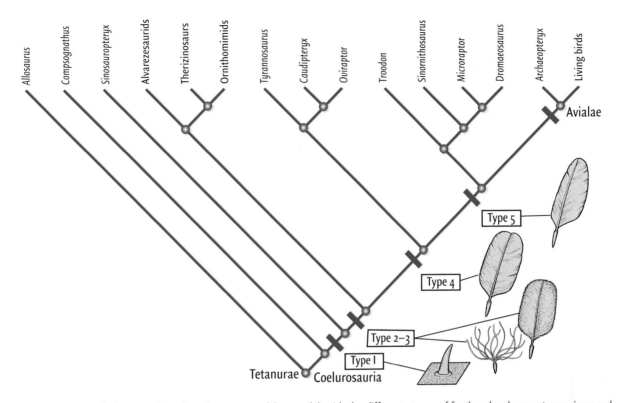

Figure 10.15. Cladogram with selected tetanurans (Theropoda) with the different stages of feather development superimposed (see text and Figure 10.9). If the authors are correct, primitive feather coats may have graced most derived tetanuran theropods.

Figure 10.16. The arboreal and cursorial hypotheses for the origin of bird flight. (a) The arboreal hypothesis, which suggests that bird flight evolved by "birds" gliding down from trees. (b) The cursorial hypothesis, which suggests that bird flight evolved by "birds" running along the ground until the animals became airborne.

Which to choose? The arboreal hypothesis is intuitively appealing, and getting airborne is easy. On the other hand, the cursorial hypothesis is strongly supported because ultimately the ancestor of birds had to have been a cursorial creature.

A problem with the cursorial hypothesis is that it has so far proven nearly insurmountable to model a cursorial theropod that developed flight by running along the ground. For this reason, an arboreal stage intermediate in the development of flight has been attractive to many scientists. Yet indications of a cursorial heritage are present in all living birds as, indeed, their limbs are little changed from the non-flying coelurosaurian condition.

Recently an interesting compromise position was proposed. Perhaps flapping wings helped early cursorial theropods to get a purchase on steep slopes, overhangs, or even tree trunks. From this it would not have been a big leap, as it were, to flapping flight. Ultimately, however, the exact scenario by which flight arose may never be known.

The story of the origin of flight reminds us of a fundamental property of evolution. Structures are not commonly invented wholesale in evolution. Evolution modifies existing structures. Here, the feathers and grasping arms of warm-blooded, non-avialan deinonychosaurs were modified – remarkably little – to permit flight. It was a breathtaking evolutionary achievement.

Summary

Birds are dinosaurs. This is not because they evolved from dinosaurs, although they did, but because they share the derived characters of Dinosauria. Living birds have a suite of highly derived anatomical features that at first glance appear to be unique. In fact, a close look at Theropoda shows that most of the supposedly uniquely avian features of living birds are actually distributed throughout theropods: cladistic analysis demonstrates that many presumed "bird" characteristics are distributed in non-bird theropods, particularly within Tetanurae and Eumaniraptora.

A small, Late Jurassic theropod, *Archaeopteryx lithographica*, shows an almost perfect intermediate mix of characters between theropods and modern birds. Although the skeleton is in many ways almost indistinguishable from deinonychosaur theropods, the animal had well-developed flight feathers, among many other "avian" characters. *Archaeopteryx* should not be considered as the ancestor of modern birds, but it shares many of the same characters as that ancestor.

The many discoveries of feathered, non-flying theropods from the Early Cretaceous Liaoning Province of China blurs the line between bird and non-bird. Clearly, although feathers are diagnostic for birds among living organisms, the Liaoning fossils demonstrate that presence of feathers does not guarantee that one is dealing with a bird. These discoveries reinforce the viewpoint that feathers had multiple uses, and were evidently invented as a form of insulation or perhaps display, only later to be co-opted for flight purposes.

An important quality of Avialae appears to be flight. The origin of flight is shrouded in mystery; birds may have evolved flight by leaping down from trees (the "arboreal hypothesis") or alternatively, may have evolved flight by fast running (the "cursorial hypothesis"). For a variety of reasons, neither hypothesis is completely satisfactory, and the evolution of flight – and thus the final step to Avialae – may have involved a mixture of fast running and leaping from high points.

Selected readings

Chiappe, L. M. and **Dyke, G. J.** 2007. The beginnings of birds: recent discoveries, ongoing arguments, and new directions. In Anderson, J. S. and Sues, H.-D. (eds.), *Major Transitions in Vertebrate Evolution*. Indiana University Press, Bloomington, IN, pp. 303–336.

Chiappe, L. M. and **Wittmer, L. M.** 2002. *Mesozoic Birds: Above the Heads of Dinosaurs*. University of California Press, Berkeley, 576pp.

Dial, K. 2003. Wing-assisted incline running and the evolution of flight. *Science*, **299**, 402–404.

Dingus, L. and **Rowe., T.** 1997. *The Mistaken Extinction*. W. H. Freeman and Company, New York, 332pp.

Gauthier, J. A. 1986. Saurischian monophyly and the origin of birds. In Padian, K. (ed.), *The Origin of Birds and the Evolution of Flight*. California Academy of Sciences Memoir no. **8**, pp. 1–56.

Hecht, M. K., Ostrom, J. H., Viohl, G. and **Wellenhofer, P.** (eds.). 1984. *The Beginnings of Birds*. Proceedings of the International *Archaeopteryx* Conference Eichstatt, Freunde des Jura–Museums Eichstatt, Willibaldsburg, 382pp.

Ji, Q., Currie, P. J., Ji, S. and **Norell, M. A.** 1998. Two feathered dinosaurs from northeastern China. *Nature*, **399**, 350–354.

Ostrom, J. H. 1974. *Archaeopteryx* and the origin of flight. *Quarterly Review of Biology*, **49**, 27–47.

Ostrom, J. H. 1976. *Archaeopteryx* and the origin of birds. *Biological Journal of the Linnaean Society*, **8**, 91–182.

Prum, R. O. and **Brush, A. H.** 2003. Which came first, the feather or the bird? *Scientific American*, **288**, 84–93.

Schweitzer, M. H., Suo, Z., Avci R. *et al.* 2007. Analyses of soft tissue from *Tyrannosaurus rex* suggest the presence of protein. *Science*, **316**, 277–280.

Sereno, P. and **Rao, C.** 1992. Early evolution of avian flight and perching: new evidence from the Lower Cretaceous of China. *Science*, **255**, 845–848.

Shipman, P. 1998. *Taking Wing*. Simon and Schuster, New York, 336pp.

Wagner, G. P. and **Gauthier, J. A.** 1999. 1, 2, 3 = 2, 3, 4: a solution to the problem of the homology of the digits in the avian hand. *Proceedings of the National Academy of Sciences*, **96**, 5111–5116.

Xu, X., Tang, Z. and **Wang, X.** 1999. A therizinosaurid dinosaur with integumentary structures from China. *Nature*, **399**, 350–354.

Xu, X., Wang, X. and **Wu, X.** 2000. A dromaeosaurid dinosaur with a filamentous integument from the Yixian Formation of China. *Nature*, **401**, 262–266.

Xu, X., Zhou, Z., Wang, X., Kuang, X., Zhang, F. and **Du, X.** 2003. Four-winged dinosaur from China. *Nature*, **421**, 335–340.

Zheng, X.-T., You, H.-L., Xu, X. and **Dong, Z.-M.** 2009. An early Cretaceous heterodontosaurid dinosaur with filamentous integumentary structures. *Nature*, **458**, 333–336. doi:10.1038/nature07856.

Zhou, Z.-H., Wang, X.-L., Zhang, F.-C. and **Xu, X.** 2000. Important features of *Caudipteryx* – evidence from two nearly complete new specimens. *Vertebrata palasiatica*, **38**, 241–254.

Topic questions

1   What are the diagnostic characters of modern birds that might be preserved in the fossil record?

2   Compare these characters to those present in non-avian theropods.

3   What characters point to eumaniraptoran theropods as good candidates for the ancestry of modern birds?

4   Why did the ancestor of birds have to have been cursorial if one goes back far enough?

5   Why is it that we no longer think feathers evolved for flight? Does this mean that they are unrelated to flight?

6   How did bird flight evolve?

7   Why are bird bones hollow and pneumatic?

8   Should *Archaeopteryx* be considered the ancestor of modern birds? It has a mix of non-avian and avian theropod anatomy, so what can be said about its relationships to the actual ancestor of all birds?

9   If we were unable to resolve the discrepancy between the paleontological and the embryological identifications of the fingers in a bird's carpometacarpus, would this in any way affect the hypothesis that birds are dinosaurs? Why?

10 If *Archaeopteryx* had never been found, would we be able to tell that birds are dinosaurs? Support your answer.

11 During bird evolution, how did the wing evolve? Was a whole new wing structure required, or were all the pieces – as well as the correct proportions – there already?

12 Why is the question, "What is a bird?" a difficult one to answer? What kinds of data would go into your answer to this question?

13 Formulate a thoughtful answer to the question "What is a bird?"

CHAPTER 11

THEROPODA III: EARLY BIRDS

CHAPTER OBJECTIVES

- Learn about the evolutionary transition from *Archaeopteryx* to modern birds

- Gain familiarity with the diversity of Mesozoic avian theropods

Mesozoic birds

Archaeopteryx, as we have seen, had many features that are far from the condition found in living birds, including teeth, an unfused hand, a bony tail, no synsacrum, and gastralia. And as it turns out, there is much about its physiology as well that would be unfamiliar to a modern ornithologist. Indeed, recent work suggests that it grew much more slowly than modern birds grow, although its growth rates appear to have been comparable to similarly sized maniraptoran theropods (Figure 11.1). So physiologically, compared to its near relatives, it was not unusual, although next to a modern bird it was quite primitive.

Archaeopteryx and its close theropod relatives had another unexpected feature as well: a well-developed sense of smell. Paleontologist Darla Zelenitsky and colleagues used CT scans to study the relative size of the olfactory bulbs in 157 non-avian theropods, fossil birds, and living birds. The relative size of the bulb was used to infer smelling capability: the proportionately larger the bulb, the better the sense of smell. While modern birds are characterized by relatively weak olfaction, their antecedents were not: the relative size of the olfactory bulbs – and the inferred olfactory acuity – increased from primitive maniraptorans through *Archaeopteryx* and its near relatives, and into early birds. Only later, when modern groups of birds became ascendent, did the sense of smell begin to diminish.

How and when did the other changes take place that distinguish living birds from *Archaeopteryx*? Here our interest will be within Avialae, the clade that includes *Archaeopteryx*, Aves (living birds), and everything in between.

The Mesozoic Avialary

Within Avialae, very close to *Archaeopteryx* is *Rahonavis* from the Late Cretaceous of Madagascar. We've opted to emphasize its avian features in tentatively considering it more derived than *Archaeopteryx*. Slightly larger than *Archaeopteryx* (the size of a crow; Figure 11.2), recent work places *Rahonavis* as a dromaeosaurid theropod, but its position above or below *Archaeopteryx* remains uncertain. It possessed an enlarged sickle-shaped claw on its feet (similar to that of dromaeosaurids and troodontids), and a long, *Archaeopteryx*-like tail. Younger than *Archaeopteryx* by 25 million years, it had forward-looking features such as pneumatic foramina

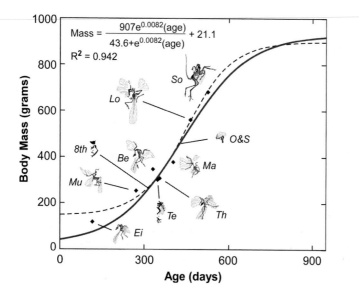

$$\text{Mass} = \frac{907e^{0.0082}(\text{age})}{43.6+e^{0.0082}(\text{age})} + 21.1$$

$$R^2 = 0.942$$

Figure 11.1. Inferred growth rates for *Archaeopteryx*. On the horizontal axis is time in days; on the vertical axis is presumed body mass. The curve is based upon 8 specimens; growth was calculated from femur length. Growth rates were calculated to be between those of living birds (which themselves span a significant range of growth rates) and living reptiles.

leading into pleurocoels in its thoracic vertebrae, which as we've seen implies unidirectional breathing and possibly a more efficient metabolism (see Box 8.1 and Chapter 12), along with a series of other bird-like characters (Figure 11.3). Time and further analyses will eventually consolidate its position.

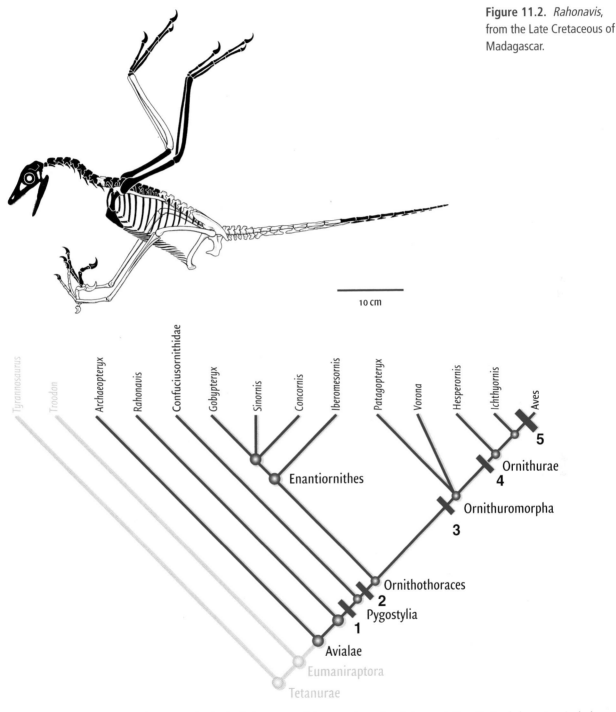

Figure 11.2. *Rahonavis,* from the Late Cretaceous of Madagascar.

10 cm

Figure 11.3. Cladogram of Mesozoic birds, depicting some of the steps in early avialan evolution. Derived characters include: at **1**, pygostyle; at **2**, reduction in number of trunk vertebrae, flexible furcula, strut-like coracoid, alula, carpometacarpus, fully folding wings; at **3**, further reduction in number of trunk vertebrae, loss of gastralia, final rotation of pubis to lie parallel with ilium and ischium; at **4**, reorientation of pubis to lie parallel to ilium and ischium, reduction of number of trunk vertebrae, decrease in size of acetabulum, patellar groove on femur; at **5**, loss of teeth.

Flight proficiency seems to have been a driving force in avialan evolution. Subsequent events included the formation of a pygostyle as well as the development of the synsacrum and other features for a rigid trunk, all of which contribute to the efficient flight that characterizes modern birds. Confuciusornithidae, a group containing *Confuciusornis* (Figure 11.4) and *Changchengornis* from the Early Cretaceous of China, is characterized by these relatively modern features (see Figure 11.3).

Thereafter, birds reduced the number of trunk vertebrae, altered the shoulder joint, and fused the digits of the hand into a carpometacarpus, among other flight-related features. All birds with these transformations are united within Ornithothoraces (*ornith* – bird; *thora* – thorax, or chest). Ornithothoracans have two main divisions: on the one hand, Enantiornithes (*enant* – opposite), and on the other Ornithuromorpha (*ornith* – bird; *uro* – tail; *morpha* – form), the lineage leading to Aves (see Figure 11.3).

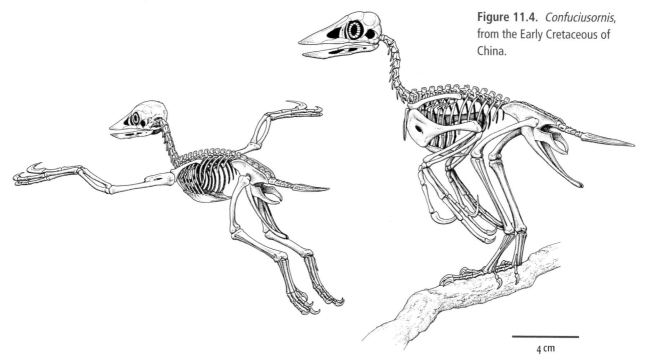

Figure 11.4. *Confuciusornis,* from the Early Cretaceous of China.

4 cm

Enantiornithes Enantiornithes ("opposite bird") are all sparrow-sized, small, arboreal birds (Figure 11.5) that had well-developed flight capabilities. They were the most diverse clade of Mesozoic birds, yet all went extinct before the close of the Era.

Enantiornithes modified the wrist joint to allow, for the first time, the wing to fold tightly against the body, and developed adaptations indicative of perching: the first toe is positioned fully opposite to the others. The perching foot is a clue that these Mesozoic birds lived in trees, reinforcing other evidence that flight was an integral part of their life habits.

Still, one would hardly call Enantiornithes modern birds – they had gastralia across their belly (a carry-over from the primitive archosaurian condition), relatively numerous back vertebrae (a number intermediate between the 13 or14 found in *Archaeopteryx* and the 4–6 found in living birds), an unfused tarsometatarsus, and an unfused pelvis.

Enantiornithes apparently had a world-wide distribution. *Nanantius* is from Australia, *Iberomesornis* hails from Spain, and *Sinornis* comes from China. Others include *Kizylkumavis* and *Sazavis* from Uzbekistan (Asia), *Alexornis* from Mexico, *Enantiornis* itself and *Avisaurus*, known from both Argentina and the USA.

Ornithomorpha Returning to the line leading to Aves, Ornithomorpha is represented by *Vorona* from the Late Cretaceous of Madagascar, *Patagopteryx* (Figure 11.6) from the Late Cretaceous of Argentina, and all remaining birds, the clade known as **Ornithurae** (*ornith* – bird; *uro* - tail).

Ornithurae is one of the most robust of all avialan clades, united by at least 15 unambiguous characters (see Figure 11.3) Not surprisingly, ornithurans include not only the closest relatives to living birds (Hesperorniformes and Ichthyorniformes), but also Aves (modern birds) as well.

Hesperornithiform (*hesper* – western) birds were a monophyletic clade of large, long-necked, flightless, diving birds that used their feet to propel themselves, much like modern loons or cormorants (Figure 11.7). They had highly reduced arms and developed powerful hindlimbs for propulsion. The hindlimbs were oriented to the side of the creature, and could not be brought under the body. For this reason, locomotion on land must have been a kind of seal-like waddling (at best).

In the water, on the other hand, hesperornithiform birds were supremely adapted for marine life. The long, flexible neck must have been useful in catching fish, a behavior indicated from coprolites preserved with their skeletons. In many respects, the group is quite close to modern birds, yet they retained teeth in their

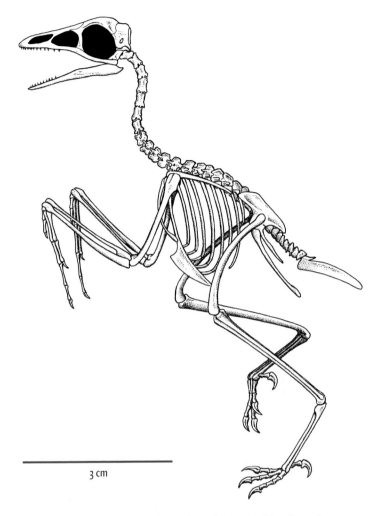

Figure 11.5. *Iberomesornis,* a member of Enantiornithes from the Early Cretaceous of Spain.

jaws. Like modern diving birds, some of the pneumaticity in the bones was lost. Presumably because strength in the chest and shoulder regions for flight was no longer needed, the furcula, coracoid, and forelimbs were highly reduced.

All of these adaptions indicate that this group had a long evolutionary history prior to their appearance in the Late Cretaceous. Hesperornithiforms include *Enaliornis* from the Early Cretaceous of England, *Hesperornis* (Figure 11.7) and its smaller relative *Baptornis*, and *Parahesperornis*, all from North America.

Closer related yet to Aves were the toothed Ichthyornithiformes (see Figure 11.3). Unlike hersperornithiforms, ichthyornithiforms were excellent flyers (Figure 11.8). *Ichthyornis*, from the Late Cretaceous of North America, had a massive keeled sternum and an extremely large deltoid crest that was probably an adaptation for powerful flight musculature. In other respects, it shared many of the adaptations of modern birds, including a shortened, fused trunk, a carpometacarpus, a pygostyle, a completely fused tarsometatarsus, and a synsacrum formed of 10 or more fused vertebrae. Found exclusively in marine deposits, ichthyornithiforms must have been rather like Mesozoic seagulls – but with teeth.

Figure 11.6. *Patagopteryx*, from the Late Cretaceous of Argentina.

Aves

Aves (modern birds) is a well-supported clade, involving as many as 11 characters of the skull, pelvis, and ankle. The Mesozoic fossil record of Aves – all of it restricted to the Late Cretaceous – is very fragmentary and scattered, although tantalizing.

The group includes screamers and waterfowl (Anceriformes), loons (Gaviiformes), and possibly shorebirds such as sandpipers, gulls, and auks (Charadriiformes), landfowl (Galliformes), wing-propelled divers such as modern petrels (Procellariiformes), and parrots (Psittaciformes). Clearly, these early records of modern birds speak, however incompletely, to the origin, initial radiation, and establishment of Aves in the closing moments of the Cretaceous.

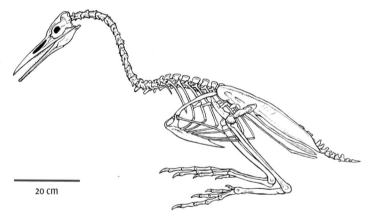

20 cm

Figure 11.7. *Hesperornis*, the diving bird from the Late Cretaceous of the USA.

Evolution of Aves

Getting to be a modern bird

For all of its limitations, the Mesozoic record provides for us insights into the transition from primitive theropods, through eumaniraptorans, to Avialae (including *Archaeopteryx*), and finally to Aves. The first part of the long evolutionary sequence – the part that ran from primitive theropods to avialans – was detailed in Chapters 9 and 10. The second part, the evolution of Aves from the primitive avialan condition, represented by *Archaeopteryx*, through the remarkable avian discoveries recently made in China, can be read from the cladogram in Figure 11.3.

Our best understanding of the sequence of evolutionary events is as follows:

1   Development of the perching adaptation in the foot, in combination with limited flapping flight capabilities.
2   Development of a pygostyle.
3   Reduction in number of trunk vertebrae; and development of a flexible furcula, strut-like coracoid, carpometacarpus, and fully folding wings.
4   Further reduction in the number of trunk vertebrae, loss of gastralia, final rotation of pubis to lie parallel with ilium and ischium.
5   Reduction of number of trunk vertebrae, decrease in size of acetabulum, formation of patellar groove, a groove at the distal end of the femur to accommodate the patella (knee cap).
6   Finally, loss of teeth (Aves).

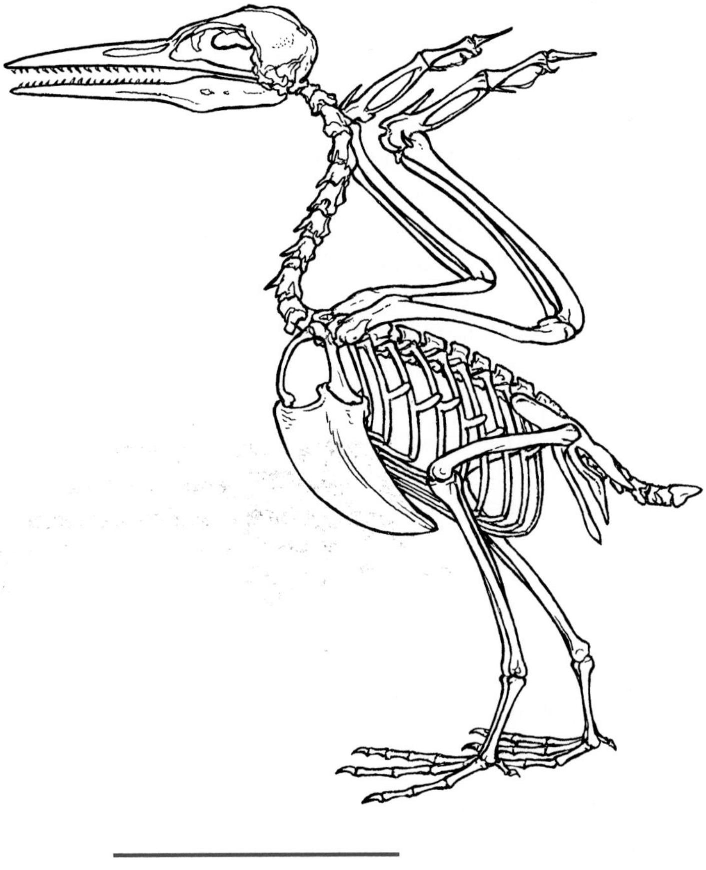

Steps 1–5 occurred in the Mesozoic;

Figure 11.8. *Ichthyornis,* a gull-like bird from the Late Cretaceous of the USA.

Step 6 may have occurred after the Mesozoic was over, because all Mesozoic avialans for which a skull is known, had teeth. Skulls are unfortunately not well known for Mesozoic Aves. Likewise, no Cenozoic toothed bird is known, although the fossil record of birds in the earliest Cenozoic is also very poor. Teeth – that final step in getting to be a modern-style bird – somehow got lost during the Mesozoic–Cenozoic transition (Box 11.1).

Cold cases

But it's by no means all figured out. Flightless birds such as emus and ostriches appear to retain in the skull primitive features whose origin may be found in Mesozoic birds. Unfortunately, the rarity of the skulls of Mesozoic birds has made connecting them with modern birds both complex and controversial. How such flightless modern birds fit into the rest of modern birds and into our understanding of Mesozoic bird history – is a story that must await another day.

Enigmata One small evolutionary radiation, known as Alvarezsauridae, from the Late Cretaceous, has been difficult to place in the phylogenetic scheme we've outlined here. This is largely due to the very dramatic and unusual specializations of their skeletons. Take *Mononykus*, for example, a Late Cretaceous theropod that evidently apparently lived in a Late Cretaceous Sahara-like sand sea (Figure 11.9).

Box 11.1 Molecular evolution and the origin of Aves

In this book, we've emphasized the fossil record as the means of telling when events occurred, mainly because this book deals with extinct organisms, the only record of which has historically been the fossil record. But in those cases in which we are dealing with living organisms, a whole different type of technique is available for study: **molecular evolution**.

Molecular evolution involves measuring the timing of molecular changes. So, for example, take two somewhat closely related living organisms, A and B. Now, choose a particular protein that they share, say, serum albumin (a protein in their blood). Their serum albumins might be quite closely related, but if some time has elapsed since A and B last shared a common ancestor, the exact molecules may have evolved and now differ slightly, in either form or composition. If we knew the rate at which the molecules diverged, we would know how distantly in the past A and B shared a single common ancestor (whose serum albumin composition and form were the ancestral ones). This very technique (and indeed this very molecule) was used in the case of humans and chimpanzees to show that the two shared a most recent common ancestor only 5 million years ago, instead of the 15 million that had been inferred from the geological record.

More recently, molecular biologists have been using a technique called **DNA hybridization**. This technique works similarly to the one described above for proteins, except that it compares two strands of DNA (instead of proteins). In the same species, the strands of DNA should be virtually identical. DNA hybridization allows molecular biologists to measure the differences between the two strands. Knowing what rate **substitutions** (or changes) occur in the DNA allows us to calculate how long ago two different creatures shared identical DNA. That number should equal the time of divergence from a common ancestor.

And what of birds? Molecular estimates of the earliest Aves have consistently been somewhat earlier than has been inferred from the fossil record. In general, the fossil record has shown that the major radiation of birds took place *after* the K/T extinction. Yet, the fossil record of birds is, as we have seen, rather spotty, and perhaps most trustworthy only in its broadest outlines.

Estimates of the radiation of Aves, based on molecular data – primarily DNA hybridization – have put the time of the radiation well within the Cretaceous, *before* the boundary. How to resolve this contradiction?

Recently, the fossil record has begun to support the molecular record… a little bit. The fossil record of modern bird groups in the Cretaceous now includes the ancestral relatives of ducks, chickens, and large, flightless birds such as ostriches and emus. Even with these new-found discoveries, however, whether or not the major radiations of birds took place before or after the K/T boundary remains unclear.

Molecular evidence has also been used in an entirely different context. A study published in April, 2008, compared proteins from 21 different living creatures, including an alligator, an ostrich, a chicken, and two extinct creatures *T. rex* and a mammoth. The *Tyrannosaurus* proteins were types of collagen extracted from an unaltered femur (see Chapter 1 and Chapter 9, footnote 3). The results were unequivocal: the proteins showed that the mammoth and an elephant were phylogenetically close and, more relevant for our story, that the *Tyrannosaurus* and the birds were close – closer to each other than any is to an alligator.

From its pelvis backward, *Mononykus* looks like a typical ho-hum theropod, with strong, elongate, well-developed hindlimbs, and a long straight tail. But the hands are fused into a short, powerful carpometacarpus, and the arms are stout and short, with a large process (the olecranon process) for developing power at the elbow joint. Among its avian-like characters is a mildly keeled sternum. Perhaps it used its shortened, yet strong, arms for burrowing.

Subsequent years have seen the discovery of other alvarezsaurids – *Parvicursor* and *Shuvuuia* from the Gobi Desert in Mongolia, and *Alvarezsaurus* and *Avisaurus* from Argentina and the USA. No feathers are preserved with any alvarezsaurid – if they were ever present.

The position of alvarezsaurids within Theropoda is still utterly unclear (Figure 11.10). Are the carpometacarpus and keeled sternum homologous with those of birds? If so, what a mess that is for the cladograms we've presented: it would call for a comparatively primitive theropod

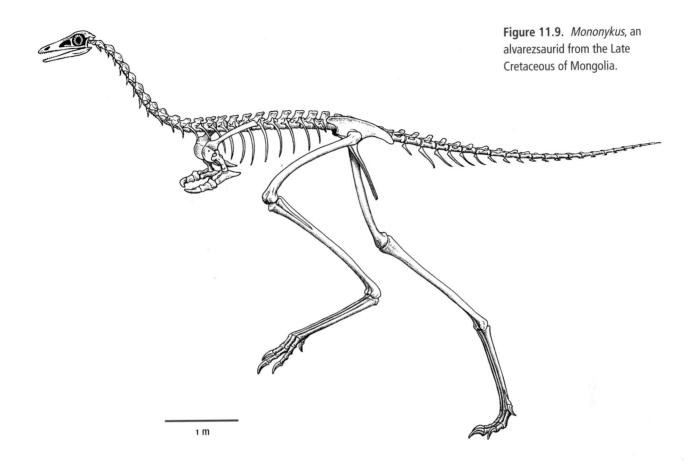

Figure 11.9. *Mononykus,* an alvarezsaurid from the Late Cretaceous of Mongolia.

1 m

evolutionary throwback (at least, from the pelvis backward) in the middle of a group of animals that are very bird-like.

If these features are not homologous, is it unparsimonious for them to have evolved twice: once in ornithothoracans (where they would be homologous with the carpometacarpus and keeled sternum of living birds) and once in some more basal eumaniraptoran, theropod lineage, in which their appearance would be utterly unique?

This story, too, must await another day for its final telling.

Figure 11.10. Two interpretations of the position of Alvarezsauridae. (a) Alvarezsaurids as birds; (b) alvarezsaurids as relatives of Ornithomimosauria.

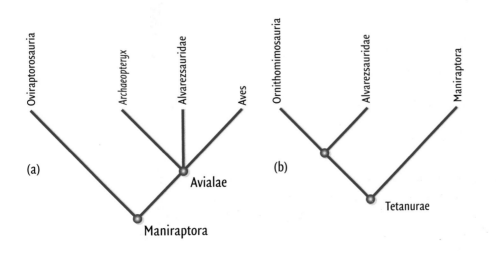

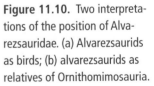

Summary

While *Archaeopteryx* was clearly important in identifying the relationship between Aves and dinosaurs, there were still a number of evolutionary steps to take before the anatomical condition seen in Aves (see Chapter 10) was achieved. Despite their rarity as fossils, the fossil record of birds indicates the general order in which these evolutionary events occurred.

The improvement of flight capability was a driving force in post-*Archaeopteryx* bird evolution. Pneumatic foramina became better developed, along with, sequentially, the pygostyle, a reduction in the number of trunk vertebrae, modifications of the shoulder, and the development of the carpometacarpus. Still, some primitive characters such as gastralia, were retained. All these features were present in Cretaceous ornithothoracan birds, including the small, comparatively common Enantiornithes, and the line leading to Aves, Ornithuromorpha.

Within Ornithuromorpha, several highly evolved birds evolved, notably the diving hesperornithiformes and the seagull-like ichthyornithiformes. These birds, for all their advancement, were not exactly like living birds, lacking a number of features diagnostic for Aves, including loss of teeth. The earliest fossil record of Aves is very fragmentary, but goes back into the Late Cretaceous.

Some forms exist that do not fit into the evolutionary scenario proposed above. These include the enigmatic alvarezsaurids, whose stout carpometacarpus, if homologous with those of birds, suggests that carpometacarpi occurred much earlier in the evolution of Theropoda that we currently believe. The possibility also exists that the carpometacarpus may have evolved twice in Theropoda.

Selected readings

Chiappe, L. M. 1995. The first 85 million years of avian evolution. *Nature*, **378**, 349–355.

Chiappe, L. M. and **Dyke, G. J.** 2002. The Mesozoic radiation of birds. *Annual Review of Ecology and Systematics*, **33**, 91–124.

Chiappe, L. M. and **Dyke, G. J.** 2007. The beginnings of birds: recent discoveries, ongoing arguments, and new directions. In Anderson, J. S. and Sues, H.-D. (eds.), *Major Transitions in Vertebrate Evolution*. Indiana University Press, Bloomington, IN, pp. 303–336.

Chiappe, L. M. and **Witmer, L. M.** (eds.) 2002. *Mesozoic Birds*. University of California Press, Berkeley, 520pp.

Clarke, J., Tambussi, C. P., Noriega, J. I., EricksonJ. M., and **Ketcham, R. A.** 2005. Definitive fossil evidence for the extant radiation of Aves in the Cretaceous. *Nature*, **433**, 305–308.

Cracraft, J. and **Clarke, J.** 2001. The basal clades of modern birds. In Gauthier, J. and Gall, L. F. (eds.), New perspectives on the origin and early evolution of birds. *Proceedings of the International Symposium in Honor of John H. Ostrom*. Peabody Museum of Natural History, Yale University, New Haven, CT, pp. 143–156.

Ericson, P. G. P., Anderson, C. L., Britton, T. *et al.* 2006. Diversification of Neoaves: integration of molecular sequence data and fossils. *Biology Letters*, **2**, 543–547.

Gauthier, J. A. and **Gall, L. F.** (eds.). (2001). *New Perspectives on the Origin and Early Evolution of Birds*. Peabody Museum of Natural History, Yale University Press, New Haven, CT, 613pp.

Topic questions

1   What were the evolutionary steps from *Archaeopteryx* to Aves?
    What were the evolutionary steps from basal tetanurans to *Archaeopteryx*?

2   Why do we link all subsequent bird evolution through *Archaeopteryx*?

3   What are the features of the rigid trunk in Aves that are lacked by pre-Aves birds?

4   What characters suggest that considerable evolution occurred before the appearance of hesperornithiform birds?

5   What are the features of ichthyornithiform birds that indicate powerful, efficient flight?

6   Choose five characters that you can see on *Mononykus*. Which are advanced, which are primitive, and at what level in the cladogram are these characters diagnostic?

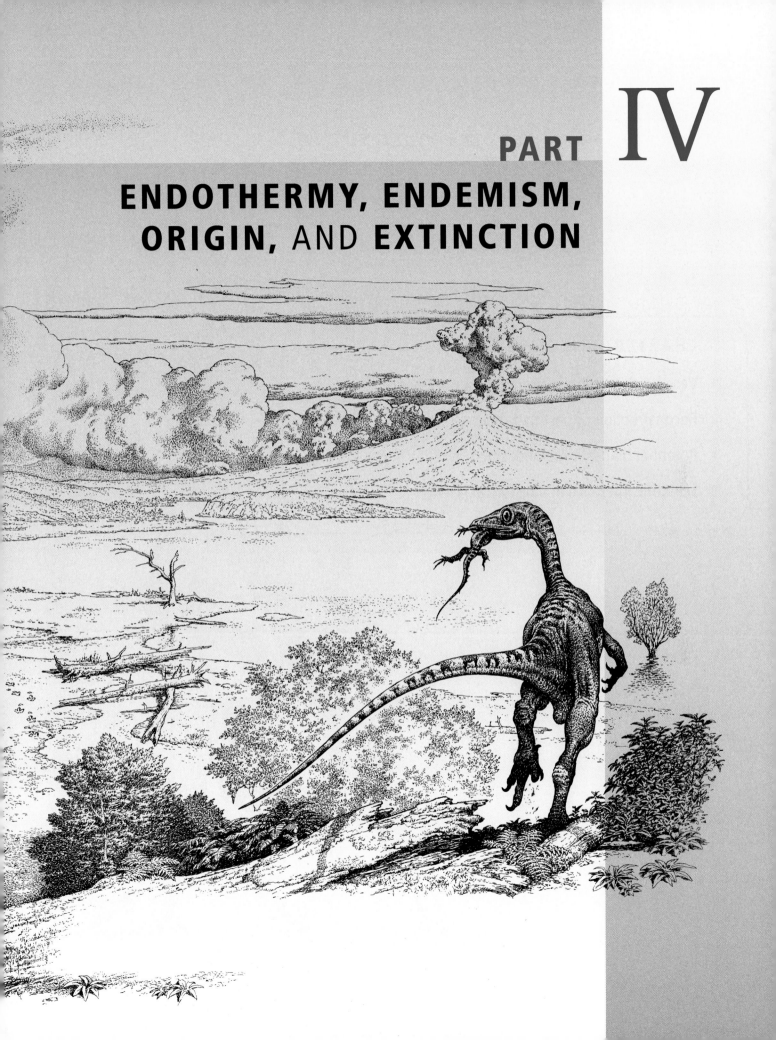

PART **IV**

ENDOTHERMY, ENDEMISM, ORIGIN, AND EXTINCTION

CHAPTER OBJECTIVES

- Vertebrate metabolic strategies
- Reconstruction of past metabolic strategies
- Potential dinosaur metabolism(s)
- Scientific inference in a historical science

CHAPTER 12

DINOSAUR THERMOREGULATION:
SOME LIKE IT HOT

The way they were

Paleontologists have been of two minds about dinosaurs. Owen originally described them as mammal-like (see Chapter 14). But by the early 1900s, the prevailing idea that the closest living relatives of dinosaurs might be crocodiles suggested to many scientists that crocodylians were a better model for dinosaur behavior. And so, for much of the twentieth century, dinosaurs were thought of as, in effect, dolled-up crocodiles. All that got revisited with the revolutions of the 1960s and early 1970s and, if you'll think back to what you've read and seen in the pages preceding this chapter, there are a lot of qualities and behaviors seen in dinosaurs that just don't say "crocodile." With all those bird-like qualities, should we think of Mesozoic dinosaurs as, perhaps, birds of a different feather?

So here is a multiple-choice question that forms the core of this chapter: Mesozoic dinosaurs were

(a)   cold-blooded, slow, and "reptilian;"
(b)   hot-blooded beasts that functioned like modern mammals and birds; or
(c)   none of the above.

Physiology: temperature talk

We begin with a bit of vocabulary and perhaps a concept or three. Physiology is the study of the various integrated functions of an animal; here we discuss the aspect of physiology known as metabolism, in this case, the chemical reactions and pathways an organism uses to obtain energy and put it to work (Box 12.1). We often hear expressions like "warm-blooded" and "cold-blooded"; but it turns out that these terms don't mean much. For example, most "cold-blooded" vertebrates have warmed blood when they are active. A more meaningful distinction is between endotherms (*endo* – inside; *therm* – heat), organisms that regulate their temperature internally, and ectotherms (*ecto* – outside), organisms that use external sources of heat to regulate their temperatures.

Considered from another perspective, in some organisms, called poikilotherms (*poikilo* – changing), temperature fluctuates, but in others, called homeotherms (*homeo* – same), the temperature remains constant. Humans are *endothermic homeotherms*: when we are unable to maintain our body temperature, we get sick. Ectotherms, such as lizards, can *tolerate* core decreases in core temperature, while endotherms must internally *regulate* their core temperatures.

Ectothermy and endothermy are two biochemically and biophysically different methods of obtaining heat. The terms poikilotherm and homeotherm, however, are endpoints in a spectrum that runs from maintaining a constant temperature to having a fluctuating temperature. While many organisms cluster at the familiar metabolic endpoints (endothermic homeotherm; ectothermic poikilotherm), many do not (Box 12.2).

In this chapter, we'll take a tour of some of the strongest arguments for and and against dinosaur endothermy. We'll start with some of the older evidence, generally marshaled in the 1970s and 1980s, and move to some more current studies. It turns out that the metabolisms of dinosaurs, as we shall see, likely did not closely match those found in living vertebrates: "(c)" is thus probably the best answer to the multiple-choice question posed above. But we can't stick a thermometer into the vent of a Mesozoic dinosaur, so how can we tell?

Box 12.1 **Chain of fuels**

We all know that somehow we get energy from the family of carbon-based molecules called carbohydrates (think "Candy bar!"). Not quite so familiar, perhaps, is how this works. Simply put, the energy stored in the bonds of carbohydrates is trarnsferred to energy stored in the bonds of a molecule called **ATP** (adenosine triphosphate). Then we – and all living organisms – access that energy by breaking the ATP molecule to produce a closely related molecule called **ADP** (adenosine diphosphate) and thereby releasing some of that energy.

Cellular respiration

In respiration, chemical bonds in carbohydrates (such as the sugar glucose) are broken via a type of reaction called **oxidation**. These reactions occur as complex, linked, series involving a number of intermediate steps. The breakdown of a single molecule of glucose (a simple carbohydrate) through this suite of reactions can produce 36 new molecules of of ATP through a series of reactions called the "citric acid cycle," so named because citric acid is produced as an intermediate step in the carbohydrate breakdown (Figure B12.1.1).

This type of metabolism, called *aerobic* (involving oxygen), however, is not 100% efficient: ATP production captures about 40%–60% of the energy of the bonds of the carbohydrates. The remainder is released as heat.

Organisms respire oxygen because energy storage as ATP involves, as we have seen, oxidation reactions. As the energy output of the organism is increased, the amount of ATP needed is increased, and hence more oxygen is consumed and more heat produced. This is why breathing, heart rate, and temperature increase when we exercise: we are using more energy, requiring more ATP to be generated, and thus we need more oxygen. And give off more heat!

There is a point, however, at which the volume of oxygen supplied by breathing is insufficient. Under such conditions, a different reaction path called *glycolysis* is followed. The process of generating energy through glycolysis is a type of metabolism called anaerobic (without oxygen).

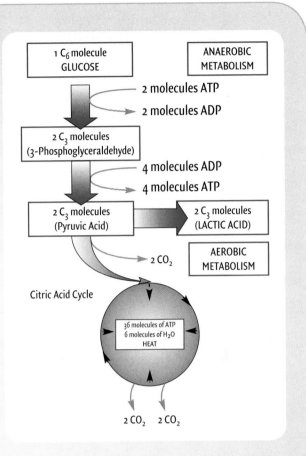

Figure B12.1.1. Cellular respiration consists of the breakdown of carbohydrates to produce energy that is stored in ATP. In this example, the 6-carbon molecule glucose is broken down. Two pathways are shown: the aerobic path, in which ATP is produced via the citric acid cycle, and the anaerobic path, in which lactic acid is ultimately produced via glycolysis.

Glycolysis bypasses the citric acid cycle, and instead directly produces two 3-carbon molecules called pyruvic acid. The pyruvic acid in turn generates lactic acid, which accumulates in the muscles and causes the familiar ache after extreme exercise (see Figure B12.1.1). After hard exercise, we breathe heavily to replenish our depleted oxygen supply, and eventually the lactic acid is removed from the muscles.

What about dinosaurs?

Given what we know now, we begin with the most obvious piece of evidence: who dinosaurs are. As we have seen, birds are dinosaurs and modern birds are surely endothermic. As we asked before (see Chapter 11), at what point during theropod evolution did "avian" endothermy evolve?

Box 12.2 Warm-bloodedness: to have and to have hot

Although endothermy is *characteristic* of birds and mammals, it is by no means restricted to these groups. For some time, physiologists have known of plants (!) that can regulate heat in a variety of ways, the most common being to decouple metabolism (described in Box 12.1) from respiration, so that energy from the breakdown of ATP is simply released as heat. Several snakes are known to generate heat while brooding eggs, although this is accomplished by muscle exertion. Certain sharks and tunas can retain heat from their core muscles by counter-current circulation, and a variety of insects, including moths, beetles, dragonflies, and bees, are known to regulate their body temperatures. Endothermy is not characteristic of these groups of organisms. Simply, it is known that some of them do maintain temperatures warmer than those of the medium (air or water) in which they are living. Indeed, it has been estimated that maintaining a temperature against an external gradient has evolved independently at least 13 different times.

This, of course, differs from the idea that endothermy is *diagnostic* of a particular group. Indeed, endothermy is characteristic of but two groups: birds and mammals.

An important clue comes with insulation. All small- to medium-sized modern endotherms are insulated with fur or feathers. There is a certain sense to this; if an ectotherm depends upon external sources for heat, why develop a layer of protection (insulation) from that external source? And can a small endotherm, constantly eating to maintain its metabolism, afford to lose heat? *Archaeopteryx*, with its plumage, is therefore usually considered to have been endothermic.[1] The discovery of non-avian, feathered (insulated) theropods from China (see Chapter 10) gives us a clue that endothermy occurred well within Coelurosauria, and perhaps at an even more basal level within Theropoda (Figure 12.1).

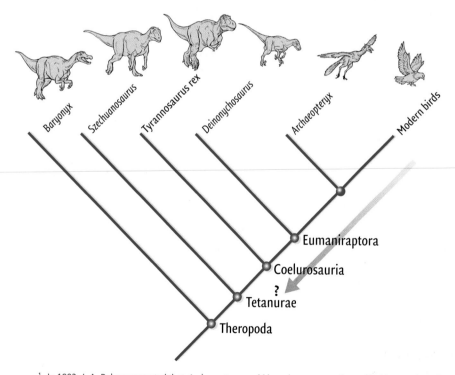

Figure 12.1. Cladogram showing the inferred "depth" within Theropoda of endothermy. While we can be certain the avialans were all endotherms, it is not clear how far back – or how deep within the cladogram – endothermy extends. With the discovery of the feathered *Yutyrannus* (Chapter 9), we now know that it goes back further than coelurosaurs.

[1] In 1992, J. A. Ruben suggested that *Archaeopteryx* could have been an ectotherm. His idea was based upon the amount of energy needed for flight, and upon the amount of energy available from an ectothermic metabolism. Since the bones of *Archaeopteryx* show limited adaptations for sustained, powered flight, Ruben argued that an ectothermic metabolism would have been more than sufficient for the kind of limited flight that apparently characterized *Archaeopteryx*. While powered flight may be possible in an ectothermic tetrapod, none (save perhaps *Archaeopteryx*) ever evolved it. Moreover, it seems to us that the presence of insulation (feathers) in *Archaeopteryx* is incompatible with an ectothermic metabolism.

Of course, using just this criterion, it's looking more and more like a lot of animals could have been endotherms. We saw that the primtive ornithischischian, *Tianyulong*, was likely feathered (see Part II; Ornitshichia and "Feathers without flight" in Chapter 10); the evidence suggests that pterosaurs were covered with a filamentous coat. On the other hand, crocodiles aren't. This distribution of coats suggests that insulation *could* be as basal as Ornithodira (Figure 12.1)!

Air sacs and unidirectional breathing in living organisms are certainly indicative of high metabolic rates and thus endothermy. As we've seen, pneumaticity – indicative of expanded air sacs – and thus unidirectional breathing, are well developed in saurischian dinosaurs as a group (Chapters 8–10). These features are indicative of highly evolved respiratory systems, and suggest higher metabolic rates than commonly associated with ectotherms, as well as incipient endothermy.

Seminal studies

Along the way, some very clever attempts were made to gain insights into the metabolisms (metabolisms, because one size need not necessarily fit all) of dinosaurs; here we review some of the most important.

Some of the earliest evidence that dinosaurs might be endothermic was marshaled by paleontologist R. T. Bakker (see Chapter 14); below we will see a few of his most persuasive arguments. These arguments are particularly interesting, because they are examples of how a creative individual, looking at the fossil record in unconventional ways, can infer things from it that far transcend the mere bones of the animals. Bakker appreciated the metabolic significance of the bird–dinosaur relationship we've just reviewed: in 1974, he and paleontologist P. M. Galton proposed that, on the basis of their presumably shared endothermic metabolism, a whole new *class* of vertebrates be recognized: Dinosauria (including birds)!

Bakker observed that all non-avian dinosaurs had a fully erect stance. Among living vertebrates, a fully erect stance occurs only in birds and mammals, both of which are endothermic. The fully erect limb position in dinosaurs was therefore suggestive to him of endothermy.

Great coincidence; but is there a causal relationship between stance and endothermy? The original idea was that the fine neuromuscular control necessary to maintain a fully erect stance would only be possible within the temperature-controlled environment afforded by an endothermic metabolism. Since Bakker's insight, further work (see also Box 4.3) suggests that when an animal with sprawling stance moves, the trunk of the organism flexes from side to side as the animal walks. Such flexion reduces the amount of air that can fill the lungs on the scrunched side (Figure 12.2), so that when the animal needs the

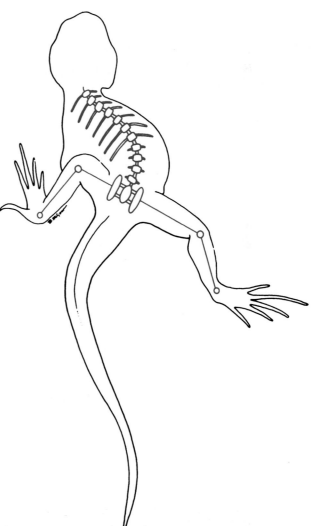

Figure 12.2. A sprawling vertebrate running quickly. The trunk alternately compresses the lung capacity on each side as the animal runs.

most air, it gets the least. The evolution of a fully erect stance may thus have been a means by which lung volume could be maximized during high-speed locomotion. Considered in this way, a fully erect posture could be a prerequisite for an endothermic metabolism.

Two other simple correlations have since been noted between anatomy and endothermy. The first is that long-leggedness is characteristic of living endotherms while living ectotherms possess relatively stubby limbs. Certainly many dinosaurs possessed rather long limbs. The second correlation between anatomy and endothermy is the observation that, among living tetrapods, the only bipeds are endotherms. A recent study reconstructed the energy required for bipeds to walk and slow run. The amount of energy required for maintaining these gaits was theoretically calculated for both large and small dinosaur bipeds, and the somewhat surprising conclusion – surprising because it ran counter to prevailing wisdom – was that the amount of energy required by large bipeds, such as hadrosaurs and tyrannosaurs, to walk and slow run was significantly greater than that which would be available to ectotherms. The data were more equivocal for smaller bipeds, but even these pushed the maximum energy available with an ectothermic metabolism to its limit. Large bipedal dinosaurs, the authors concluded, must thus have been endotherms.

Box 12.3 In the tracks of dinosaurs

Trackways, the most tangible record of locomotor behavior, provide evidence for one aspect of an animal's walking and running capabilities, and the only independent test of anatomical reconstructions. When footprints are arranged into alternating left–right–left–right patterns, they demonstrate that all dinosaurs walked with a fully erect posture. But how can trackways also give us an indication of locomotor speed?

We begin with stride length; that is, the distance from the planting of a foot on the ground to its being planted again. When animals walk slowly, they take short strides, and when animals are walking quickly or running, they take considerably longer strides. This much is intuitive for anyone trying to catch a bus about to pull away from the curb. Now, consider the situation when you are being chased by something smaller than you. The creature chasing you must take long strides for its size, and more of them too, just to keep up. So there is clearly a size effect during walking and running, and these will likely be different for different kinds of animals under consideration.

How, then, to relate stride, body size, and locomotor speed? British biomechanist R. M. Alexander provided an elegant solution to this problem by considering **dynamic similarity**. Dynamic similarity is a kind of conversion factor:

it "pretends" that all animals are the same size and that they are moving their limbs at the same rate. With these adjustments for size and footfall, it doesn't matter if you're a small or large human, a dog, or a dinosaur. All will be traveling with "dynamic similarity"; only speed will vary. That variable Alexander terms "dimensionless speed."[1] It is dimensionless speed that has a direct relationship with relative stride length. Stride length, of course, can be measured from trackways, which in turns allows us, for the first time, to calculate locomotor speed in dinosaurs.

To see how all this works, let's use Alexander's example of the trackway of a large theropod from the Late Cretaceous of Queensland, Australia. The tracks are 64 cm long, which Alexander, from other equally sized theropods, estimated must have come from a theropod with a leg length of about 2.56 m. The stride length of these tracks is 3.31 m, so the relative stride length (stride length:leg length) is 1.3. The dimensionless speed for a relative stride length of 1.3 is 0.4. And from all these measures, this Australian theropod must have been traveling reasonably quickly, at about 2.0 m/s, or 7.2 km/h.

As complicated as this approach appears, it represents the best method for estimating the actual speeds implied by

---

[1] Dimensionless speed may appear oxymoronic, but in fact is equivalent to real speed divided by the square root of the product of leg length and gravitational acceleration.

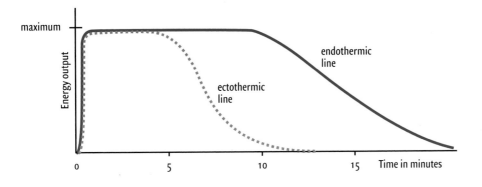

Figure 12.3. Energy output versus time in ectotherms and endotherms. The curves show that the muscles of both endotherms and ectotherms achieve their maximum energy output virtually instantaneously. In general, however, an endotherm *sustains* maximum energy output for more than twice as long as an ectotherm does.

Legs and lungs Endotherms produce more energy than ectotherms; it takes longer, during heavy use, for the muscles of endotherms to enter an anaerobic physiological state (characterized by lactic acid production). This means that, generally, endothermic tetrapods are capable of higher levels of activity sustained over longer periods of time than are ectothermic tetrapods (Figure 12.3).

trackways. But what about the fastest speeds a dinosaur might have been capable of? In 1982, R. A. Thulborn of the University of Queensland developed a method by which absolute locomotor abilities could be calculated. Thulborn's work relied heavily upon Alexander's slightly earlier studies on speed estimates from footprints and the relationship between body size, stride length, and locomotor speed among living animals. For both approaches, Thulborn determined that relative stride length has a direct relationship with locomotor speeds at different kinds of gaits (for example, walking, running, trotting, galloping). Explicitly (for the quantitatively oriented among you):

$$\text{Locomotor velocity} = 0.25(\text{gravitational acceleration})^{0.5}$$
$$\times (\text{estimated stride length})^{1.67}$$
$$\times (\text{hindlimb height})^{-1.17}$$

Thulborn used this equation to estimate a variety of running speeds for more than 60 dinosaur species. The first group of estimates were for the walk/run transition, where stride length is approximately two to three times the length of the hindlimb. A potentially more important estimate – especially for dinosaurs fleeing certain death or pursuing that all-important meal – is maximum speed, which Thulborn calculated using maximum relative stride lengths (which range from 3.0 to 4.0) and the rate of striding, called limb cadence (estimated at $3.0 \times$ hindlimb length$^{-0.63}$).

Although we report some of these speeds elsewhere in this book, it is of some value to summarize the overall disposition of Thulborn's estimates. For small bipedal dinosaurs – which would include certain theropods and ornithopods – all appear capable of running at speeds of up to 40 km/h. Ornithomimids, the fastest of the fast, may have sprinted up to 60 km/h. The large ornithopods and theropods were most commonly walkers or slow trotters, probably averaging no more than 20 km/h. Thus the galloping, sprinting *Tyrannosaurus*, however attractive the image, did not impress Thulborn (or us) as likely.

Then there were the quadrupeds. Stegosaurs and ankylosaurs walked at a no more than a pokey 6 to 8 km/h. Sauropods likely moved at 12 to 17 km/h. And ceratopsians – galloping along full throttle like enraged rhinos? Thulborn estimated that they were capable of trotting at up to 25 km/h.

Are these estimates accepted uncritically by all? P. Dodson has argued that these calculations would suggest that humans can run as quickly as 23 km/h, which in life they cannot. So it is possible that these calculations overestimate the speeds at which dinosaurs could run. On the other hand, anatomically, humans are not dinosaurs and these calculations could be valid for dinosaurs but be inapplicable to humans. At a minimum, they give some indication of the relative speeds of dinosaurs; for example, how quickly *T. rex* might have run in comparison with *Triceratops*.

A variety of small- to medium-sized bipedal dinosaurs such as dromaeosaurids and ornithomimids are characterized by gracile bones, in which the thigh is short relative to the length of the calf. This, in turn, suggests high levels of sustained running – behavior generally not characteristic of modern ectotherms.

And what of the larger dinosaurs, especially those that were not bipedal? Here the issue becomes murkier. The walking speeds of all tetrapods can be calculated from a combination of footprint spacing (stride length) and the length of the hindlimb (Box 12.3). But, of course, the walking that produced most trackways was generally not full-tilt running.

Could quadrupedal dinosaurs have run like the fastest mammals today? Ancestry gives a hint. In mammals, a fully erect posture evolved in a quadrupedal ancestor; however, in dinosaurs the fully erect posture evolved in a biped. Quadrupedal dinosaurs are thought to have evolved their four-legged stance secondarily (see Chapters 5 and 6) and thus the front limbs of dinosaur quadrupeds look, and likely functioned, differently from those of mammals (see Figure 6.24).

Of predators and their prey Perhaps Bakker's most brilliant insight was his inference of metabolism from the numbers of predators and prey found in the fossil record. He began with a simple reality of energy budgets, based in modern animals: *endothermy is much more costly in terms of energy use than ectothermy*. He estimated that it costs 10–30 times as much energy to maintain an endothermic metabolism as to maintain an ectothermic metabolism, in part because so much energy is expended on maintaining a constant body temperature.

Given that fact, Bakker reasoned, if predators are endothermic, they should require more energy than if they were ectotherms, and this should be in turn reflected in the weight proportions of predators to prey, or **predator : prey biomass ratios**. He estimated that predator : prey

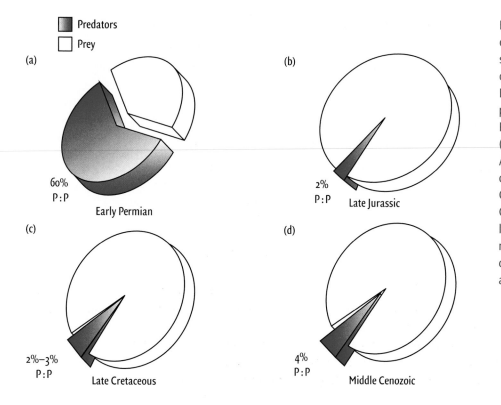

Figure 12.4. The proportions of predators to prey (P : P) in selected faunas in the history of life, as reconstructed by R. T. Bakker. Predators are shaded; prey are unshaded. (a) Early Permian of New Mexico; (b) Late Jurassic of North America; (c) Late Cretaceous of North America; (d) Middle Cenozoic of North America. The Cenozoic fauna (d) is mammalian and obviously endothermic, providing clear guidelines on what predator : prey ratios are in an endothermic fauna.

biomass ratios for *ectothermic* organisms are around 40%, while predator:prey biomass ratios for *endothermic* organisms are 1%–3%. Here, then, was an order of magnitude difference in the biomass ratios, which ought to be recognizable in ancient populations. Now, by counting specimens of predators and presumed prey in major museums and by estimating the specimens' living weights, Bakker was armed with a tool from modern ecosystems that he believed could reveal the energetic requirements of ancient ecosystems.

His results seemed unequivocal: among the dinosaurs, the predator:prey biomass ratios were very low, ranging from 2% to 4%. He interpreted this low number to indicate that predators and prey in dinosaur-based food chains were endothermic (Figure 12.4).

This study, for all its originality, had some problems. The assumption that there is an order of magnitude difference between ectothermic predator:prey biomass ratios and endothermic predator:prey biomass ratios – may not hold in all cases. Bakker's assumption that prey are approximately the same size as the predators is clearly not correct (consider a bear eating a salmon), and has drastic effects on the resultant ratio. Most significantly, the predator:prey biomass calculation assumes that all deaths are the result of predation; that there can be no mortality due to other causes. This is simply not the case in modern ecosystems.

There are serious and legitimate problems with using fossils. Most obvious are difficulties in estimating dinosaur weights (Box 12.4). Moreover, the preservation of dinosaur material is subject to a variety of biases. How can one ever be sure that the proportions of the living community are represented? Because we can't, paleontologists commonly talk about fossil *assemblages*, which, as we've seen (Chapter 2), may have nothing to do with the proportions of the same animals in the living *community* in which those animals lived.

Finally, Bakker obtained his data by counting specimens in museum collections, specimens that were likely collected because they were rare or particularly well preserved. Museum collections thus tend to represent assemblages of well-preserved organisms with a higher percentage of rare organisms than was present in the original fauna.

Several other studies have since attempted to duplicate Bakker's study, with mixed and/or equivocal results. Ultimately, too much uncertainty for the results to be definitive was introduced through the brilliant, but flawed, idea of predator:prey biomass ratios. But Bakker evidence and advocacy had kindled the flame of endothermic dinosaurs, and they would never be thought of as glam crocodiles again!

Hearts and minds All living endotherms possess four-chambered hearts. The four-chambered heart system, in which the oxygenated blood is completely separated from the deoxygenated blood, may be a prerequisite for endothermy. Endothermy requires relatively high blood pressures in order to perfuse complex, delicate organs such as the brain with a constant supply of oxygenated blood. Such high blood pressures, however, would "blow out" the alveoli in the lungs. For this reason, mammals and birds separate their blood into two distinct circulatory systems: the blood for the lungs (pulmonary circuit) and the blood for the body (systemic circuit). The two separate circuits require a four-chambered heart – a pump that can completely separate the circuits.

Would such a heart be possible in dinosaurs? The nearest living relatives of dinosaurs, birds and crocodiles, possess four-chambered hearts; thus it is likely that a heart with a double-pumping system was present in basal Dinosauria.

This idea was strongly reinforced by the discovery in 2000 of what was controversially inferred to be a four-chambered heart with an aorta. The "heart" was preserved as an iron-stone mass within the thoracic cavity of the basal ornithopod *Thescelosaurus*, and identified

Box 12.4 Weighing in

Back in the day it used to be "easy" to estimate the weight of a dinosaur. Just as American paleontologist E. H. Colbert did in the early 1960s, you made a scale model of the dinosaur, and then calculated its displacement in water (Figure B12.4.1).

That displacement was then (a) multiplied by the size of the scale model (for example, if the model were 1/32 of the original, the weight of the displaced water would have to be multiplied by 32) and (b) further modified by some amount using a fudge factor corresponding to the density of tetrapods.

But what is the density of a tetrapod? Based upon studies with a baby crocodile, Colbert determined that baby crocodiles, at least, have a specific gravity of 0.89. Studies with a large lizard (*Heloderma*) showed that the specific gravity of that lizard was 0.81. Among mammals, it would not be surprising to find the density of a whale differing from that of a cheetah, which might in turn differ from that of a gazelle. Full disclosure: there is no uniform density shared by all tetrapods! Oh well.

What if you had a big thigh bone, say, but not much else? Your plans are not quite DOA, because there turns out to be a relationship between limb cross-sectional area and weight. This relationship makes intuitive sense, because obviously, as a terrestrial beast becomes larger, the size (including cross-sectional area) of its limbs, to support its ever-increasing weight, must also increase. The question is, does it increase in the same manner for all tetrapods? If so, a single equation could apply to all. Surprise: it does *not*!

As noted by J. O. Farlow, weight is dependent upon muscle mass and muscle mass is really a consequence of behavior. Therefore weight estimates of dinosaurs are in part dependent upon presumed behavior. For example, reconstructing the weight of a bear would involve assumptions of muscle bulk and gut mass very different from those used in reconstructing the weight of a deer (Figure B12.4.2).

Indeed, the cross-sectional area of their limb bones may be identical, but they may weigh very different amounts. Moreover, our knowledge of dinosaurian muscles and muscle mass is incomplete. So the cross-sectional area method, although convenient and used by a number of workers, has the potential for serious mis-estimations of dinosaur weights.

In short, while the displacement method could be more accurate than limb cross-sectional calculations, it is still dependent upon inferred muscle mass (and thus behavior), and at best is a terribly crude estimate. In these days of CT scans, computers, and the information highway, can't we do better?

The answer is, of course we can, as recently demonstrated by a group of scientists at the University of Manchester. These scientists began with complete skeletons – all the better to estimate your weight, my dear! – and digitally reconstructed the fleshed-out bodies of five dinosaurs (three large-bodied tetanuran theropods, an ornithomimid, and a duck-billed dinosaur) using laser scanning techniques (Figure B12.4.3).

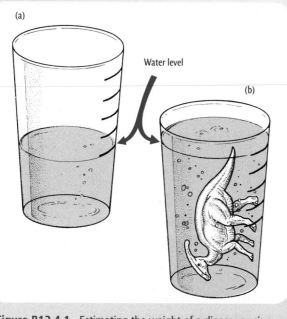

Figure B12.4.1. Estimating the weight of a dinosaur using displacement. For explanation of (a) and (b), see the text.

Figure B12.4.2. Estimating the weight of a dinosaur by comparing the cross-sectional areas of bones.

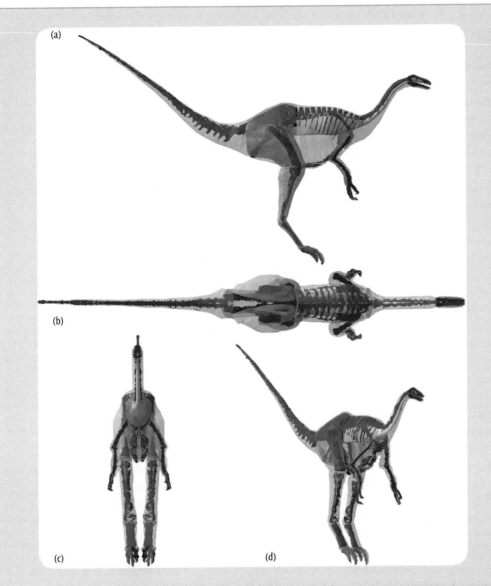

Figure B12.4.3. The "best estimate," fleshed-out reconstruction of the ornithomimid *Struthiomimus* in (A) right lateral, (B) dorsal, (C) cranial and (D) oblique right craniolateral views (not to scale), using laser scanning technology.

(a)

(b)

(c)   (d)

Among the most extraordinary features of these models were reconstruction of real internal anatomy, such air sacs, features that doubtless took up considerable space in non-avian dinosaurs but would cause the density of their owners to diverge significantly from that hopeful (but meaningless) quantity, the "specific gravity of tetrapods." The use of computers allowed the researchers to make innumerable other decisions as well; generally, how much "meat" to put on any given series of bones. Different parts of the body were accorded different densities, in accordance with their biology.

Ultimately, the mass estimates for these dinosaurs varied significantly, "emphasizing the high levels of uncertainty inevitable in such reconstructions," as the researchers put it. Still, some "best estimates" were produced: 7655, 6071, and 6177 kg (= 8.44, 6.69, and 6.81 tons, respectively) for the three large-bodied theropods, 423 kg (= 932 lbs) for the ornitho-

mimid, and 813 kg (= 1792 lbs.) for the duckbill. Science is supposed to be testable; how could these scientists figure out whether or not their final numbers had any meaning?

One test was to use the method on the skeleton of a living ostrich. If the method had any validity, it should predict from the skeleton the correct weight of the modern bird, in which the actual mass is known. In fact, the ostrich's predicted weight closely matched that of the estimate calculated from the skeleton, suggesting the the method has some validity.

Ultimately, the most striking thing about the research was what it said about the weight distribution in these creatures. In all cases, the center of mass was consistently located low on the body, in front of the hip. This reinforces the idea, based upon the design of the bones, that these animals all tended to keep the backbone parallel, or nearly so, with the ground (as we saw in Chapter 4).

Box 12.5 **Dinosaur smarts**

How can we measure the intelligence of dinosaurs?[1] The short answer is "Not easily." However, it is clear that, at a very crude level, there is a correlation between intelligence and brain:body weight ratios. Brain:body weight ratios are used because they allow the comparison of two differently sized animals (that is, brain:body weight ratios allow comparison of chihuahua and St Bernard dogs). The correlation suggests that, in a general way, the larger the brain:body weight ratio, the smarter the organism. Indeed, mammals have higher brain:body weight ratios than fish and are generally considered to be more intelligent.

But how smart could a very large dinosaur with a minuscule brain be (for example, see Box 5.2)?

It is well known that organisms change proportions as they increase in size; this is **allometry**. And it turns out that brain:body weight ratios follow allometric principles as well: brains do not increase in size proportionally to the rest of the animal. For example, the brain of a 0.5 m rattlesnake is proportionally larger than the brain of a 3 m anaconda. Does this mean that the anaconda is significantly stupider than the rattler? Obviously not. So, when considering how big or small a brain is in an animal, there has to be a way to compensate meaningfully for size. A quantitative method of doing this was first proposed by psychologist H. J. Jerison, who, in the early 1970s, developed a measure called the "encephalization quotient" (EQ). Jerison constructed an "expected" brain:body weight ratio for various groups of living vertebrates (reptiles, mammals, birds) by measuring many brain:body weight ratios among these animals. Jerison noted that, on the basis of EQ, living vertebrates cluster into two groups, endotherms and ectotherms. The endotherms show greater encephalization (higher EQs) and the ectotherms showed lower encephalization (lower EQs). Thus, for Jerison, living endotherms and ectotherms could be distinguished by brain size. Having constructed a range of expected brain:body weight ratios, he could account for size in different organisms (and accommodate what might at first seem like an extraordinarily large or small brain). Noting that some organisms still didn't exactly fit in his ectotherm or endotherm group (by virtue of having brains either larger or smaller than expected), he measured the amount of deviation, and then termed this EQ.

Paleontologist J. A. Hopson, now knowing what he could expect for living vertebrates, measured how much the

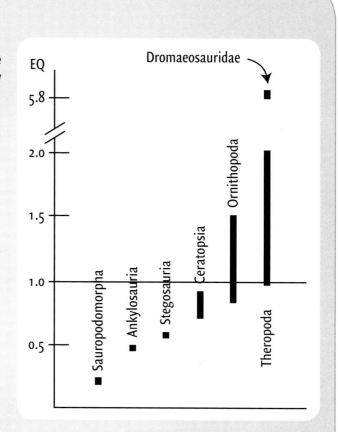

Figure B12.5.1. EQs (Encephalization Quotients) of dinosaurs compared. The line at 1.0 represents the crocodilian "norm," and suggests that many groups of dinosaurs had larger brains than would be predicted from a conventional reptilian model (the crocodile). Note also the break between 2.0 and 5.8; if these measures mean anything, apparently coelurosaurs significantly outdistanced other dinosaurs in brain power. (Data from Hopson, 1980.)

estimated brain:body weight ratio EQ of extinct vertebrates deviated from the expected brain:body weight ratios of their living counterparts. Figure B12.5.1 shows the EQs for several major groups of dinosaurs as reconstructed by Hopson. Because dinosaurs are "reptiles," he measured the deviation of various groups relative to a "reptilian" norm, in this case a living crocodile. Significantly, many ornithopods and theropods show a brain:body weight ratio that is significantly larger than would be expected if a modern reptilian level of intelligence is being considered.

using computed tomography (a CT scan). Doubters doubted; advocates advocated; and the issue remains unresolved.

In the late 1970s, attempts were made to assess the intelligence of dinosaurs using the encephalization quotient or EQ (Box 12.5). The idea was that living endotherms (birds and mammals) have significantly higher EQs than do living ectotherms (reptiles and amphibians), presumably because their more refined levels of neuromuscular control require an endothermic metabolism. EQ was reconstructed in dinosaurs using brain endocasts, internal casts of the braincases of dinosaurs (see Box 9.1). Based upon EQ, coelurosaurs were likely as active as many birds and mammals, while large theropods and ornithopods were somewhat less active than birds and mammals, but more active than typical living reptiles. Using EQ as representative of activity levels, other dinosaurs appear to have been in the range of living reptiles.

The nose shows Endothermy requires the lungs to replenish their air (ventilate) at a high rate. And high rates of ventilation lead to water loss, unless something is done to prevent it. What modern mammals and birds do is to grow convoluted sheets of delicate, tissue-covered bone, called respiratory turbinates, in the nasal cavaties. The mucus-covered surfaces of the turbinates pull moisture out of the air before it leaves the nose, thus conserving water (Figure 12.5).

Biologist John Ruben wondered if the fact that modern endotherms require a means to conserve moisture during respiration might provide a clue about dinosaur metabolism. Although a number appear to have had olfactory turbinates, indicative of a well-developed sense of smell, apparently none – as far as we currently know – had respiratory turbinates to allow them to recoup respired moisture. Considered exclusively on this basis, Ruben concluded that dinosaurs could not have been endothermic in the way that most mammals and birds are today.

Where the wild things are The distribution of dinosaurs around the globe far exceeds the current distribution of modern ectothermic vertebrates, which are generally not found above and below, respectively, latitudes 45° north and 45° south. Large modern ectotherms rarely occur above latitude 20° north and below latitude 20° south (Figure 12.6).

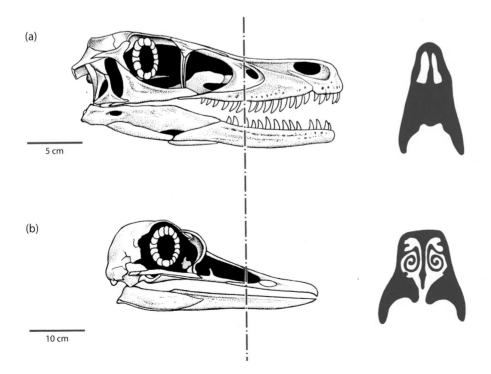

(a)

5 cm

(b)

10 cm

Figure 12.5. Cross-sections (solid shading) through the nasal regions of (a) an extinct dinosaur (*Velociraptor*) and (b) a living bird (*Rhea*); skulls and positions of the cross-sections are shown to the left. The nasal cavity of the bird shows convoluted respiratory turbinates, while that of the dinosaur does not.

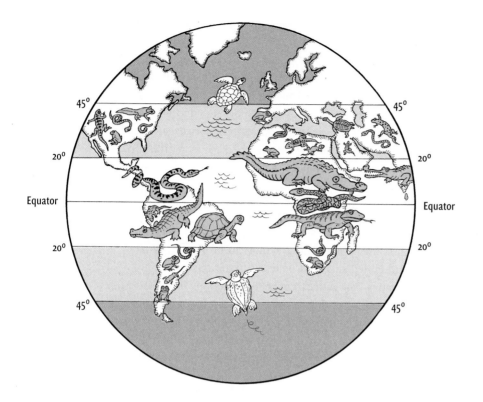

Figure 12.6. The latitudinal distribution of ectothermic tetrapods on Earth today. The larger terrestrial tetrapods do not get much beyond about latitude 20° north and south. These include large snakes and lizards, crocodilians, and tortoises.

Correcting for continental movements, Cretaceous dinosaur-bearing deposits have been found close to latitudes 80° north and 80° south of the equator. Both the northern and southern sites experienced extended periods of darkness, and we can be reasonably sure that, at least occasionally, temperatures in winter fell below freezing.

The Arctic assemblage, from North America, includes hadrosaurids, ceratopsids, tyrannosaurids, and troodontids. The Antarctic dinosaur assemblage, from Australia, includes a large theropod and some basal euornithopods (including many juveniles). Along with these dinosaurs are fish, turtles, pterosaurs, plesiosaurs, birds (known solely from feathers), and, incredibly, a improbable late-surviving temnospondyl (an amphibian group that apparently went extinct in the Early Jurassic everywhere else in the world; see Figure 13.6).

This mix of animals is not particularly easy to interpret in terms of endothermy or ectothermy. Several members of the southern dinosaur assemblage had large brains and well-developed vision, potentially useful during periods of extended darkness. Others, however, were not so well equipped. Burrowing may have been an option for some, but not for all.

In the case of the North American faunas, only *Troodon* is of a size that could make burrowing feasible. Migration was potentially a solution to inclement winter weather, although the dinosaurs would have had to migrate for tremendous distances before temperatures warmed sufficiently.

Newer approaches

Haversian bone Bones grow by remodeling, which involves the resorption (or dissolution) of bone first laid down – primary bone – and redeposition of a kind of bone called secondary bone. Secondary bone is deposited in the form of a series of vascular canals called Haversian canals, and resorption and redeposition of secondary bone can occur repeatedly during remodeling. When remodeling occurs, a type of Haversian bone known as **dense secondary Haversian bone** is formed. This bone has a distinctive look about it (Figure 12.7), visible in modern as well as fossil bone (Figure 12.8).[2]

Dense secondary Haversian bone is found in many mammals and birds – all, of course, endotherms. Among *extinct* vertebrates, dense secondary Haversian bone has been observed in dinosaurs, pterosaurs, and therapsids (including Mesozoic and Cenozoic mammals). Given the distribution of dense secondary Haversian bone in vertebrates, it is not too great a leap to suppose that dinosaurs, too, must have been endotherms.

Although this idea was initially promising, dense secondary Haversian bone forms due to a variety of factors, only one of which is endothermy. Secondary Haversian canals are known to be correlated with size and age, and possibly with the type of bone being replaced, the amount of mechanical stress undergone by the bone, and **nutrient turnover** (the metabolic interaction between soft tissue and developing bony tissue). The last two, however, are still related to metabolism, and as we shall see (below) bone histology, the study of bone as a type of living tissue, has played an important role in understanding dinosaur metabolism.

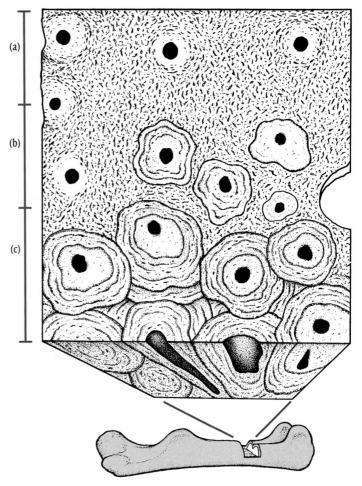

Figure 12.7. Primary bone in the process of being replaced by Haversian bone in the leg of a hadrosaurid. Longitudinal canals (at the top of the figure) in primary lamellar bone (a) are resorbed (b) and then reconstituted as Haversian bone (c).

LAGs Concentric growth rings have been observed in the bones and teeth of dinosaurs. Among modern tetrapods, such growth rings are typically found in ectotherms, where they are believed to represent seasonal cycles. During times of slowed metabolism (such as dry seasons in the tropics, or cold seasons in more temperate latitudes), growth is stymied – hence the term "**lines of arrested growth**," or LAGs (Figure 12.9).

So bone records a pattern of ring-like deposits representing annual cycles of growth and stasis. Among living homeothermic endotherms, on the other hand, such patterns are rarer, because the relatively constant, elevated metabolic rates ensure growth at a constant rate.

Most dinosaurs – with the exceptions of some theropods, hypsilophodontids, iguanodontids, and sauropods – show well-developed LAGs, suggesting to researchers that growth in dinosaurs might have been more susceptible to external climatic influences than had been predicted by the simple homeothermic endothermic view of dinosaur metabolism. In two small flyers and one large flightless enantiornithine (*Patagopteryx*) bird, the presence of LAGs led to

---

[2] Fossil bone can preserve fine anatomical details. To see these, a thin slice of bone can be cut, mounted on a glass slide, and ground down to ~30 microns, so thin that light can be transmitted through it. The slide with the slice of bone can then be studied under a microscope.

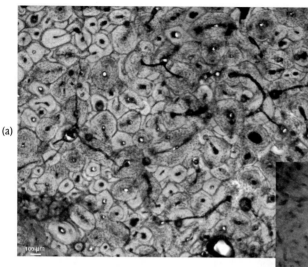

(a)

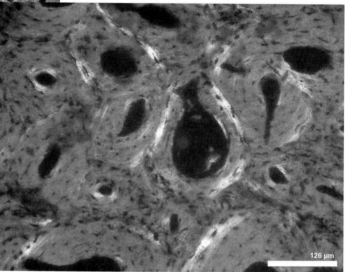

(b)

Figure 12.8. (a) Dense Haversian bone in *Tyrannosaurus*. (b) Magnified view of dense Haversian bone in *Archaeomimus*.

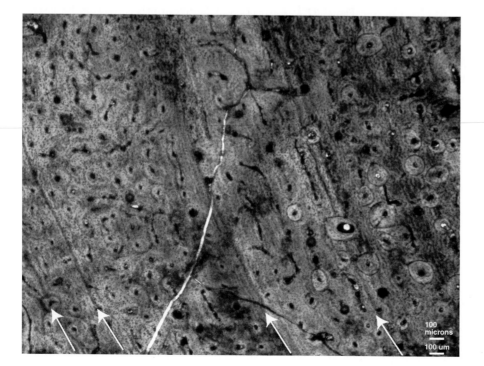

Figure 12.9. Lines of arrested growth (LAGs) in a *Tyrannosaurus* fibula. Arrows indicate LAGs.

the conclusion that the early birds' metabolism(s) were subject to seasonal growth, even though the birds clearly bore feathers. The presence of LAGs in these early birds could mean that, ultimately, they had not quite attained the level of endothermy seen in living birds (see below).

Enantiornithine bone histology contrasts with that of ornithurine birds (for example, *Hesperornis* and *Ichthyornis*). In these, the bone tissue looks very similar to that in modern birds. So too is the bone tissue of the primitive, Early Cretaceous *Confuciusornis*.

LAGs as time markers But are the LAGs truly seasonal? Our best evidence suggests that indeed they are. Of course, there is no numerical way to date them (see Chapter 2), so the evidence is circumstantial. Yet, it comes from several sources and those sources all produce a coherent story. For one thing, comparable lines known to form annually are found in a wide variety of groups closely and not-so-closely related to Dinosauria. Moreover, the amount of bone that forms between the lines is consistent with annual (seasonal) cycles. Finally, the spacing of the lines shows a steady decrease, consistent with a decrease in growth rate as the animals get older. Indeed, LAGs have never been found to come from anything other than annual growth.

And so if we accept that LAGs are annual signals, they suddenly give us insights into a heretofore murky aspect of dinosaur paleobiology: how fast they grew, and how long they lived.

Growth How, then, does one evaluate the age of a dinosaur? Paleohistologists A. de Ricqlès, J. Horner, and K. Padian describe their technique:

1 Count LAGs, which yield a number presumably corresponding to the age of the dinosaur specimen;
2 Measure bone thickness between each LAG, and divide that thickness by the number of days/year, which should produce the amount of bone deposited per day; and
3 Measure total bone thickness, and develop an estimate of how much time it would reasonably take to form.

From those steps, they infer the age of the particular specimen in question.

But how about entire growth trajectories, that is, the pattern of growth through an animal's life? To take a familiar example, human growth trajectories aren't linear: we start with rapid growth, decreasing gradually until the onset of sexual maturity, after which growth, if it occurs at all, slows dramatically. By the time most humans have completed the first quarter of their lives, growth has stopped. By contrast, many animals, such as crocodiles, have continuous growth throughout their lives (although growth rates are fastest when they are juveniles; Figure 12.10).

Two approaches can be used to learn about the growth of a dinosaur. The lengths of the long bones – femur or tibia, generally – can be measured as representative of growth; alternatively, the animal's weight can be estimated (see Box 12.4) and considered as representative of its growth. Figure 12.11 shows estimated growth curves for the sauropod *Janenschia*, for some tyrannosaurs; a growth curve for *Archaeopteryx* is shown in Figure 11.1.

So in the end, what does all this say about the subject of this chapter, dinosaur endothermy? It has become well established that metabolic rates are well correlated with maximal growth rates. Fast growth rates – such as are found in modern birds – are highly suggestive of an endothermic metabolism. Figure 12.12 contrasts the growth rates of dinosaurs with those of some familiar groups of living animals.

While some larger dinosaurs may have undergone very rapid growth to attain large size (Figure 12.13), dinosaurs as a group, based upon their growth rates, were generally on

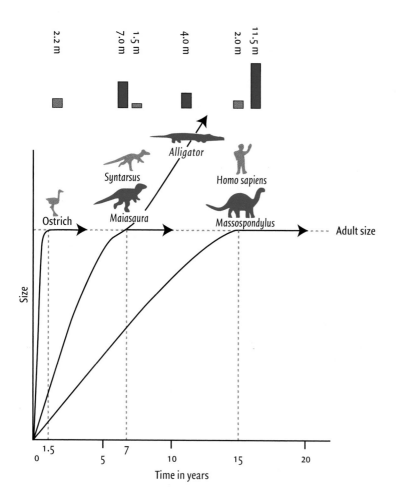

Figure 12.10. Estimated growth rates of some dinosaurs, *Alligator*, and a human. Unlike the other tetrapods presented, *Alligator* grows continuously throughout its life; hence, it has no fixed "adult size." Sexual maturity usually comes within six to eight years. (Estimates for *Syntarsus* and *Massospondylus* from the work of A. Chinsamy; estimates for *Maiasaura* from the work of J. Peterson and J. Horner.)

the low side of modern endothermic growth rates. This in turn suggests that their metabolism, although properly termed endothermic, must have been slower than that found in, for example, modern birds.

Stable isotopes Incredibly, fossil vertebrates carry around their own thermometers. These come in the form of **stable istopes**; that is, isotopes that, unlike their unstable brethren, do *not* spontaneously decay. Of particular interest to us are the istopes of the element oxygen. There are three: ^{16}O, by far the most common,[3] ^{17}O, and ^{18}O. The last, ^{18}O, is particularly interesting, because its proportion to ^{16}O varies as temperature varies. Therefore, if a substance contains oxygen, one can learn something about the temperature at which that substance formed by the ratio ^{18}O:^{16}O. In the case of bone and teeth, the oxygen is contained in phosphate (PO_4) that forms part of the mineral matter in the bone. Thus, knowing the oxygen isotopic composition of the bone, one can learn something of the temperature at which the bone formed. A number of different approaches have been tried including core-to-extremities measurements, direct temperature estimates using isotope ratios, and most recently, "clumped isotope geothermometry."

Core-to-extremities If dinosaurs were poikilothermic, reasoned paleobiologist R. Barrick and geochemist W. Showers, there should be a large temperature difference between parts of the

---

[3] The isotope ^{16}O comprises 99.763%, ^{17}O 0.0375%, and ^{18}O 0.1905% of total atmospheric oxygen.

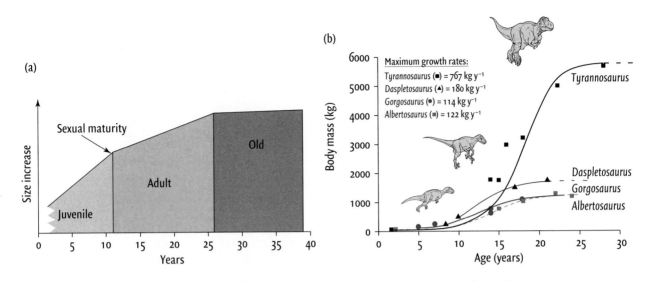

Figure 12.11. Inferred growth curves for (a) the sauropod *Janenschia*; and (b) tyrannosaurs. While age (in years) is on the horizontal axis in both curves, the vertical axis of the sauropod curve is based upon inferred size, while the vertical axis of the tyrannosaur curve is based upon inferred weight. *T. rex*, strikingly heavier than other tyrannosaurids, evidently underwent a striking growth spurt between 10 and 25 years; other tyrannosaurids had proportionately smaller growth spurts at the same time interval or slightly earlier.

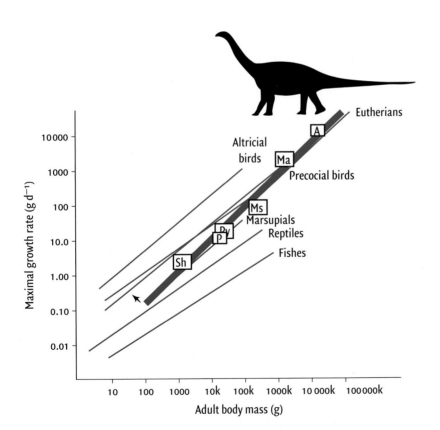

Figure 12.12. Maximum composite growth rates of various groups of animals compared. Composite rates constructed from growth rates of several animals in each group. The slope of the composite dinosaur growth rate (based upon data from 6 dinosaurs) is slightly steeper than the others, and higher than all except birds and placental mammals. The difference in slope between dinosaurs and other other vertebrate groups in the graph suggests that dinosaurs grew in different ways from these groups. Redrawn from Erickson, G. M. 2005. Assessing dinosaur growth patterns: a microscopic revolution: *Trends in Ecology and Evolution*, **20**, 677–684.

skeleton located deep within the animal (that is, ribs and trunk vertebrae) and those located toward the exterior of the animal (that is, limbs and tails; Figure 12.14).

If, however, dinosaurs were homeothermic, there should be little temperature difference between bones deep within the animal and those more external, because the body would be maintaining its fluids at a constant temperature. The difference in temperatures – or lack thereof – should be reflected in the proportions of ^{18}O to ^{16}O.

Barrick and Showers showed that bones from the cores of some of the dinosaurs tested (*Tyrannosaurus*, *Hypacrosaurus*, *Montanoceratops*, and a juvenile *Achelousaurus*) showed little temperature variation, suggesting that they were formed under homeothermic conditions. The small euornithopod *Orodromeus* and a nodosaurid ankylosaur that they tested, on the other hand, had an isotopic variation (hence, an inferred temperature variation) that pushed the limits of conventional homeothermy.

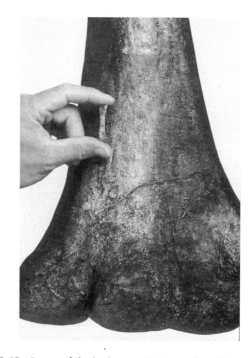

Figure 12.13. Femur of the hadrosaur *Maiasaura* hatchling compared to that of an adult. Note the size of the human hand.

The Jurassic dinosaurs *Ceratosaurus* and *Allosaurus* all showed an ectotherm-like variability in their core regions (the pelvis, in this case). In general, the extremites of these dinosaurs fell within 4 °C of the cores (Figure 12.15). Barrick and Showers concluded that virtually all of these dinosaurs were homeotherms that experienced some "regional heterothermy."

Direct temperature measurements As we have seen, Barrick and Showers' work depended upon isotope ratios, but there was a catch. The calculation of temperature from isotopic ratios

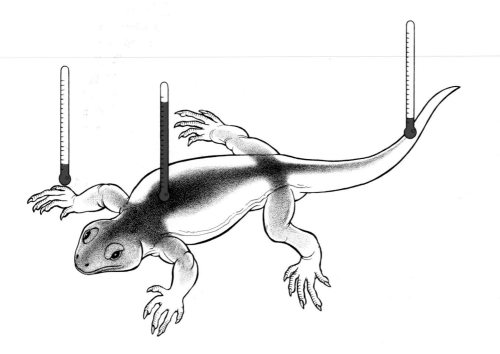

Figure 12.14. Core-to-extremity temperatures in a poikilothermic ectotherm. Because this tetrapod's temperature fluctuates with the ambient temperature, when it's cold outside, its extremities are much colder than its core.

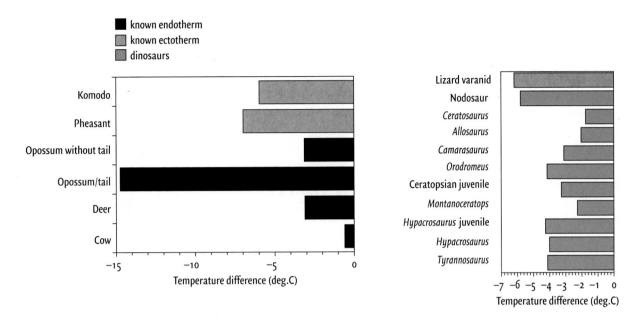

Figure 12.15. Estimated maximum temperature variations between bones located in the core of the body and those located at the extremities in living and extinct vertebrates, reconstructed with the use of oxygen isotopes. Living vertebrates are represented by the Komodo dragon (an ectothermic lizard) and a selection of mammals (endotherms). Note that the greatest variation between core and extremities occurs in the opossum (conventionally considered to be an endotherm), reinforcing the idea that even endotherms can undergo significant fluctuations between core and extremities (in this case the long tail). The researchers concluded that all dinosaurs tested, except the ankylosaur, matched the definition of homeotherms.

requires knowing the isotopic composition of the water in which the chemistry took place; in this case, the temperature of the water that these extinct organisms drank. But how could one ever know that? Surely that water was as long gone as the dinosaurs being tested! Barrick and Showers didn't need to worry about this because they were comparing *relative* temperatures – variations between cores and extremities.

A number of later studies by a variety of paleontologists and geochemists attempted, however, to obtain the actual temperatures at which various dinosaurs functioned. This was done by calibrating the estimated temperatures of known ectotherms to obtain those of dinosaurs. So, for example, crocodiles were an important feature of many dinosaur ecosystems. Knowing the temperatures at which modern crocodiles function, one could calibrate the estimated isotopic composition of the water, until the calculated temperature of fossil crocodiles was within a couple of degrees of the known temperatures of modern crocodiles. Once the isotopic composition of the water was properly estimated, the real temperatures of dinosaurs could then be determined.

The results of these types of studies have been consistent. For Cretaceous theropods, sauropods, ornithopods, and ceratopsians, the very considerable body of data collected suggests that these animals maintained body temperatures of between 30 and 37 °C, well within the range of modern endotherms, at least.

Clumped isotopes The Achilles heel, as we have seen, of direct isotope-based temperature measurements is that that isotopic composition of the water that the animals drank has to be estimated. Very recently, however, a new technique – clumped isotopes – has been applied to the problem of the body temperatures at which ancient bone and teeth formed. Clumped isotopes refers to the attraction – the "clumping" – of ^{18}O and ^{13}C at different temperatures. The amount of clumping is temperature-dependent, and so with this technique, by measuring the clumping, there is no need to estimate the isotopic composition of the water. Geochemist R. A. Eagle and colleagues applied this technique to the teeth of five Jurassic sauropod specimens, and determined that these animals operated at temperatures between 31 and 38 $^{\circ}$C. This temperature is again suggestive of an endothermic metabolism.

Different strokes for different folks?

Some paleontologists suggest that some dinosaurs – particularly large ornithopods and theropods – maintained something close to endothermic homeothermy as fast-growing juveniles, but may have became closer to homeothermic ectotherms as adults. While this may be true, the elevated temperatures consistently recorded by geochemists suggest that these animals maintained a temperature different from that of the air that surrounded them: the hallmark of an endothermic metabolism. The recent discovery of feathers in large, as well as small theropods, supports this view.

Similarly, some scientists have argued that large sauropods could have relied upon a strategy called gigantothermy: small surface:volume ratios (resulting from large size) retained core heat, allowing sauropods to maintain a homeothermic metabolism without the metabolic cost of being truly endothermic. Yet, the results of the clumped isotopic studies suggest that sauropod body temperatures may have been too *cool* for a passive strategy like gigantothermy. To those researchers, sauropods must have actively maintained their temperatures: they must have been true endotherms. Considerable evidence suggests that small- to medium-sized theropods, and perhaps similarly sized ornithopods, were likely homeothermic endotherms throughout their active lives; not perhaps to the extent that a modern bird is, but perhaps somewhat like the kind of endothermy exhibited by animals like an opossum.

What is unimpeachable at this point is that dinosaurs were not ectotherms in the crocodilian sense. They were something else, and it is reasonable to suppose that different endothermic strategies could have been adoped by different dinosaurs, including, of course, birds. Our best call is that dinosaurian physiology was likely a complex mix of various endothermic strategies, relating to size, behavior, and, potentially, environment.

Summary

Physiologists avoid terms like "warm-blooded" and "cold-blooded" and instead replace them by more meaningful terms like endothermy and ectothermy, which describe the heat source, and homeotherm and poikilotherm, which describe the degree to which temperatures fluctuate in organisms. All of these terms should be thought of as endpoints on spectra of metabolic strategies.

Anatomical indicators in dinosaurs suggestive of an endothermic metabolism include: the erect stance, which among living vertebrates is exclusively posessed by endotherms; the inferred presence of a four-chambered heart, necessary to sustain the higher blood pressures associated with an endothermic metabolism; and the relatively high EQs of some dinosaurs.

Each of these indicators,however, is inconclusive: the erect stance argument has been criticized as being purely coincidental (and not causal) and the high EQs of some dinosaurs are matched by strikingly low EQs in others. Moreover, the absence in non-avian dinosaurs of respiratory turbinates has suggested that they perhaps didn't maintain the high rates of ventilation seen in living endotherms.

The presence in dinosaurs of Haversian canals appears to suggest endothermy. These, coupled with LAGs, suggest rapid growth rates that today are known only in endotherms. These suggest that, as juveniles at least, many dinosaurs may have possessed endothermic metabolisms.

Because endothermy is, in terms of energy, quite costly to maintain, it was suggested that the ratio predators : prey in endothermic eocsystems ought to be significantly smaller than the ratio predators : prey in ectothermic ecosystems. Several attempts were made to calculate such ratios for dinosaurs. None ultimately proved definitive for a variety of reasons, including the unreliability of museum collections as accurate indicators of ancient communities, the fact that endothermic predators sometimes eat ectothermic prey (and vice versa), and the fact that the limiting factor on prey populations is not generally predation.

The existence of polar-dwelling dinosaurs has been interpreted as suggestive of an endothermic metabolism, since today large ectotherms don't get much above 20° N or below 20° S latitude. Yet, a high-latitude temnospondyl amphibian also preserved suggests that, for whatever the reasons, the past distribution(s) of ectotherms was greater than today.

With insulatory coatings potentially going as far back as Ornithodira, there is considerable evidence that endothermy may go back equally far within archosaurs. Regardless, as we saw in Chapter 11, *Archaeopteryx* and its avialan cohort enjoyed slower growth rates that we see in modern birds, suggesting that the insulation in basal ornithodires may be indicative of early flirtations with endothermy.

^{18}O:^{16}O ratios are temperature-sensitive, and have been used as a kind of paleothermometer in well-preserved fossil bone. The idea was that ectotherms would show a greater range of temperature fluctuations from core to extremities than endotherms. In fact, dinosaurs (and even some mammals) produced a somewhat mixed signal: while some dinosaurs, such as hadrosaurs, showed little temperature variation, ankylosaurs and two of the large theropods tested showed ectotherm-like variations. Reinforcing the point that endothermy and ectothermy are actually endpoints in a spectrum and that metabolisms among vertebrates are not easily predictable, an opossum, a living marsupial (mammal), also showed the kind of variation expected in an ectotherm. Clumped isotopic studies, although in the early phases, suggest endothermy in those dinosaurs in which they have been attempted.

While it is clear that the old "cold-blooded" lizard or crocodile model of dinosaur metabolism is defunct, the record suggests that dinosaurs likely enjoyed a range of metabolic strategies, most or all of them rooted in some kind of at least primitive type of endothermy.

Selected readings

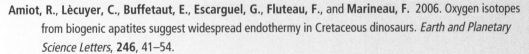

Amiot, R., Lècuyer, C., Buffetaut, E., Escarguel, G., Fluteau, F., and **Marineau, F.** 2006. Oxygen isotopes from biogenic apatites suggest widespread endothermy in Cretaceous dinosaurs. *Earth and Planetary Science Letters*, **246**, 41–54.

Bakker, R. T. 1975. Dinosaur renaissance. *Scientific American*, **232**, 58–78.

Bakker, R. T. 1986. *The Dinosaur Heresies*. William Morrow and Company, New York, 481pp.

Bakker, R. T. and **Galton, P. M.** 1974. Dinosaur monophyly and a new class of Vertebrates. *Nature*, **248**, 168–172.

Barrick, R. E., Stoskopf, M. K. and **Showers, W. J.** 1997. Oxygen isotopes in dinosaur bone. In Farlow, J. O. and Brett-Surman, M. K. (eds.), *The Complete Dinosaur*. Indiana University Press, Bloomington, IN, pp. 474–490.

de Ricqlès, A., Horner, J. R. and Padian, K. 2006. The interpretation of dinosaur growth patterns. *Trends in Ecology and Evolution*, **21**, 596–597.

Desmond, A. 1975. *The Hot-Blooded Dinosaurs*. The Dial Press, New York, 238pp.

Eagle, R. A., Tütken, T., Martin, T. S. *et al.* 2011. Dinosaur body temperatures determined from isotopic (13C-18O) ordering in fossil biominerals. *Science*, **333**, 443–445.

Erickson, G. M. 2005. Assessing dinosaur growth patterns: a microscopic revolution: *Trends in Ecology and Evolution*, **20**, 677–684.

Farlow, J. O. 1990. Dinosaur energetics and thermal biology. In Weishampel, D. B., Dodson, P. and Osmólska, **H.** (eds.), The Dinosauria. University of California Press, Berkeley, pp. 43–55.

Fricke, H. and Rogers, R. 2000. Multiple taxon–multiple locality approach to providing oxygen isotope evidence for warm-blooded theropod dinosaurs. *Geology*, **28**, 799–802.

Horner, J. R., de Ricqlès, A. and Padian, K. 2000. Long bone histology of the hadrosaurid dinosaur *Maiasaura peeblesorum*: growth dynamics and physiology based upon an ontogenetic series of skeletal elements. *Journal of Vertebrate Paleontology*, **20**, 115–129.

Reid, R. E. H. 1997. Dinosaurian physiology: the case for "intermediate" dinosaurs. In Farlow, J. O. and Brett-Surman, M. K. (eds.), *The Complete Dinosaur*. Indiana University Press, Bloomington, IN, pp. 449–473.

Rich, P. V., Rich, T. H., Wagstaff, B. E. 1988. Evidence for low temperatures and biologic diversity in Cretaceous high latitudes of Australia. *Science*, **242**, 1403–1406.

Spotila, J. R., O'Connor, M. P., Dodson, P. and Paladino, F. V. 1991. Hot and cold running dinosaurs: body size, metabolism, and migration. *Modern Geology*, **16**, 203–227.

Thomas, R. D. K. and Olson, E. C. (eds.). 1980. *A Cold Look at the Warm-Blooded Dinosaurs*. AAAS Selected Symposium no. 28, 514pp.

Topic questions

1   What are meant by the terms endothermy, ectothermy, poikilothermy, and homeothermy?

2   What are remodeled bone, Haversian canals, LAGs, and respiratory turbinates?

3   Was a four-chambered heart evidence for endothermic dinosaurs? What was it used to support?

4   Give some anatomical evidence that dinosaurs were endothermic. Critique it.

5   What is a predator:prey biomass ratio? How were these used in assessing dinosaur metabolism?

6   Critique the predator:prey biomass ratios.

7   How would the distribution of animals on Earth have anything to do with their metabolisms? What kinds of strategies were available to dinosaurs for protection against long, cold winters?

8   What is ^{18}O? What is meant by the statement that "Fossil vertebrates carry around their own paleothermometers?"

9   Why would comparison of core temperatures to those of the extremities be useful in determining whether an animal had an endothermic or an ectothermic metabolism?

10   How are rates of growth calibrated in fossil dinosaurs?

CHAPTER 13

THE FLOWERING OF THE MESOZOIC

CHAPTER OBJECTIVES

- Introduce large-scale patterns of dinosaur evolution

- Develop a deeper understanding of climate in the Mesozoic Era

- Introduce a few important Mesozoic plants

- Introduce dinosaur–plant co-evolution

Dinosaurs in the Mesozoic Era

Throughout much of this book, we have considered dinosaurs as individuals; who they were, what they did, and how they did it. Now we'll step back and take a look at the large-scale ebb and flow of dinosaur evolution. Before we can do that, though, we need to think about what might be *missing*.

Preservation

Table 13.1 shows the distribution of dinosaurs among the continents through time. The paucity of dinosaur remains from Australia and Antarctica, however, is surely more a question of *local* preservation and inhospitable conditions today for finding and collecting fossils than defining where dinosaurs actually lived. Into this mix must be factored *geological* preservation; some time intervals simply contain more rocks than others. For example, the terrestrial Middle Jurassic is not well represented by rocks, with the result that it artificially appears to have been a time of very low tetrapod diversity (Box 13.1). The Late Cretaceous is rather the opposite, with the happy result that we have a rich record of Late Cretaceous dinosaurs. Several methods of estimating the completeness of fossil preservation have been developed to mitigate these problems (Box 13.2).

Dinosaurs through time

We can think of these geographical and temporal distributions as pages in a notebook, in which each succeeding page represents a new time interval, with new continental arrangements, and new and different assemblages of dinosaurs populating the continents. Considered in this way, the sequence of dinosaurs through time is like a grand pageant through Earth history, in which

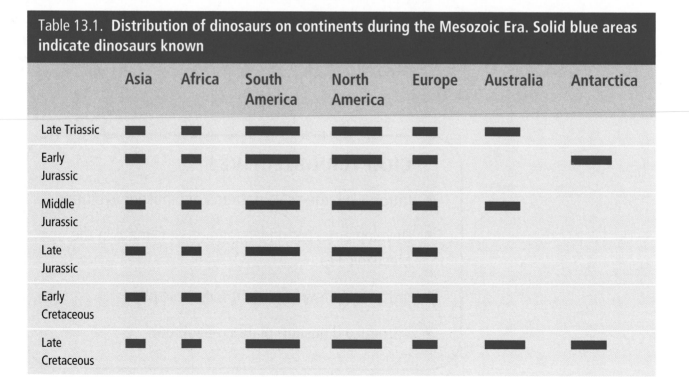

Table 13.1. Distribution of dinosaurs on continents during the Mesozoic Era. Solid blue areas indicate dinosaurs known

	Asia	Africa	South America	North America	Europe	Australia	Antarctica
Late Triassic	▬	▬	▬▬	▬▬	▬	▬▬	
Early Jurassic	▬	▬	▬▬	▬▬	▬		▬▬
Middle Jurassic	▬	▬	▬▬	▬▬	▬	▬▬	
Late Jurassic	▬	▬	▬▬	▬▬	▬		
Early Cretaceous	▬	▬	▬▬	▬▬	▬		
Late Cretaceous	▬	▬	▬▬	▬▬	▬	▬▬	▬▬

Box 13.1 **The shape of tetrapod diversity**

For more than 25 years, the University of Bristol's M. J. Benton has been compiling a comprehensive list of the fates of tetrapod families through time. We see several interesing features of the curve that results from this compilation. Note the drop in families during Middle Jurassic time (Figure B13.1.1). This, as we have seen, is an **artifact**; that is, a specious result. This particular one comes from the lack of find localities more than from a true lack of families during the Middle Jurassic. Then, notice the huge rise in families during the Tertiary. Some of this may be real, and perhaps attributable in part to Tertiary birds and mammals (both of whom are very diverse groups), but some of it might be another artifact, due to what is called the "**pull of the Recent**." The pull of the **Recent** is the inescapable fact that, as we get closer and closer to the Recent, fossil biotas become better and better known. This is because more sediments are preserved as we get closer and closer to the Recent, and a greater amount of sedimentary rocks preserved means more fossils. The big spike at the end of the Jurassic is the Morrison Formation of the US Western Interior, a unit that preserved an extraordinary wealth of fossils.

So a curve like Benton's requires skill to understand and to factor out the artifacts. Nonetheless, we can see that, generally, dinosaurian diversity increased throughout their stay on Earth and, as they progressed through the Cretaceous,

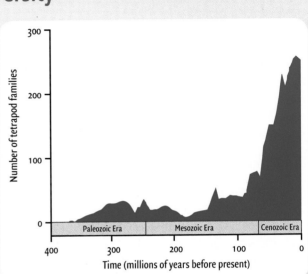

Figure B13.1.1. M. J. Benton's estimate of vertebrate diversity through time. On the *x*-axis is time; on the *y*-axis is diversity as measured in numbers of tetrapod families.

dinosaurs continued to diversify. The increase in diversity shown in Benton's diagram may reflect the increasing global endemism of the terrestrial biota, itself driven by the increasing isolation of the continental plates.

each interval of time has a characteristic fauna that gives that time a characteristic quality (Figure 13.1).

Now let's look at this information in a more quantitative way: the diversity, that is, the number of different *types*, of dinosaur genera over time (Figure 13.2). This tracks large-scale, global fluxes in dinosaurs through the approximately 163 million years that they were on Earth.

In the beginning … the Late Triassic (228–200 Ma)

Figure 13.2 shows that dinosaurs radiated quickly in the Late Triassic. Exactly how dinosaurs came to be the dominant terrestrial vertebrates in the Late Triassic remains tantalizingly shrouded in the mists of antiquity. However, our best data suggest that dinosaurs likely moved quickly into a world *abandoned* by other vertebrates rather than possessing superior adaptations that somehow allowed them to *out-compete* pre-existing tetrapods (mainly therapsids and primitive archosaurs; see Figures 13.3–13.7) and take charge.

These potentially dramatic ecological steps are not so easily revisited, because the diversity of early dinosaurs is small, and the times of their appearances are not particularly well known. The earliest dinosaurs known are *Herrerasaurus* and *Eoraptor* from the Ischigualasto Formation

Box 13.2 Counting dinosaurs

Since 1990, there has been almost a doubling (85%) in the known number of dinosaur genera. So our understanding of even who dinosaurs were is changing so much, it's well to think about ways to estimate the total number of genera, that is, the generic diversity of dinosaurs, that ever existed. In this box, we introduce several of the ways that diversity can be estimated.

Cladistic estimates

Although we have heretofore emphasized the use of cladograms for reconstructing evolutionary relationships, they can also portray the relative sequence of the appearance of organisms. For animals – such as dinosaurs – with a fossil record this relative sequence from the cladogram can be compared with the real sequence of appearance that comes from the geological record of the same dinosaurs. The cladogram-based sequence ought to compare well with the sequence of stratigraphic occurrence of the fossils themselves.

In addition, the combination of phylogeny and stratigraphy has a lot to say about presence and meaning of gaps in the fossil record. That is, even if an ancestor is not preserved, the cladogram allows us to infer when it must have existed. To understand this, we turn to an example.

Suppose Dinosaur X and Dinosaur Y were each other's closest relative; thus they share a unique common ancestor. If Dinosaur X is known from rocks dated at 100 Ma and Dinosaur Y came from 125 Ma rocks, then this ancestor had to be at least 125 million years old (that is, the age of the older of the two dinosaur

species). And if this is true, then there must be some not-yet-sampled history between this ancestor and Dinosaur X – to the tune of 25 million years – all because of phylogenetic continuity calibrated through the use of stratigraphy. Such a 25 million year gap can be referred to as a **minimal divergence time (MDT)**; it can be calculated for any two taxa so long as their phylogenetic relationships and stratigraphic occurrences are known and is an estimate of the completeness of the fossil record. Lineages must have existed that have so far not left us a physical record (through fossils) of their existence; these are called **ghost lineages**. MDTs are measures of their duration (Figure B13.2.1).

Ceratopsia counted

It has sometimes been claimed that ceratopsians have one of the best fossil records among all dinosaur groups. How can we test this? There are 32 ceratopsian species known, less than are found in theropods, sauropodomorphs, and ornithopods, yet more than in ankylosaurs, stegosaurs, and pachycephalosaurs. Averaged over their total time on Earth, ceratopsians apparently produced new species at the rate of one every 1.9 million years, as compared with a high of one new species per 1.4 million years for sauropodomorphs and a low of one per 5.6 million years for stegosaurs. By this reckoning, ceratopsians had relatively high rates of speciation.

To estimate the total diversity of a group, however, we turn to ghost lineages. For ceratopsians, MDT values range from 0 to nearly 30 million years, with an average of just over

of Argentina, reliably dated to about 230 Ma. Indeed, phylogenetic and biogeographical perspectives point to South America as the cradle of Dinosauria (see Chapter 15).

What we can be sure of is that Late Triassic terrestrial vertebrate faunas were not dinosaur-dominated; rather they were an eclectic mixture, including therapsids (advanced synapsids; Figure 13.3), Earth-bound archosaurs (Figure 13.4), primitive turtles (Figure 13.5), some crocodile-like amphibians called temnospondyls (Figure 13.6), and pterosaurs (Figure 13.7). Oh yes, and the very earliest mammals, tiny, shrew-sized, insectivorous creatures, were present (Figure 13.3d). As it turned out, their appearance on Earth was approximately coincident with – or even slightly preceded – that of dinosaurs.

Contintental distributions and the Late Triassic fauna What kinds of evolutionary forces might have been driving the distinctive Late Triassic faunas? The very distributions of the continents likely play a role in the composition of global faunas.

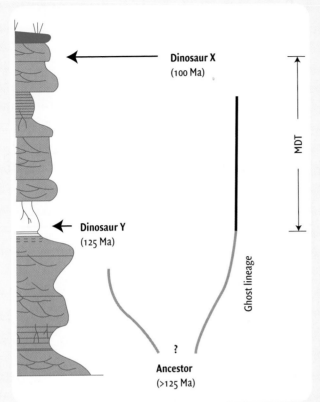

Figure B13.2.1. Ghost lineages and MDTs. Dinosaur X and Dinosaur Y are preserved 25 million years apart. If they are closely related, they both share a common ancestor that is at least as old as Dinosaur Y. Thus an estimate of the minimum divergence time (MDT) of the two lineages is as old as, or older than, Dinosaur Y (125 Ma). The record of that divergence is unpreserved and is therefore called a ghost lineage (curved gray lines on the figure).

5 million years. These are among the smallest MDTs for all Dinosauria, which suggests that the fossil record of this group is comparatively pretty well represented. Furthermore, actual ceratopsian species counts are nearly 70% of the total after ghost lineages have been added to the diversity total. On these measures, ceratopsians do indeed have one of the best records of all of the major dinosaur groups.

Other ways

Sophisticated statistical treatments have opened up other ways to count dinosaurs as well. Dodson (1990) used a published compilation, and simply counted. Fastovsky *et al.* (2004) used an updated version of the compilation used by Dodson (1990), and applied a statistical technique called rarefaction to the data. This technique allowed them to compare different sized samples to determine whether the diversity of dinosaurs actually changed through time, or whether just the samples varied because of preservation.

Another interesting statistical approach was applied by Wang and Dodson (2006), who developed a method for estimating the number of fossils for particular groups *that have yet to be discovered*! Here they introduced a metric called "coverage," which statistically assesses exactly how closely the *known diversity* from a locality (that is, what has been collected) conforms to its *actual diversity* (that is, a complete inventory of what theoretically ought to be preserved in the locality). Using the coverage metric, Wang and Dodson were able to take the total number of currently known dinosaur genera – 527 genera as of the year 2006 – and estimate the total number of dinosaur genera that ever existed: approximately 1,850 genera.

Consider this example from the modern world: the large herbivore fauna of Africa is rather different from that of North America. And both differ from that of India. There are no physical connections among these continents that would allow the fauna of one to spread to the other. Each of these faunas – in fact, the ecosystems of which they are a part – has developed in relative isolation, and is therefore distinct. This type of distinctness is called endemism. A region that is populated by distinct faunas unique to it is said to show *high endemism*. High endemism is caused by evolution on widely separated continents, because there is no opportunity for faunal interchange.

Aternatively, if faunas of two continents appear very similar to each other, then it is likely that some land connection was present to allow the fauna of one continent to disperse to the other continent. Thus we can imagine a region characterized by *low endemism*; because the continents are closely allied with each other, and there are extensive opportunities for faunal interchange.

Figure 13.1. The great historical pageant of dinosaurs through time. The paleoenvironments of each time interval – and the dinosaurs that populated them – all have different qualities that characterized each successive ecosystem.

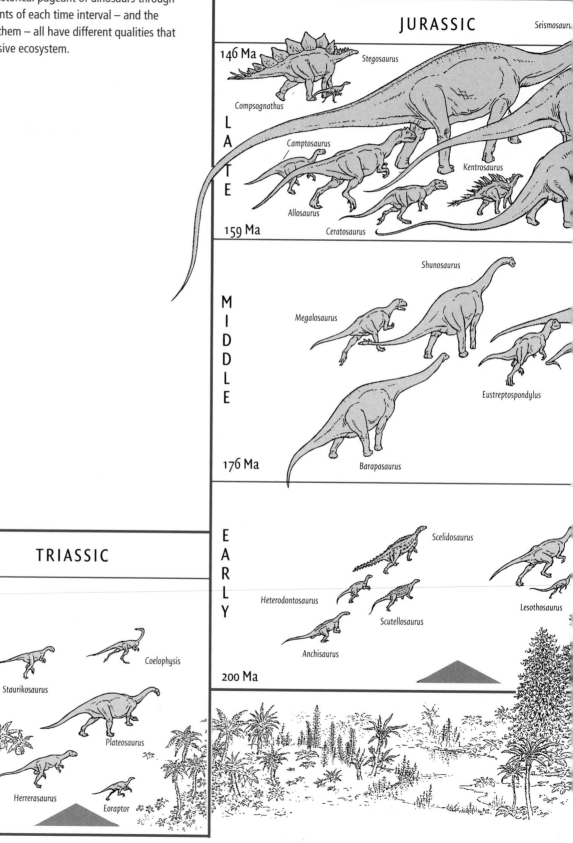

Figure 13.4. Assorted primitive archosaurs. (a) A phytosaur (*Rutiodon*); (b) an aetosaur (*Stagonolepis*); (c) a rauisuchian (*Postosuchus*); and (d) a primitive crocodile (*Protosuchus*).

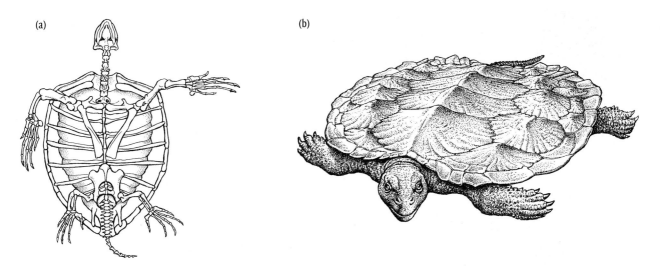

Figure 13.5. A tale of two turtles. (a) The skeleton of a modern specimen "turned turtle" with the shell bones removed; (b) the primitive Triassic turtle *Proganochelys*.

Figure 13.6. A temnospondyl (an archaic amphibian) grabbing a snack.

Figure 13.7. The primitive rhamphorynchid pterosaur *Dimorphodon.*

It is now clear that, during the Triassic and Early Jurassic, global terrestrial vertebrate faunas were characterized by unusually low endemism. The supercontinent of Pangaea still existed during this time, and land connections were more or less continuous among all of today's continents (see Figure 2.5). The interesting mixtures of faunas outlined here look similar on a global scale during these time intervals. While Pangaea remained united, land connections existed and endemism was low. Here, then, is an excellent example of large-scale, non-biological events driving and modifying large-scale patterns of biological evolution.

Jurassic (200–146 Ma)

The Early Jurassic (200–176 Ma) was the first time on Earth when dinosaurs truly began to dominate terrestrial vertebrate faunas. Many of the players in the terrestrial game were now dinosaurs, although they continued to share the limelight with some relict non-amniotes (see Chapter 4), a few of the very highly derived, mammal-like therapsids (including some puny mammals), turtles, pterosaurs, and the newly evolved crocodilians. Interestingly, the Early Jurassic faunas retained some of the low endemism that characterized the Late Triassic world. The unzipping of Pangaea was in its very earliest stages, and it had not gone on so long that the fragmentation of the continents was yet reflected through increased global endemism. *That* was to await the Late Jurassic.

The Middle Jurassic (176–161 Ma) has historically been an enigmatic time in the history of terrestrial vertebrates. As noted earlier, Middle Jurassic terrestrial sediments are quite uncommon. When we look at the total diversity of tetrapods through time (Box 13.1), the curve all but bottoms out during the Middle Jurassic. Did vertebrates undergo massive extinctions at the end of the Early Jurassic? Probably not. More likely the curve is simply reflecting the serendipitous absence of terrestrial Middle Jurassic sediments on Earth. Without a good sedimentary record to preserve them, we can have little knowledge of the faunas that came and went during that time interval.

Regardless, the Middle Jurassic must have been an important time in the history of dinosaurs. With the dismemberment of Pangaea well underway by this time, dinosaurs had diversified, and endemism was on the rise. Many of the non-dinosaurian tetrapods that characterized earlier faunas – advanced therapsids, for example – were largely out of the picture. The insignificant exception to this, of course, were mammals, hanging on by the skin of their multi-cusped, tightly occluding teeth. The Middle Jurassic must have been a kind of pivot point in the history of dinosaurs, because it was then that most of the major dinosaur groups – sauropods, large theropods, thyreophorans, and ornithopods – assumed their familiar forms and consolidated their hold on terrestrial ecosystems. It's a shame that we cannot know more of this crucial time.

By the Late Jurassic (161–146 Ma; see Figure 2.6), global climates had stabilized and were generally warmer and more equable (less seasonal) than they presently are (see Chapter 2). Polar ice, if present, was reduced. Sea levels were higher than today. Dinosaur faunas were more endemic than ever before.

The Late Jurassic has been called the Golden Age of Dinosaurs.[1] Many of the dinosaurs that we know and love were Late Jurassic in age. Somehow that special Late Jurassic blend of equable climates, small brains, and massive size epitomizes dinosaur stereotypes and exerts a fascination on humans. Many *were* large – gigantic sauropods (*Brachiosaurus*, *Diplodocus*, *Camarasaurus*, among others) as well as theropods that reached upward of 16 m – but many were not (for example, *Compsognathus*). It was during the Late Jurassic that the first known "bird" (*Archaeopteryx*) appeared. Moreover, this was the time of stegosaurs, ornithopods, and even a few ankylosaurs. By Late Jurassic time, dinosaurs had consolidated their dominance of terrestrial vertebrate faunas.

---

[1] If only because Late Cretaceous dinosaurs weren't fully appreciated in the late 1800s when the expression "dinosaur" was coined!

Cretaceous (146–65.5 Ma)

The Early Cretaceous (146–100 Ma) was a time of enhanced global tectonic activity. With this came increased continental separation, as well as greater amounts of CO_2 in the atmosphere, producing "greenhouse" climates. Climates from the Early through mid Cretaceous (to about 96 Ma) were therefore warmer and more equable than today (see Chapter 2).

Who enjoyed these balmy conditions? Representatives of groups including all of our old friends from earlier times as well as a number of new groups of dinosaurs made their appearance. The Early Cretaceous marks the rise of the largest representatives of ornithopods. Ankylosaurs also became a significant presence among herbivores of the Early Cretaceous times, as did the earliest ceratopsians.

Moreover, the balance of the faunas seems to have changed. During the Late Jurassic, sauropods and stegosaurs were the major large herbivores, with ornithopods represented primarily by smaller members of the group. Now, in the Early Cretaceous (and, in fact, throughout the Cretaceous), ornithopods first make their mark. Sauropods and stegosaurs were still present, but the significance of these groups, particularly stegosaurs, seems to have been greatly reduced. Was the spectacular Cretaceous ascendency of ornithopods due to the feeding innovations developed by the group? The parallel success of ceratopsians in Late Cretaceous time, and the independent invention by that group of similar feeding innovations, suggest that sophisticated chewing didn't hurt. Then, too, the Early Cretaceous witnessed a revolution in carnivorous theropods, most notably, the deinonychosaurs of both North America and Asia.

During the Late Cretaceous (100–65.5 Ma; see Figure 2.7), never before in their history had Dinosauria been so diverse, so numerous, and so incredible. Something happened; there was a bump in diversity without parallel in dinosaur history. The Late Cretaceous boasted the beefiest terrestrial carnivores in the history of the world (tyrannosaurids), a host of sickle-clawed brainy (*and* brawny) killers worthy of any nightmare, and herds of horned herbivores, honking hadrosaurids, ankylosaurs, and dome-heads.

Climate seems not to have affected diversity. In fact, although diversity increased, from the mid-Cretaceous time onward, seasonality gently increased. This occurred at the same time as a marine regression, which undoubtedly played a role in the destabilization of climate. At the very end of the Mesozoic, there is no evidence for a sudden drop in temperatures, or any significant modification of climate; there was, however, a temporary increase in global temperatures just before the Cretaceous–Tertiary boundary. This has been attributed to the effects of Deccan volcanism (see Chapter 16).

Endemism Southern continents tended to maintain the veteran Old Guard: a large variety of sauropods, some ornithopods and ankylosaurs, and, in South America, the abelisaurid theropods. In northern continents, however, new, very different faunas appeared. Among herbivores, sauropods were still present, although rare at higher latitudes. Lots of new creatures roamed, including potentially migrating herds of pachycephalosaurs, ceratopsids, and hadrosaurids.

Finally there is the magnificent diversity of Late Cretaceous theropods. Nothing shaped quite like tyrannosaurids had ever been seen, or has existed since. Yet, the Late Cretaceous theropod story might be better told in the diversity of smaller forms: oviraptorosaurs, alvarezsaurids, dromaeosaurids, troodontids, and therizinosaurs.

Across the Bering Straits? North America and Asia share a rich Late Cretaceous record, including ceratopsians, tyrannosaurids, and ornithomimosaurs. Because of this, there may have been multiple migrations of herds of dinosaurs across a Bering land bridge throughout much of the Late Cretaceous (see Figure 6.31), just as humans are thought to have migrated to North America from Asia many tens of millions of years later.

Out with a whimper or a bang?

And then, in the earliest Cenozoic, it was over. Just like that. While one of the great enigmas of dinosaur paleontology has historically been how the animals went extinct, the problem has begun to yield to concentrated study over the past 30 years. We'll save that story for Chapter 16.

After the ball is over With the end of the Cretaceous, non-avian dinosaurs disappeared from Earth forever, and it definitely *was* the end of an Era. Mammals, well entrenched as the dominant terrestrial vertebrates in the Cenozoic, would be no more likely to give up their place in Tertiary ecosystems to dinosaurs than dinosaurs had been likely throughout the 163 million years of their incumbency to give up *their* place to mammals!

Or so the story is told by mammals (us!). But there is one small point for consideration: in today's world, some 65 million years after the dinosaurs supposedly became extinct, there are just under 5,500 species of mammals. By contrast, there are somewhere around 10,000 species of birds alive today. Can it really be said that we've left the Age of Dinosaurs and entered the Age of Mammals? A less-biased observer than a mammal might conclude that we're still very much within the Age of Dinosaurs!

Plants and dinosaurian herbivores

As in most extant terrestrial mammalian communities, the majority of dinosaurs were herbivorous. If dinosaurs were numerous enough, and their impact on terrestrial ecosystems was important enough, there ought to be some relationship between herbivorous dinosaur evolution and plants.

Plants

Most paleobotanists – people who study extinct plants – recognize two major groupings of Mesozoic plants. The first is a non-monophyletic cluster of plants including ferns, lycopods, and sphenopsids (Figure 13.8). All of these plants tend to be low-growing and primitive, but, like most land plants, they are vascular; that is, they possess specialized tissues that conduct water and nutrients throughout the plant.

The second major grouping of plants consists of gymnosperms and angiosperms. Together these two groups are united by the diagnostic character of possessing a seed (Figure 13.9). Seeds are ultimately nutrient-bearing pods apparently developed for the dissemination of gametes. Gymnosperms are today best known as pines and cypress, and a lesser-known but Mesozoic-ly important group known as cycadophytes: plants with a pineapple-like stem and bunches of leaves springing out of their tops. Angiosperms today consist of magnolias, maples, grasses, roses, and orchids, among many other groups (Figure 13.9).

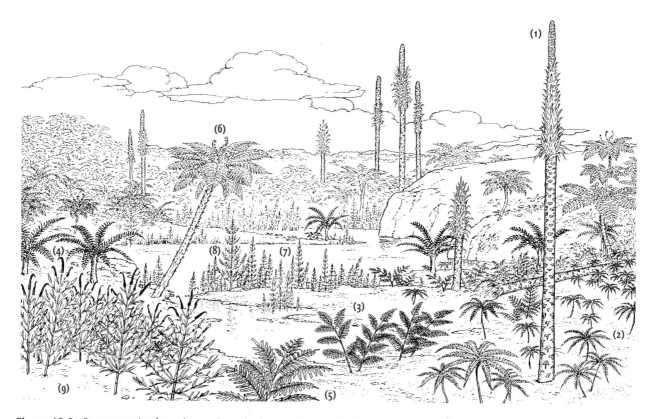

Figure 13.8. Representative ferns, lycopods, and sphenopsids from the Mesozoic. Club moss: (1) *Pleuromeia* (Early Triassic). Ferns: (2) *Matonidium* (Jurassic–Cretaceous); (3) *Onychiopsis* (Jurassic–Early Cretaceous); (4) *Anomopteris* (Middle–Late Triassic); (5) Osmundaceae (Late Paleozoic–Recent). (6) Tree fern (Jurassic). Sphenopsids: (7) *Equisetum* (Late Paleozoic–Recent); (8) *Neocalamites* (Triassic–Lower Jurassic); (9) *Schizoneura* (Late Paleozoic–Jurassic).

Several qualities distinguish these plant groups. In general, Mesozoic gymnosperms tended to be of three types: conifers, cycadophytes, and ginkgoes. Conifers – epitomized, for example, by pines – were very tall and woody plants. They had relatively little nutritive value pound for pound, possessing coarse thick bark and cellulose-rich leaves. The modern representatives of these plants tend to secrete a variety of ill-tasting or poisonous compounds as a strategy to discourage their consumption; there is no reason to suppose that their Mesozoic counterparts were any different.

Cycadophytes, on the other hand, tended to be fleshier and softer, with perhaps more nutritive value. Ginkgoes would also have been plants available for dinosaur consumption, and circumstantial evidence suggests they too were eaten by Mesozoic herbivorous dinosaurs (Figure 13.9).

Flowering plants evolved an entirely different approach to life from gymnosperms. Far from discouraging herbivores from consuming them, they evolved a variety of strategies to actively court their consumption by herbivores: bright tasty flowers, fruits with tough seeds that can survive a a trip through a digestive tract. Consumption by herbivores in the case of angiosperms appears to be a strategy for seed dispersal, not the destruction of the plant.

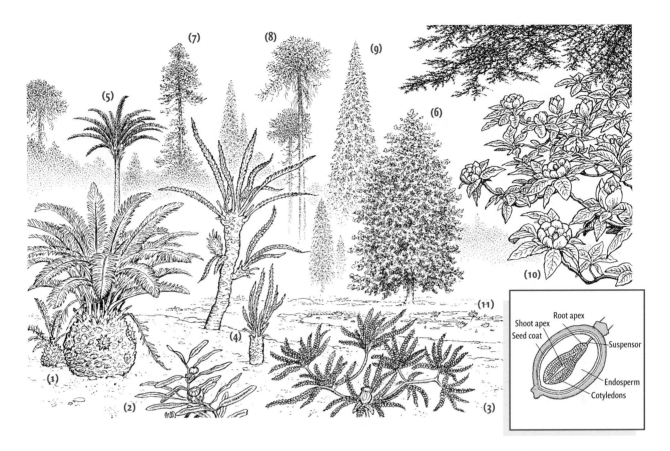

Figure 13.9. Representative cycads, ginkgoes, gymnosperms, and angiosperms from the Mesozoic. (1) Cycadeoids or bennettitaleans, as they are sometimes called (Triassic–Cretaceous); (2 and 5) *Williamsoniella* spp. (Triassic–Jurassic); (3) *Wielandiella* (Jurassic); (4) *Williamsoniella sewardiana* (Jurassic). Ginkgo: (6) *Ginkgoites* (Triassic–Recent). Conifers: (7) *Sequoia* (mid Cretaceous–Recent); (8) *Araucaria* (Late Triassic–Recent); (9) *Pagiophyllum* (Triassic–Cretaceous). Angiosperms: (10) Magnoliaceae (magnolias; still small-flowered in the early days of their appearance on Earth; Cretaceous?–Recent); (11) Nymphaeaceae (water lilies; Late Cretaceous–Recent). *Inset:* Seed (dicot) in cross-section. The cotyledons, shoot apex, root apex, and suspensor are all parts of the embryonic plant. The endosperm is a food source for the embryo as it develops, and the seed coat protects the embryo and its food source.

Dinosaurs and plants

Our analysis is built around Figure 13.10, which compares the record of Late Triassic through Late Cretaceous plant diversity with that of dinosaurian herbivores. The lower part of the figure gives approximations of the global composition of plants through the time of the dinosaurs. The upper part of the figure is divided into various groups of herbivorous dinosaurs.

Plants In terms of plants, Figure 13.10 shows some key patterns. Lycopods, seed ferns, sphenopsids, and ferns decrease in global abundance during the Late Triassic interval. From then until the end of the Mesozoic, they constitute a roughly constant proportion of the world's floras. Not so with the gymnosperms, which dramatically increase their proportion of the total global flora during the Late Triassic. And it is clear that much of that increase is taken up by conifers, which consitute around 50% of the world's total floras throughout the rest of the Mesozoic.

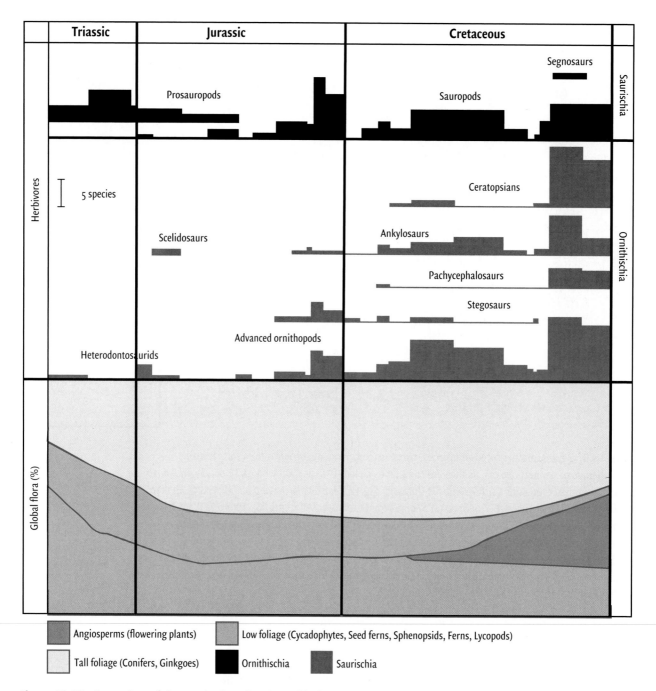

Figure 13.10. Comparison of changes in plant diversity and herbivorous dinosaur diversity during the Late Triassic through Late Cretaceous time interval. Upper part of the diagram shows diversity of major groups of herbivorous dinosaurs through time. Comparison between this diagram and that shown in Figure 13.2 suggests that one of the most important things driving dinosaur evolution and diversity was the development of new (or improved) ways to exploit the environments in which they lived (see the text).

Our best guess is that angiosperms first evolved in the very early part of the Cretaceous; however, it was during mid-Cretaceous times that they underwent a tremendous evolutionary burst. The uniquely efficient angiosperm seed dispersal mechanisms afforded by flowers were (and are) unparalleled in the botanical world, and consequently flowering plants have blossomed as no other group of plants has.

Co-evolution What is the relationship between dinosaurian herbivores and plants? It cannot be purely by chance that the rise of tall coniferous forests is coincident with the appearance on Earth of the world's first tall herbivores: prosauropods (and later sauropods). Here we see the mark of co-evolution, the evolution of one group affecting – and even effecting – the evolution of another. In this case, it's plants and dinosaurs: were those tall prosauropods favored by natural selection that could take advantage of comparatively succulent leaves at the tops of conifers? Alternatively, were conifers that were particularly tall favored by natural selection in response to the increasing height of prosauropods? Which is cause and which is effect is something we'll likely never know.

The figure also reveals another compelling relationship. The rise of the angiosperms occurs at approximately the same time as several major radiations of dinosaur groups. Were these groups – ceratopsians, pachycephalosaurs, hadrosaurids, and late-evolved ankylosaurs – groups that somehow took advantage of angiosperms as a food source and diversified? Is this a clue to what these dinosaurs were eating (Box 13.3)?

It appears that, during the middle of the Mesozoic, few vertebrates fed very *selectively* upon the relatively slow-growing conifers, cycads, and ginkgoes that formed the majority of terrestrial floras. Instead, it has been suggested, dinosaur feeding consisted of low browsing and was rather generalized (similar to the way a lawn-mower "grazes" over whatever is in its path). Because so many of these Mesozoic herbivores were also very large and may have lived in large herds, they likely cleared expansive areas, trampling, mangling, uprooting, and otherwise disturbing areas that otherwise might be colonized by plants.

Such low-level, generalized feeding and disturbances of habitats tended to emphasize fast growth in plants, but discouraged the establishment of seed-dispersal relationships between plants and animals. Thus the picture of Mesozoic plant–herbivore interactions appears to be one in which (a) plants produced vast quantities of offspring to ensure the survival of the family line into the next generation and (b) herbivores took advantage of the rapidly and abundantly reproducing resource base to maintain their large populations of large individuals. Plant–herbivore co-evolution during the Mesozoic appears to have been based on habitat disturbance, generalized feeding, and rapid growth and turnover among plants.

To date, "mummified" remains[2] of hadrosaurids (*Edmontosaurus* and *Corythosaurus*) do not show the remains of angiosperms in the digestive tract, but rather the remains of coniferous plants. Late Cretaceous coprolites, reliably attributed by size to either ceratopsids or hadrosaurids, contained conifer fragments as well. If angiosperms were fueling this dinosaur radiation, where are the angiosperm pieces that we might hope to find?

Yet dinosaur chewing efficiency increased markedly through the latter part of the Mesozoic. This is not to say that non-chewing dinosaurs were in a state of decline; as Figure 13.10 shows, sauropods – for whom chewing was a minimalist artform – were successful throughout the Cretaceous. Moreover, animals that indulged in rudimentary chewing – such as ankylosaurs and pachycephalosaurs – underwent strong evolutionary bursts during the latter part of the Cretaceous. Still, ceratopsids and hadrosaurids – groups that elevated chewing to new heights - are characteristic of the Late Cretaceous radiation. Did advanced chewing mechanisms allow hadrosaurids and ceratopsids to take advantage of food resources not heretofore available to other dinosaurs?

---

[2] These rare and spectacular fossils are not truly mummified, which would mean that their original tissue is preserved (as is the case with the famous mummies of ancient Egypt. Rather, in this case, the original animal dried out (like a mummy) and was buried by sediment intact. Then, post-burial, all the original organic tissue was replaced by minerals resulting in, as discussed in Chapter 1, a perfect natural forgery of the original desiccated carcass (including soft tissue such as skin impressions, tendons, and stomach contents).

Box 13.3 Dinosaurs invent flowering plants

In his popular book *Dinosaur Heresies*, R. T. Bakker proposed that dinosaurs "invented" flowering plants. The germ behind Bakker's hypothesis is that Late Jurassic herbivores, epitomized by sauropods, were essentially high-browsers, while Cretaceous herbivores, epitomized by ornithopods, ankylosaurs, and ceratopsians, were largely low-browsers. Bakker argued that Cretaceous low-browsers put tremendous selective pressures on existing plants, so that survival could occur only in those plants that could disseminate quickly, grow quickly, and reproduce quickly. Angiosperms, he argued, are uniquely equipped with those capabilities. In his scenario, Bakker has Cretaceous low-browsing dinosaurs eating virtually all the low shrubbery, and plants responding by developing a means by which animals simply couldn't keep up with the growth, reproduction, and dissemination of the plants.

How likely is this? We are not sure. Troubling, of course, is the strongly North American and Asian cast of this hypothesis; southern latitude faunas seem to have had just the faunal compositions that Bakker claims would *not* have put intense selective pressure on contemporary low-growth floras. Yet, angiosperms were radiating world-wide by Late Cretaceous time. Still, what is significant about Bakker's hypothesis is that in it, he, as well as others, recognizes and attempts to define the co-evolution between dinosaurs and plants.

We would be remiss if we did not mention the sloth-like therizinosaurs – that strange Late Cretaceous theropod foray into herbivory (see Figure 9.31). By the chewing standards of their ornithopod brethren, these animals were mighty primitive. Yet, was there something about the rise of angiosperms that fueled their unusual radiation too?

Indeed, the Late Cretaceous could also lay claim to being the "Golden Age of Dinosaurs"; as most were herbivorous, it seems that they themselves flowered during the flowering of the Mesozoic.

Summary

Here we look at the overall sweep of non-avian dinosaur evolution. Factoring in time intervals of a poor geological record, in which preservation is artifically low, dinosaurs as a group increased markedly in number and diversity, particularly during the Late Jurassic-through-latest Cretaceous time interval. This increase is attributable to ceratopsian and ornithopod herbivores, and theropods.

The global pattern of dinosaur evolution from the Late Triassic to the Late Cretaceous is one of generally increasing endemism, likely attributable to the increasing separation of continental masses. Late Triassic and Early Jurassic dinosaur faunas shared their terrestrial world with a variety of other vertebrates; and global vertebrate faunas were relatively homogeneous. Distinct among all the Late Triassic and Early Jurassic vertebrates, however, dinosaurian herbivores were the first to be able to reach, and thus add to their diets, tall foliage.

By Middle Jurassic time, dinosaurs likely consolidated their dominance in the terrestrial realm, even though terrestrial deposits from this time interval are comparatively rare. This fact is especially unfortunate for understanding the details of dinosaur evolution, since many of the groups that became so abundant and diverse in the Cretaceous had their roots in the Middle Jurassic. The Late Jurassic has been called the "Golden Age of Dinosaurs," with the abundance of many familiar forms including very large theropods and sauropods.

The Cretaceous was a truly astounding time in dinosaur evolution. Aside from the wholesale dominance of new forms (in particular, ornithopods and ceratopsians, as well as a wide range of theropods), many of the spectacular adaptations that we've seen, such as advanced chewing, evolved in the Cretaceous. A driving force in all this evolutionary ferment may have been the rise of flowering plants; yet what we know of dinosaur diets suggests that the fibrous gymnosperms constituted the bulk of the nutrition. Processing such plants may have been the driving force in the development of sophisticated modes of chewing. It can certainly be said that, as plants evolved effective methods for dispersal and colonization, dinosaurs apparently hitched a ride, increasing markedly in number and diversity as they took advantage of the radiation of vascular plants.

The end of the Cretaceous (explored in greater detail in Chapter 15) was of course the end of non-avian dinosaurs. Overall diversity trends show no gradual decrease from some previous high point; rather, non-avian dinosaurs increase in diversity throughout the Mesozoic and then abruptly, at 65.5 Ma, disappear from the fossil record.

Selected readings

Bakker, R. T. 1986. *The Dinosaur Heresies*. William Morrow and Company, New York, 481pp.

Chin, K. 1997. What did dinosaurs eat? Coprolites and other direct evidence of dinosaur diets. In Farlow, J. and Brett-Surman, M. K. (eds.), *The Complete Dinosaur*. Indiana University Press, Bloomington, IN, pp. 371–382.

Dodson, P. 1990. Counting dinosaurs. *Proceedings of the National Academy of Sciences*, **87**, 7608–7612.

Fastovsky, D. E. 2000. Dinosaur architectural adaptations for a gymnosperm-dominated world. In Gastaldo, R. A. and DiMichele, W. A. (eds.), *Phanerozic Terrestrial Ecosystems*. The Paleontological Society Papers, vol. **6**, pp. 183–207.

Fastovsky, D. E., Huang, Y., Hsu, J., Martin-McNaughton, J., Sheehan, P. M. and Weishampel, D. B. 2004. The shape of Mesozoic dinosaur richness. *Geology*, **32**, 877–880.

Tiffney, B. H. 1989. Plant life in the age of dinosaurs. *Short Courses in Paleontology*, **2**, 34–47.

Tiffney, B. H. 1997. Land plants as food and habitat in the age of dinosaurs. In Farlow, J. and Brett-Surman, M. K. (eds.), *The Complete Dinosaur*. Indiana University Press, Bloomington, IN, pp. 352–370.

Wang, S. C. and Dodson, P. 2006. Estimating the diversity of dinosaurs. *Proceedings of the National Academy of Sciences*, **103**, 13601–13605.

Weishampel, D. B. and Norman, D. B. 1989. Vertebrate herbivory in the Mesozoic: jaws, plants, and evolutionary metrics. *Geological Society of America Special Paper* no. **238**, pp. 87–100.

Wing, S. and Tiffney, B. H. 1987. The reciprocal interaction of angiosperm evolution and tetrapod herbivory. *Review of Palaeobotany and Palynology*, **50**, 179–210.

Topic questions

1   What is meant by the words conifer, co-evolution, cycadophyte, angiosperm, diversity, and endemism?

2   What is the general pattern of dinosaur diversity through time? When were dinosaurs at their most diverse? When were they at their least diverse?

3   On what continents are the most dinosaurs found? Why might this be?

4   Describe climatic conditions throughout the Mesozoic.

5   Describe the degree to which the continents were separated throughout the Mesozoic. When were the continents most like they are now? When were they least like they are now?

6   What is the relationship between endemism and the distribution of continents? Why is this so?

7   Describe the general outlines of plant evolution through the Mesozoic.

8   What is the general relationship between dinosaur diversity and plant evolution through the Mesozoic?

9   What is the relationship between herbivore chewing specializations and plant diversity in the Mesozoic?

10 Drawing on material from other chapters, can you think of highly evolved behaviors that appear to be related to plant diversity?

11 What kinds of evolutionary changes characterize Theropoda in the context of plant diversity and increased ornithischian diversity in the Mesozoic?

CHAPTER 14

A HISTORY
OF PALEONTOLOGY
THROUGH IDEAS

CHAPTER OBJECTIVES

- Outline the history of paleontological thought

- Understand relationships to larger intellectual movements

- Introduce the stories of some famous paleontologists

- Provide a historical context for the subjects discussed in this book

The idea of ideas

Ernest Rutherford[1] once infamously remarked, "In science there is only physics; all the rest is stamp collecting." And what could be more like stamp collecting than paleontology, that endless litany of names, dates, and locations?

Paleontology *would* be stamp collecting, if it weren't for the ideas – the creativity – that grew with the field. The history of paleontology, therefore, is really the history of the ideas that forged the discipline.

In the beginning

Tradition usually identifies the beginning of dinosaur paleontology as 1822, when Mary Ann Mantell, wife of English physician Gideon Mantell, found large teeth along a Sussex country lane while her husband was busily tending patients (Figure 14.1). Gideon was something of a fossil collector, and the discovery baffled him, because the teeth looked very much like those of the living herbivorous lizard *Iguana*, but were ominously much, much bigger (Figure 14.2).

But of course the Mantells weren't the first humans to see dinosaur fossils; however, they may have been the first to interpret them meaningfully in a Western scientific context. Fossils of all types must have been remarked upon for as long as there have been humans.

For example, Adrienne Mayor, classical folklorist and historian of science, has reconstructed the origin of the legend of griffins, sharp-beaked, winged, four-legged creatures whose mythology was known across all of Europe and Asia (Figure 14.3). Her idea is that traders along ancient gold-trading caravan routes stretching from Europe through central Asia encountered abundant, beautifully preserved fossils of *Protoceratops* (see Chapter 6), whose strange (to

Figure 14.1. Gideon Mantell (1790–1852), the man who first recognized non-avian dinosaurs for what they were.

them) combination of beak, frill, and limbs were explained as the mythical griffin's beak, wings, and legs. The richness of the Asian deposits was revealed more than a thousand years later in the American Museum of Natural History's Central Asiatic fossil Expeditions of the 1920s (Box 14.1) and ancient traders, Mayor suggests, could hardly have failed to notice the bones of strange, articulated, bird-like creatures emerging from weathering desert sands. Mayor hypothesizes that the griffin legend spread from central Asia along trade routes to Europe.

---

[1] Nobel Prize-winning New Zealand physicist, 1871–1937; pioneer in radiation and radioactivity.

Figure 14.2. Mantell's *Iguanodon* teeth.

Figure 14.3. A griffin.

Seventeenth and eighteenth centuries

The griffin legend, then, is one explanation, outside of a scientific context, for the observation of dinosaur material. But, with a few exceptions, the birth of the Western scientific tradition is generally reckoned to have occurred in association with the Enlightenment, the seventeenth- and eighteenth-century intellectual revolution devoted to the ability of reason and observations to reveal truth. The Enlightenment brought with it a number of scientific conclusions important to our story, including:

- The Earth is not static, that is, it has changed through time.
- The Earth of is of great antiquity (its age was not well understood until the middle of the twentieth century).

Box 14.1 Indiana Jones and the Central Asiatic Expedition of the American Museum of Natural History

He stands in the middle of the remote, rugged, Mongolian desert: high leather riding boots, riding pants, broad-brimmed felt hat, leather-holstered sidearm hanging from a glittering ammunition belt. He carries a rifle and knows how to use it. Nobody else dresses like him, but then nobody else is the leader of the American Museum's Central Asiatic Expeditions to Mongolia (a place which, at the time of the expeditions, the 1920s, could have been the Moon). He is Roy Chapman Andrews, who 50 years later will be the inspiration, it is most plausibly rumored, for Indiana Jones (Figure B14.1.1).

Andrews always knew that he was a man with a destiny. Although he began his career at the American Museum of Natural History (AMNH) modestly (he scrubbed floors), training in mammalogy (an M.A.), sheer will, charisma, and a *very* good idea carried him the distance. He had traveled extensively, spoke several Asian languages more or less fluently (at a time when very few Westerners did), and had fabulous contacts in Beijing (then called Peking).

His idea was simple: to run an expedition to what was then known as Outer Mongolia and to see what he could see. Andrews' timing was superb: the Director of the AMNH, the powerful Henry Fairfield Osborn, had concluded that the cradle of humanity was located in Outer Mongolia, and so Andrews was effectively offering Osborn the opportunity to prove his thesis right (the possibility that Osborn could be *wrong* did not seem to be of concern). The logistics of the expedition were extravagant: Dodge cars, resupplied by a caravan of camels, would bear the brunt of the expedition (Figure 1.7). The expedition itself would consist of a range of earth scientists – paleontologists, geologists, and geographers – to explore the Gobi Desert, the huge desert that forms the vast southern section of Mongolia (then called "Outer" Mongolia, as if to emphasize its remoteness) and northern China.

The journey was not without its risks. The Gobi Desert is a place of temperature extremes, beset by relentless strong winds. Politically, at the time, the region was in an uproar. China, the base of operations, was torn by civil strife. And in 1922, the year of the first of three expeditions, a revolution shook Mongolia. Moreover, only one fossil, a rhinoceros tooth, had ever been found in Mongolia.

As it turned out, the Central Asiatic Expeditions were an unqualified success. Although Osborn's theory was not supported, Andrews brought back a wealth of fossils, including abundant dinosaur material, that made Osborn's error easy to forget. Among the most famous dinosaur finds of his

Figure B14.1.1. Roy Chapman Andrews (1884–1960), explorer, adventurer, and leader of what he called "The New Conquest of Central Asia."

expedition, for example, were *Protoceratops* (the species name of this famous dinosaur is *andrewsi*) and eggs – the first time that dinosaur eggs were ever found. Other incredible finds included *Velociraptor* and a group of tiny Mesozoic mammals (still the rarest of the rare). Andrews and his field parties also found the largest land mammal and the largest carnivorous land mammal of all time (both Cenozoic in age). Other fossils were obtained whose significance was not completely understood. For example, it was only in 1992 that a specimen of *Mononykus*, collected by Andrews' scientists in the 1920s, was finally correctly identified. All in all, it was quite a haul.

Andrews and his parties survived the Mongolian revolution of 1922, but eventually the expeditions came to an end when the political situation in China became too unstable and travel too dangerous. Andrews, himself, eventually went on to get the job held much earlier by Osborn: that is, Director of the AMNH. He assured his place in history, however, by leading the Central Asiatic Expeditions.

- The sequence of the rock record reveals the history of the Earth.
- Fossils are the remains of once-living organisms.
- Organisms on Earth were not static, they too had clearly changed through time, some in startling ways.

It was during this time that we have the earliest description of a dinosaur fossil, in this case the lower end of a theropod thigh (likely *Megalosaurus*) from Oxfordshire, England. As the bone was large, it was interpreted by the Reverend Dr Robert Plot in 1677 to have been the end of a thigh bone of an antediluvian (pre-Biblical Flood) giant – man or beast (Figure 14.4).[2]

The nineteenth century through the mid-twentieth century

Mantell's discovery turned him into the world's first true dinosaur junkie. While there is no space to recount the details, his is the remarkable story of a man consumed by a passion for paleontology so great that he vanquished the skepticism of the greatest anatomist of his time (see Box 14.2), built a museum, and wrote the first description (and guided the first reconstruction) of a dinosaur – *Iguanodon*. It was a passion so all-consuming that it ultimately cost Mantell his livelihood and his marriage.

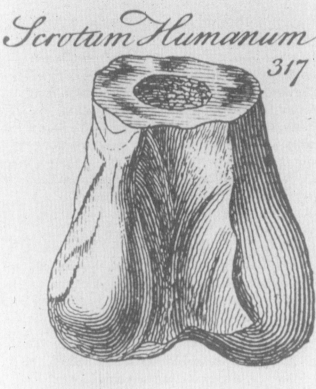

Figure 14.4. Robert Plot's drawing of the lower end of a *Megalosaurus* (?) thigh bone. Dr Plot was the first Professor of Chemistry at the University of Oxford. For explanation of the Latin inscription, see footnote 2.

Dinosaurs in the Victorian Age

Perhaps it had to do with the Victorian penchant for collections and museums, perhaps it was just the novelty of the beasts being uncovered, but Victorian England was dino crazy. In 1824, the natural historian William Buckland (1784–1856) described a jaw fragment with a single recurved, serrated tooth as *Megalosaurus*. This was the first named dinosaur, now known to be a theropod, but at the time Buckland thought it was just a rather large lizard.

By 1842, enough of dinosaurs was known for the rising young English anatomist Richard Owen (Box 14.2) to invent a new term: Dinosauria (*deino* – terrible; *sauros* – lizard). The

---

[2] Stranger still, in 1763, Richard Brooke drew the specimen in a publication on the uses of various natural objects (including fossils) in medicine. The specimen appeared to Brooke to preserve a giant's testicles; hence, his Latin description identified the fossil as "*scrotum humanum.*" It has been suggested that, tongue firmly in cheek, the rule of priority in the Linnaean classification (see Box 4.2) dictates that this first dinosaur bone should be referred to a genus *Scrotum*, species *humanum*.

Box 14.2 Sir Richard Owen: brilliance and darkness

Richard Owen (Figure B14.2.1) was the dean of natural historians in Victorian times, that iconic age of natural history. In his day, he was among the most powerful and influential scientists in England. His personality was at once brilliant, irascible, politically astute, ruthless, and condescending, and it would not be going too far to call him a liar. He was, to say the least, a man of contradictions.

Owen looked the part. He was tall and gaunt with high cheekbones and, as he grew older, strangely bulging eyes. Cloaked, hands resting gently upon a skull, he looked like he came directly from Central Casting for the part of a Victorian serial killer.

Owen was born in 1804, trained to be a physician, and early on demonstrated a penchant for anatomy. Bill Bryson inimitably describes a memorable event from the life of the young Owen as he copped corpses for dissection:

> Once while carrying the head of a black African sailor [that Owen had severed from the corpse for study], … Owen slipped on a wet cobble and watched in horror as the head bounced away from him down the lane and through the open doorway of a cottage, where it came to rest in the front parlor. What the occupants had to say upon finding an unattached head rolling to a halt at their feet can only be imagined. (Bryson, 2003, p. 87)

By the age of 21, Owen was hired by the Royal College of Surgeons in London to assist in the curation of the Hunterian Collection, a collection of biological oddities and medical curiosities amassed by John Hunter, a famous London surgeon. Hunter's notes had been destroyed in a fire,

Figure B14.2.1. Sir Richard Owen (1804–1892), eventually of the British Museum of Natural History, the brilliant nineteeth-century English anatomist and the father of the term "Dinosauria."

charter members of the group were *Iguanodon* (an ornithopod), *Megalosaurus* (a theropod), and *Hylaeosaurus* (an ankylosaur). Presciently, Owen's initial idea of Dinosauria was that its members were endotherms like mammals and birds, a conclusion based upon the now-discredited idea that Mesozoic air was somehow thinner than modern air: right idea, wrong reasons! It seems that Owen balked at the idea, which had currency in certain circles at that time, that organic evolution (as it was understood before Darwin) was a kind of linear process that ran from quite simple to more complex. Owen thought that by demonstrating that an ancient group of organisms had modern levels of complexity, he would successfully undermine the notion of evolution as it was then understood.

Victorians immortalized their conception of dinosaurs with a variety of images and sculptures. The dinosaurs were reconstructed as large, heavy-set quadrupeds, the most famous of which were created life-sized in plaster and tile by an English sculptor, Benjamin Waterhouse

and so the daunting job was to organize, identify, and catalog disorganized drawers of biological detritus. Owen proved to be particularly adept, using clever inferences and his growing knowledge of comparative anatomy to identify and catalog specimens for which there was no recorded information. As his reputation grew, the medical career receded quietly into the distance.

Instead, he became a lecturer in comparative anatomy, and began to publish scholarly tomes on organisms ranging from the living chambered cephalopod *Nautilus* to the first description of the newly discovered *Archaeopteryx*. It was Owen who first described the exotic South American fossils that Charles Darwin brought back with him from his voyage on the *Beagle*, and, naturally enough, it was Owen who made the connection between the still-fragmental and isolated bits of fossil material that at the time constituted all of Dinosauria.

Owen was undaunted by any anatomy. A single bone fragment from New Zealand led him to the then-outrageous conclusion that flightless, ostrich-like giant birds lived there at one time. He later named the animal *Diornis*, a name that still applies to the large flightless ostrich relatives that populated New Zealand during pre-Columbian times. He described a new genus of ape, first discovered in 1847: *Gorilla*. At the height of his powers, the breadth of his knowledge of comparative anatomy was likely unequaled.

Comparative anatomy, which might have led him to an appreciation of the ideas of his contemporary Charles Darwin, never led him to embrace evolution by natural selection. Instead, he identified forms as divinely created "archetypes," from which came a variety of predetermined variations. But with Darwinian evolution still controversial, thanks in part to Owen's objections, his political star ascended along with his academic star. He gave regular, popular lectures (some attended by members of the royal family) at the Royal College of Surgeons.

But along with the rise of Owen's powers and reputation came a rise in some unfortunate personality quirks. He was, not to put too fine a point upon it, unpleasant. He was arrogant and condescending to presumed inferiors; a remarkably inclusive category. Charles Darwin, famously tolerant, disliked him enough to remark upon the fact in his autobiography. Owen dissembled, claiming for himself honors and positions that he didn't actually hold. He barred talented contemporaries from access to specimens that would have allowed them to carry out their science. Ever jealous of his place in history, he reserved some of his most finely honed vituperation for Gideon Mantell, who had made the mistake of discovering and describing the first dinosaurs, and then actually attempting to claim credit for his own accomplishments! Owen even went so far as to write an anonymous, scathing obituary of Mantell when the man finally had the good grace to die. The occasion of Owen's receiving the Royal Medal – the Royal Society's highest honor – was marred by the discovery that Owen claimed credit for someone else's discovery. Owen's very real, legitimate accomplishments juxtaposed against his personal meaness appear almost pathological.

Eventually, people began to catch on to Owen's meaner side, and his star fell into eclipse. Owen was discredited at both the Zoological Society of London and the Royal Society, and eventually took a job as superintendent of the natural history collections in the British Museum. He had a stunning vision for the collections – that they would be available to any and all who cared to look, and advocated their separation from the rest of the British Museum. He died in 1892. The Natural History Museum, today a public museum much as Owen had envisioned it, finally separated from the British Museum in 1963.

Hawkins (1807–1889), on the occasion of the opening of the Crystal Palace in 1854. The Crystal Palace was the brainchild of Prince Albert, Queen Victoria's Royal Consort, and was a kind of permanent monument to England's scientific and technological prowess. Waterhouse Hawkins posed his bestiary frolicking in a park in the Crystal Palace grounds (Figure 14.5), where, remarkably, they can be seen to this day.[3] It rarely gets more surreal than it did on New Year's Eve, 1853, when Owen and Waterhouse Hawkins hosted a dinner for England's best and brightest *inside* the unfinished sculpture of *Iguanodon* (Figure 14.6).

---

[3] Not so Waterhouse Hawkins' efforts in Central Park, New York. There, the sculptor – or rather, the cost of his project – ran afoul of Tammany Hall, the "Boss" Tweed-led political machine that dominated New York City politics for much of the latter half of the nineteenth century. One spring day in 1871, Waterhouse Hawkins showed up at his studio to find his near-finished sculptures smashed beyond repair. The bits were buried in Central Park, where they are thought to remain to this day.

Box 14.3 Dinosaur wars in the nineteenth century: boxer versus puncher

One of the strangest episodes in the history of paleontology was the extraordinarily nasty and personal rivalry between late nineteenth-century paleontologists Edward Drinker Cope and Othniel Charles Marsh (Figure B14.3.1). In many respects, it was a boxer versus puncher confrontation: the mercurial, brilliant, highly strung Cope versus the steady, plodding, bureaucratic Marsh. Their rivalry resulted in what has been called the "Golden Age of Paleontology," a time when the richness of the dinosaur faunas from western North America first became apparent – when the likes of *Allosaurus*, *Apatosaurus*, and *Stegosaurus* were first uncovered and brought to the world's attention. But the controversy had its down side too. Who were these men, and why were they at each other's throats?

Cope was a prodigy; one of the very few in the history of paleontology. By the age of 18, he had published a paper on salamander classification. By 24, he became a Professor of Zoology at Haverford College, Philadelphia. Blessed with independent means, within four years he had moved into "retirement" (at the grand old age of 28) to be near Cretaceous fossil quarries in New Jersey. He quickly became closely associated with the Philadelphia Academy of Sciences, where he amassed a tremendous collection of fossil bones which he named and rushed into print at a phenomenal rate (during his life he published over 1,400 works). He was capable of tremendous insight, made his share of mistakes, and was girded with the kind of pride that did not admit to errors.

Marsh, nine years older than Cope, was rather the opposite, with the exception that he, too, eventually rushed his discoveries into print almost as fast as he made them (some thought faster) and that he, too, did not dwell upon his mistakes. Marsh's own career started off inauspiciously; with no particular direction, he reasoned that if he performed well at school he could obtain financial support from a rich uncle, George Peabody. This turned out to be perhaps the most significant insight in Marsh's life: Marsh persuaded Peabody to underwrite a natural history museum at Yale (which to this day exists as the Yale Peabody Museum of Natural History), and, while he (Peabody) was at it, an endowed chair for Marsh at the Museum.

The careers of the two paleontologists moved in parallel; Marsh slowly publishing but acquiring prestige and rank, while Cope frenetically published paper after paper. At first, there was no obvious acrimony, but this changed when Marsh apparently hijacked one of Cope's New Jersey collectors right out from under him. Suddenly, the fossils started going to Marsh instead of Cope. Then, in 1870, Cope showed Marsh a reconstruction of a plesiosaur, a long-necked, flippered, marine reptile. The fossil was unusual to say the least, and Cope proclaimed his findings in the *Transactions of the American Philosophical Society*. Marsh detected at least part of the reason why the fossil was so unusual: the head was on the wrong end (the vertebrae were reversed). Moreover, he had the bad manners to point this out. Cope, while admitting no error, attempted to buy up all the copies of the journal. Marsh kept his.

Cope sought revenge in the form of correcting something that Marsh had done. The rivalry ignited, and the battle between the two spilled out into the great western fossil deposits of the Morrison Formation. Both hired collectors to obtain fossils, the collectors ran armed camps (for protection against each other's poaching), and, between about 1870 and 1890, east-bound trains continually ran plaster jackets back to New Haven (Connecticut) and Philadelphia. There Marsh and Cope rushed their discoveries into print, usually with new names. The competition between the two was fierce, as each sought to out-science the other. Discoveries (and replies) were published in newspapers as well as scholarly journals, lending a carnival atmosphere to the debate. Because Philadelphia and New Haven were not that far apart by rail, it was possible for

From 1842 onward, membership of Owen's Dinosauria grew by leaps and bounds. Much of the attention was devoted to basic collecting and description, asking questions like, "What is this creature? A new genus? A species of an existing genus? Maybe even a new family?" This otherwise-healthy penchant for discovery, description, and naming reached absurd levels during the latter half of the century, fueled by the extraordinarily rich fossil beds of the North American West and the very strange competition between Yale's O. C. Marsh and the Philadelphia Academy's E. D. Cope (Box 14.3).

Figure B14.3.1. The two paleontologists responsible for the Great North American Dinosaur Rush of the late nineteenth century. (a) Edward Drinker Cope (1840–1897) of the Philadelphia Academy of Sciences; and (b) Othniel Charles Marsh (1831–1899) of the Yale Peabody Museum of Natural History.

one of the men to hear the other lecture on a new discovery, and then rush home that night and describe it and claim it for himself. Because many of the fossils in their collections were similar, it was easy to do and each accused the other of it.

Both Cope and Marsh eventually aged and, in Cope's case, his private finances dwindled. Moreover, a new generation of paleontologists arose that rejected the Cope–Marsh approach, believing, not unreasonably, that it had caused more harm than good. Both men ended their lives with somewhat tarnished reputations. History has viewed the thing a bit more dispassionately, and it is fair to state that the result ultimately was an extraordinary number of spectacular finds and a nomenclature nightmare that has taken much of the past 100 years to disentangle (see Desmond, 1975; Ottaviani *et al.*, 2005).

Within 30 years of the establishment of Dinosauria, there was a revolution in scientists' conceptions of how dinosaurs looked, driven by remarkable finds such as a complete hadrosaurid from New Jersey (1858) and 33 complete *Iguanodon* skeletons, recovered from coal layers outside of the town of Bernissart, Belgium (1877–1878; Box 14.4). In the hands of imaginative, skilled paleontologists such as J. Leidy (the hadrosaurid) and L. Dollo (the *Iguanodon* specimens), dinosaurs were transformed from overfed, bear-like lumbering lizards to something more terrifying, unimaginable, and wonderful than anybody could have invented.

Figure 14.5. Sculptor Water-house Hawkins' life-sized dinosaurs on the Crystal Palace grounds, Sydenham, south London, England, still enchanting visitors after 160 years.

Dinosaurs divided And they were *different*. Not just from living animals, but also from each other. This was duly noted by Harry Govier Seeley, vertebrate paleontologist at Cambridge University, and Friedrich von Huene, dean of German dinosaur paleontology at the University of Tübingen, both of whom recognized the fundamental division in Dinosauria between Ornithischia and Saurischia (Figure 14.7).

That dinosaurs had two different types of pelvis implied to Seeley that the ancestry of Ornithischia and Saurischia was to be found separately and more deeply among primitive archosaurs, within a now-abandoned group called "Thecodontia" (see below; see also Chapters 4, 10, and 13). Therefore, Seeley's dinosaurs were not, had he known the term, monophyletic (Figure 14.8).

The perception that dinosaurs were at least diphyletic (that is, having two separate origins) continued well into the twentieth

Figure 14.6. Contemporary lithograph of the New Year's Eve, 1853, dinner inside of Waterhouse Hawkins' model of *Iguanodon*.

Box 14.4 **Louis Dollo and the beasts of Bernissart**

Louis Antoine Marie Joseph Dollo, a Belgian paleontologist with a name almost as luxuriant as his moustache, gave us our first true picture of dinosaurs, through an incredible preservation of articulated *Iguanodon* skeletons in Belgium (Figure B14.4.1). Born in Lille, France, in 1857, Dollo first pursued a career in civil engineering, but soon was hired by the *Museé Royal d'Histoire Naturelle* in Brussels, Belgium. Here he was in charge of the study and museum exhibition of these specimens.

In 1878, commercial coal miners identified fossil bone some 322 m (1,056 feet) below ground, which was immediately brought to the attention of the Brussels museum and to Dollo in particular. This occurrence of bone turned into a treasure trove of more than 30 articulated skeletons of the Early Cretaceous ornithopod called *Iguanodon* (Figure B14.4.2).

Dollo's research on *Iguanodon* was unlike contemporary approaches, which tended to ask questions about to which taxon the material belonged and how it would be classified. Instead, thanks in part to the exquisite preservation of the Bernissart material, he devoted himself to understanding the anatomy and function of these extinct forms in ways that had not been possible before. He sorted the *Iguanodon* material into two species by successively eliminating different sources of skeletal variation. He used the disparity between forelimb and hindlimb length, the development of ossified tendons across the back, and footprints to establish bipedality in *Iguanodon*. And he outlined new approaches to reconstructing the jaw systems of numerous dinosaurs including *Iguanodon*, putting them into their comparative context with living vertebrates. In doing so, Dollo turned paleontological attention to what he called "ethological paleontology" – the study of behavior and environment of extinct organisms – which Othenio Abel, a German paleontologist, termed paleobiology in 1912.

Dollo also worked on the other fossil forms from Bernissart, including its turtles, crocodilians, and amphibians. When not busy with research on the riches of Bernissart, he conducted research on a number of new Late Cretaceous dinosaurs

Figure B14.4.1. Louis Dollo (1857–1931), the Belgian paleontologist of the Musée Royale de Sciences Naturelles Belgiques, who, along with Joseph Leidy, Professor of Anatomy at the University of Pennsylvania, first understood the shapes of dinosaurs.

and mosasaurs from Belgium and elsewhere in the European lowlands, and on Antarctic fishes, among modern organisms.

Other than his work on *Iguanodon,* best known is Dollo's Law of Irreversible Evolution. This biological principle, which Dollo formulated in 1893, argued that evolution is not a reversible process. That is, structures eliminated during the course of evolution cannot themselves reappear in the same form within a given lineage of organisms.

Dollo died in his adopted home of Brussels in 1931.

Figure B14.4.2. Several death-posed *Iguanodon*, the great beast of Bernissart, Belgium.

(a)   (b)

Figure 14.7. (a) Cambridge University's Harry G. Seeley (1839–1909) and (b) Friedrich von Huene (1875–1969), University of Tübingen.

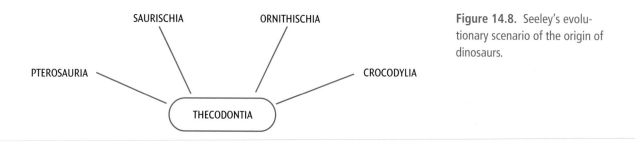

Figure 14.8. Seeley's evolutionary scenario of the origin of dinosaurs.

century. Most paleontologists, until even the early 1980s, thought that dinosaurs had at least two and likely three or four, separate origins within "thecodonts." Certainly, saurischians and ornithischians must have had separate origins; after all, their hip structure was different. And among saurischians, surely sauropods and theropods had separate origins; after all, they *look* so different. And finally, among ornithischians, ankylosaur ancestry was also often sought separately within some thecodontian group.

And what about Owen's bold suggestion that these dinosaurs were endothermic? Within ten or so years – and despite some early advocacy of it by other natural scientists – it was largely forgotten, the victim of the "fact" that dinosaurs were reptiles, and reptiles are cold-blooded.[4]

---

[4] A few paleontologists, notably G. R. Wieland of Yale University, shared a vision of some kind of dinosaur homeothermy.

As late as 1953, Roy Chapman Andrews evocatively described *T. rex's* meal in cold-blooded – literally and figuratively – terms:

> Then it [*Tyrannosaurus*] settles to the feast. Huge chunks of warm flesh, torn from the Duckbill's body, slide down the cave-like throat ... The King's stomach is full to bursting. Walking slowly to the jungle, he stretches out beneath a palm tree ... For days, or perhaps a week, he lies motionless in a death-like sleep. When his stomach is empty, he gets to his feet and goes to kill again. That is his life – killing, eating, and sleeping.[5]

Ironically – for how our views have changed! – this description would have appeared stranger to a nineteeth-century paleontologist like Cope or Marsh than to most paleontologists 70 years later. In retrospect, it seems puzzling that thoughtful scientists could have so meticulously described the bones and studied the relationships, yet with hardly any thought assumed "reptilian" ectothermy for dinosaurs for so long. Yet a look at publications through this period suggests that dinosaur metabolism rarely crossed their minds. Such is the strength of ideas.

Dinosaurs in the first half of the twentieth century

The first 60 years of the twentieth century brought about an expansion and consolidation of our basic understanding of dinosaurs and their diversity. Collecting, describing, and naming were the game, and our understanding of fundamental dinosaur morphology and diversity was dragged into a modern framework. In North America in the early years of the twentieth century, spectacular collections were made by Charles H. Sternberg along the Red Deer River in Alberta, Canada (Box 14.5). There he and his crews floated along the river in a mobile field camp, swatting mosquitoes and harvesting Upper Cretaceous dinosaurs from the sandstones and mudstones exposed in its banks. No less impressive were the efforts of the American Museum of Natural History's redoubtable Barnum Brown (Box 14.6), who, one summer in 1902, unearthed a large theropod that his sponsor, H. F. Osborne, dubbed *Tyrannosaurus rex*. For exotic and ill fated, however, none of this held a candle to Tendaguru, the richly fossiliferous series of excavations in the then German colony of "German East Africa" (Tanzania) from which the world set eyes on the full magnificence of *Brachiosaurus* (Box 14.7). Dinosaur discoveries continued at a rapid rate, new names proliferated, skilled descriptions of the new material were written, but, from the standpoint of *ideas*, the field had largely stagnated.

These discoveries were carefully collected and described by a host of extraordinarily fine, dedicated paleontologists, including (in addition to those mentioned above) E. H. Colbert, W. Granger, C. W. Gilmore, J. B. Hatcher, L. M. Lambe, A. F. de Lapparent, R. S. Lull, W. D. Matthew, A. K. Rozhdestvensky, R. M. Sternberg, and C. C. Young. Each of these remarkable men made important contributions, and the full story of each would fill a book as long as this. It's really a shame that space keeps us from highlighting their lives and work. Yet no account of dinosaur paleontologists should omit the brilliant Franz Baron Nopcsa – paleontologist, Albanian nationalist, polyglot, and spy for the Austro-Hungarian Empire in World War I (Box 14.8).

---

[5] Andrews, R.C. 1953. *All About Dinosaurs*. Random House, New York, pp. 64–67.

Box 14.5 **Rollin' on the river**

The great, long-lived dynasty of fossil collectors was surely the Sternbergs, *père et fils*. Of these, the father, Charles Hazelius (1850–1943 (C.H.)), is perhaps the most highly regarded; yet, between him and his three sons, George F. (1883–1969), Charles M. (1885–1981 (C.M.)); and Levi (1894–1976), much of the last third of the nineteenth and the first half of the twentieth centuries was occupied with fossil collecting.

C.H., whose life-long piety translated into an interest in natural history and fossils, got his start in 1876, when he attempted to join O. C. Marsh's collecting teams in the west. When that didn't materialize, he turned to E. D. Cope, who sent him $300 and put him to work. Sternberg worked for Cope until 1897 (Cope died while Sternberg was in the field); but, during his long and productive life, C.H. systematically

mined the Pierre Shale in Montana and Wyoming (Upper Cretaceous mosasaurs, plesiosaurs, and ammonites; see Figure 16.9), the Niobrara Chalk in Kansas (Upper Cretaceous marine creatures such as turtles, mosasaurs, plesiosaurs, fish, and even pterosaurs), Tertiary-aged strata of Kansas and Oregon (a variety of mammals), and the Permian of Texas (recovering the first specimens of the primitive synapsid *Dimetrodon* (see Figure 4.10) and the temnospondyl *Eryops*). It was a stunning haul.

But C.H.'s reputation – and those of his sons, who accompanied him on these expeditions – was cemented by his raft-based explorations along the Red Deer River of Alberta, Canada (Figure B14.5.1). He writes matter-of-a-factly in his account of this work:

Figure B14.5.1. Charles H. Sternberg (1850–1943), professional dinosaur collector, and his crews floating along the banks of the Red Deer River, Alberta, Canada.

The second part of the twentieth century to today

The 1960s and early 1970s are rightly known for social revolution, but they also spawned a revolution in paleontology. This was when the field of paleobiology was invented, a field that attempted to unravel the *biology* of fossil ecosystems.

Revolutions in dinosaur paleobiology

Yale University paleontologist J. H. Ostrom's 1969 description of *Deinonychus anthirropus* was the seminal event (Figure 14.9). Here was a predatory dinosaur (see Figures 9.6 and 9.21) obviously built for extremely high levels of activity. Its skeletal design simply made no sense

We reached Drumheller [*Alberta, Canada*], where we purchased … a five-horsepower motor boat; we also built a flat boat 12′ by 28′, upon the deck of which we pitched two tents, one for sleeping purposes, the other for a kitchen … We threw a rope to Charlie [*C.M.*] in his motor boat, which he fastened to a post on the small stern deck. A kindly hand pushed us off into the stream, where Charlie got up power and towed us into the current … Our motor boat, under Charlie's management, went 'chug, chug' down the river at the rate of 5 miles per hour. (Sternberg, 1985, p. 45)

Therein began one of the legendary collecting expeditions of all time. As the flatboat floated/motored down the river, they periodically put in at various promising exposures along the banks, set up camps on land, and excavated. The collection was formidable: the ceratopsians *Centrosaurus, Styracosaurus, Chasmosaurus*; the theropod *Gorgosaurus*; and the hadrosaurid *Lambeosaurus* (called "*Stephanosaurus*" by Sternberg). Along the way he met Barnum Brown (see Box 14.6), himself rafting and collecting.

C.H. was never an academic, and published relatively few scientific papers given the magnitude and importance of the fossils he collected. Yet he had a literary bent, and published an autobiography as well as an account of his fossil expeditions (see Selected readings). The latter, although originally self-published in 1917, became something of a classic and went to three editions.

The Sternberg family developed an important relationship with Canadian paleontology. Their collections remained in Canada and, in the end, while George became the curator of what was eventually named the Sternberg Museum at Fort Hays State University (Kansas), C.M. was hired on the staff of the Geological Survey of Canada, eventually ending up at the National Museum in Ottawa, while Levi was on the staff of the Royal Ontario Museum in Toronto. All taken, the Sternbergs represented two generations of lives devoted to paleontology.

C.H., as he aged, returned to Canada to be with his sons. His reaction to his adopted home sums the man up:

Though the ties of nearly a lifetime, that bound me to many a dear friend … must be severed, though I must leave the protecting folds of my father's flag and mine, and live under a flag that has waved a thousand years – under a monarch, in fact – I, a republican of republicans! Think of it! After three years' of residence in the beautiful city of Ottawa … after four seasons of work among buried dinosaurs and three winters spent in the laboratory of the Victoria Memorial Museum, I am free to confess … so far as personal liberty is concerned, that I was under the employ of his Royal Majesty George the Fifth of England … I have learned … that a man is as much a man amidst the snows of the Lady of the North, under the Union Jack, as under my own beloved Stars and Stripes. (Sternberg, 1985, pp. 7–8)

considered otherwise. Ostrom doubted that such levels of activity were likely in an animal with the metabolism of a crocodile, and he argued for the possiblity that *Deinonychus* might have been an endotherm.

Five years later Ostrom published an exacting study of the earliest-known bird, *Archaeopteryx*. Reviving the ideas of T. H. Huxley, a contemporary of Charles Darwin, Ostrom concluded that a close relationship between dinosaurs and birds was inescapable. "All available evidence," he wrote, "indicates unequivocally that *Archaeopteryx* evolved from a small coelurosaurian dinosaur and that modern birds are surviving dinosaur descendants." This also suggested that dinosaur physiology should be considered more along bird than crocodilian lines. With these two papers, Ostrom rewrote the book on both dinosaur physiology *and* bird origins!

That same year, Robert T. Bakker, Ostrom's student, and Peter M. Galton, a paleontologist from the University of Bridgeport, Connecticut, proposed to remove dinosaurs from "Reptilia"

Box 14.6 "Mr Bones"

There have been dinosaur collectors; there have even been extraordinary dinosaur collectors, and then, in a league quite by himself, there is the legendary Barnum Brown (Figure B14.6.1). Born in 1873, and named after the then-popular circus showman P. T. Barnum, Brown had an extraordinarily long and stunningly productive career. He virtually single-handedly turned the American Museum of Natural History (AMNH) from a place with not a dinosaur on the premises to perhaps the world's greatest dinosaur collection. Its great hall of Cretaceous dinosaurs has been described as a monument to his accomplishments. Brown's lifetime spans Victorian to modern paleontology, and it is fair to say that Brown did his fair share to contribute to that growth.

Brown was a collector and, in the end, his most memorable contribution was collection and not scientific description. But what a collector! He began his collections in the fossil grounds of Wyoming originally prospected by Marsh. Initially he met with little success (Marsh's collectors had done their jobs very well indeed), but toward the end of the season he discovered the still-productive Bone Cabin Quarry, a site so rich that local ranchers built an entire cabin out of fossil bones. After three years of collecting, 35 tons of fossil bones were sent back to the AMNH, including what eventually became the largest mounted specimen of its time, the AMNH's magnificent "brontosaurus" mount (see Figure B8.2.1a).

Meanwhile, in the early 1900s, Brown began to prospect in the now-legendary Hell Creek badlands of eastern Montana. There in 1902 he found the first of two magnificently preserved *Tyrannosaurus rex*. The thing was preserved in a calcite-hardened sandstone concretion, and, aside from the dynamite necessary to free it from the hillside in which it was found, he had to cut a road to carry the massive blocks out to the nearest railroad for shipment to New York (Figure B14.6.2).

Figure B14.6.2. A 1985 photograph of the site of Barnum Brown's first *T. rex* discovery. The arrow points to the remains of the wagon trail cut by Barnum Brown into the hillside to remove the massive blocks containing the fossil.

Brown's third major venue for fossil collecting was the Red Deer River in Alberta, Canada. There, following the style of C. H. Sternberg (see Box 14.5), he fitted a barge with a canvas tent and prospected along the shores of the river. Here he collected trainloads of fossils, including the beautiful hadrosaur specimens for which the AMNH is justifiably renowned.

With these collections, Brown's reputation was thoroughly cemented. Besides possessing a truly remarkable intuition for finding spectacular fossils (he was said to be able to "smell" them), Brown was a fossil collector with *style*: he always stayed in the best accommodations when traveling, dressed impeccably even while carrying out the dirtiest fieldwork, and was often seen in tailored suits and in an opulent full-length fur-trimmed coat. The result was that his reputation preceded him, and often crowds pressed him as he arrived in the field. Indeed, he was rumored to be quite the ladies' man.

By the early 1930s, Brown had cut a funding deal with the Sinclair Oil Company (whose logo, not coincidently, was, and is, a green sauropod) and expanded his collecting efforts to fossils other than dinosaurs and locales other than western North America. He traveled by virtually any conveyance available, and eventually ended up everywhere in the world, save Japan, Australia, Madagascar, and the South Sea Islands. His finds were numerous and varied: mummified musk ox, fossils of every stripe, and even the first Folsom projectile point, a find that indicated that humans were in North America far earlier than predicted. His discovery in 1934 of the Jurassic Howe Quarry bonebed was another highwater mark in a career full of them; more than 20 dinosaurs repesented by 4,000 bones.

World War II and age slowed the pace of his collecting down. Still he led tours at the AMNH, and periodically collected for the institution. He died a week before his ninetieth birthday, in the midst of planning a collecting trip to the Isle of Wight.

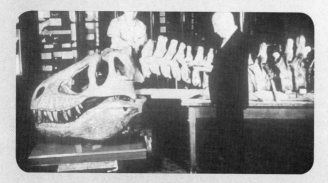

Figure B14.6.1. Barnum Brown (1873–1963), collector with the American Museum of Natural History, with his most famous discovery: *Tyrannosaurus rex. Tyrannosaurus* is the fossil laid out behind Brown.

Box 14.7 Tendaguru!

Tendaguru, located in the hinterland of Tanzania on the eastern coast of Africa and today monotonously formed of broad plateaus blanketed by dense torn trees and tall grass thick with tse-tse flies, was formerly the site of perhaps the greatest paleontological expedition ever assembled, and much – thousands of millennia – before that the place where dinosaurs came to die.

Let's go back to 1907, when Tanzania was part of German East Africa. This was the era of massive western European colonialism in Africa. With the widespread colonialism came scientists. And to then German East Africa came paleontologists in search of fossils.

The fossil wealth of Tendaguru was first discovered in 1907 by an engineer working for the Lindi Prospecting Company (established 1903). Word spread quickly, ultimately to Professor Eberhard Fraas, a vertebrate paleontologist from the Staatliches Museum für Naturkunde in Stuttgart, who happened to be visiting the region. So excited was he at the prospect of collecting dinosaurs after his visit to Tendaguru that he took specimens back to Stuttgart (including what was eventually to be called *Janenschia*) and more especially started drumming up interest among other German researchers to continue field work in the area.

It was Wilhelm von Branca, director of the Humboldt Museum für Naturkunde in Berlin, who was the first to seize upon the opportunity presented to him by Fraas. Yet before mounting an expedition of the kind demanded by Tendaguru, Branca had to tackle the problem of its financial backing. By seeking support from a great many sources, he received more than 200,000 deutschmarks – a fortune for the time – from the Akademie der Wissenschaften in Berlin, the Gesellschaft Naturforschender Freunde, the city of Berlin, the German Imperial Government, and almost a hundred private citizens.

With money, material, and supplies in hand, the Humboldt Museum expedition set off for Tendaguru in 1909. For the next four field seasons, it was bonanza time. Under the leadership of moustached and jaunty Werner Janensch (Figure B14.7.1) for three of these seasons (Hans Reck took charge in the fourth season), these years were to possibly the greatest dinosaur collecting effort in the history of paleontology. The first season involved nearly 200 workers, mostly natives, laboring in the hot sun as they dug huge bones out of the ground. During the second season, there were 400 workers and in the third and fourth seasons 500 workers. By the end of

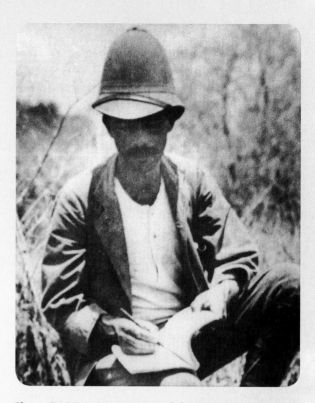

Figure B14.7.1. Werner Janensch (1878–1969), paleontologist at the Humboldt Museum, and the driving force behind the extraordinarily successful excavations at Tendagaru, Tanzania.

the expedition's efforts, some 10 km² of area was covered with huge pits, attesting to the diligence and hard work of these laborers.

But there's more. Many of these native workers brought their families with them, transforming the dinosaur quarries at Tendaguru into a populous village of upward of 900 people. With all these people, water and food was a severe problem. Not available locally, water had to be brought in, carried on the heads and backs of porters. And with the vast quantities of food that had to be obtained for workers and their families, and the pay for work carried out in the field, it is not surprising that the funds amassed by Branca disappeared at a great rate.

Still, the rewards were great indeed. Over the first three seasons, some 4,300 jackets were carried back to the seaport of Lindi – a four-day walk away and a trip made 5,400 times there and back by native workers, each with the fossils balanced on his or her head – to be shipped from there to Berlin.

and establish them and birds as a new Linnaean class of vertebrates, Dinosauria (Figure 14.10). The basis for this proposal was the "key advancements of endothermy and high exercise metabolism." This radical proposal fired imaginations and incensed detractors. In the end, however, it didn't stick because the authors failed to sufficiently demonstrate any relationship between dinosaurs and birds other than the assertion that they shared an endothermic metabolism – a point that was heatedly debated.

Endothermy The debate over dinosaur endothermy climaxed with the 1980 publication of an American Association for the Advancement of Science special volume that covered the 1978 proceedings of a symposium devoted solely to dinosaur thermoregulation. In general, the contributions in the book struck a compromise between Bakker's concept of dinosaurs having bird/mammalian levels of endothermy and the then-prevalent view of dinosaurs as oversized, uninspired crocodiles. Still, most authors seemed to lean toward, at a minimum, some kind of homeothermy (see Chapter 12).

Figure 14.9. John H. Ostrom (1928–2005), Yale University, the paleontologist whose ideas ignited the modern era of dinosaur research.

Box 14.8 Franz Baron Nopcsa: Transylvanian dinosaurs and espionage

There was never anyone quite like him before and it is very unlikely that his kind will be seen again. Franz Baron Nopcsa (Figure B14.8.1) was one of the first paleontologists who saw to it that dinosaurs were interpreted in their full biological context. For this, he is generally regarded as the founder of the field of paleobiology. From him, we've learned about the unusual dinosaur fauna from Transylvania, that part of western Romania where his noble family's estate was located. This skeletal material formed the mainstay of Nopcsa's research, including soft tissue reconstruction and its relevance to jaw mechanisms, paleoecological reconstructions of the region as a Late Cretaceous island, and the evolution of the dwarf dinosaurs that lived on the island. He also published extensively on the early evolution of birds from small predatory dinosaurs, the origin of new evolutionary conditions due to disease, the pituitary gland and large size in dinosaurs, and the relationship between bone histology, growth rates, and thermoregulation among dinosaurs. A polyglot, Nopcsa published not only in German, but also regularly in English, Hungarian, and French.

Remarkable though these achievements were, they were conducted against a background of scientific and political involvement in the founding of the state of Albania. Nopcsa was captivated by the geography and people of this stark, yet beautiful land of the western Balkans. He began working there in 1906 and by the end of his career had published some still-current monographs on the geography, geology, and ethnography of Albania and its people.

By 1912, Austria–Hungary was exceedingly worried about safe travel to the Mediterranean and saw Nopcsa's knowledge of the geography of the area to be tactically important to them. So Nopcsa became a spy during the first and second Balkan wars, the first of which was to establish the new country of Albania from the dying Ottoman Empire in Europe, and the second was to prevent neighboring states from absorbing its territory. At that time, the new country needed a king, so Nopcsa volunteered for the job, suggesting to the Austro-Hungarian army chief of staff that he would fund the war effort with money he would obtain from marrying the daughter of some American millionaire and would pledge

Figure B14.8.1. Franz Baron Nopcsa (1877–1933), Transylvanian nobleman, patriot, spy, and brilliant paleontologist.

Albania as an ally to the Empire in exchange for recognition as King. As far as is known, his proposal received no response, although he continued his spy work in Romania during World War I.

Despite his international activities, Nopcsa was a private man. He lived most of his life in Vienna, except for two years as Director of the Hungarian Geological Survey in Budapest. Living with him was his secretary, friend, and lover, Bajazid Elmas Doda, an Albanian he met in 1906. Transylvania was ceded to Romania after World War I and the Nopcsa estate was lost. Thereafter, Nopcsa's mental health declined and early in the morning of 15 April, 1933, he dosed Bajazid's tea with sleeping powder and then shot him. Going into his work room, Nopcsa wrote a suicide note and then killed himself.

While modern research on dinosaur metabolism has lost some of its contentiousness, it continues apace. Areas of study proving to provide important insights are dinosaur developmental biology as tracked through histology, and stable isotope geochemistry.

Phylogenetic systematics enters the fray

Amid all of this intellectual ferment, yet another revolution was not-so-quietly taking place. This was the *cladistic revolution* (see Chapter 3). The idea was not so new (although not nearly as old as that of endothermic dinosurs); the basics had first been articulated by a German entymologist, Willi Hennig, in 1950 (Figure 14.11). English translations of Hennig's ideas appeared in 1966 and again in 1979. Hennig's great insight was, as we've seen in Chapter 3, to develop a *scientific (testable) method* whereby relationship can be inferred from anatomy.

The method was initially not widely appreciated, and the results were a bit shocking to people trained in the traditional Linnaean classification system (see, for example, Box 4.2), with the result that between 1966 and 1990 (or thereabouts), this approach engendered considerable controversy. Nonetheless, when the computer algorithms were developed that allowed cladograms to be generated from large and complex datasets, cladograms became a ubiquitous and powerful tool for deciphering the relationships of both living and extinct organisms (including dinosaurs).

Starting in 1984, cladistic analysis exploded onto the dinosaurian systematic scene. The results were four-fold:

1 The disbanding of "Thecodontia."
2 The origin of dinosaurs.
3 The internal pattern of relationships within ornithischian and saurischian clades.
4 The relationships of birds to dinosaurs.

Figure 14.10. (a) Robert T. Bakker, then at Yale University, and (b) Peter M. Galton, University of Bridgeport, who first proposed that Linnaeus' venerable "Class" Aves be subsumed within a larger, new "Class Dinosauria."

In all four aspects, the changes wrought by cladistic analyses in our understanding of archosaurs in general, and dinosaurs in particular, were nothing short of revolutionary.

Thecodontia "Thecodontia" (*theco* – socket; *dont* – tooth; teeth set in sockets) was a term invented in 1859 by Richard Owen to group primitive archosaurs (see Figure 13.4). Thinking cladistically, however, "Thecodontia" was diagnosed using the same diagnostic characters as Archosauria; therefore, it must include *all* archosaurs – not just primitive ones. So the group "Thecodontia" was superfluous when there was already a group called Archosauria. Cladistic analysis, therefore, propelled the abandonment of what had for 120 years been considered a very important group of animals.

Dinosaur origins Recall that Seeley divided dinosaurs into two groups – ornithischians and saurischians (on the basis of their pelvic anatomy) – and that he viewed these as having separate origins within "Thecodontia." The elegant 1986 cladistic work of Jaques A. Gauthier (Figure 14.12), now at Yale University, however, provided ample corroboration of a monophyletic Dinosauria, identifying upward of 10 derived features uniting all dinosaurs with each other (see Chapter 4). Since then, numerous cladistic analyses of both new and old taxa have confirmed that *dinosaurs share a single, most recent common ancestor, itself a dinosaur.* Ornithischia and Saurischia – each monophyletic – are more closely related to each other than they are to anything else.

Ornithischian and saurischian relationships There are some other differences between the precladistic view of dinosaurs and a more modern one. For example, before cladograms, all large, carnivorous dinosaurs were grouped together within the evocatively named Carnosauria; Coelurosauria was kept for the smaller forms. Cladistic analysis, however, paints a very different picture: one in which several different lineages of large-sized theropods, with their short arms and huge heads, evolved *independently*. The large theropod *Carnotaurus* is as far from other large (tetanuran) theropods as it is possible for one theropod to be from another. And Coelurosauria, once the exclusive domain of small, light-bodied forms, is now home to the mighty *Tyrannosaurus rex* as well. The cladograms tell us that when theropods grow very large, they independently tend to take a similar form: large head, small arms, typically with

Figure 14.11. Willi Hennig (1913–1976) of the Deutsches Entomologisches Institut, the German entomologist who was the father of cladistic analysis (phylogenetic systematics).

Figure 14.12. Jacques A. Gauthier, then at the California Academy of Sciences, who carried out the seminal cladistic studies of bird origins and dinosaur monophyly.

reduced numbers of fingers, and long tails. What was it about their behavior or basic design that produced that shape?

Cladograms also revealed the important links between ceratopsians and pachycephalosaurs (once thought to be a kind of ornithopod), confirmed suspected links between ankylosaurs and stegosaurs, and situated the enigmatic therizinosaurs (once thought to be troodontids, or ornithischians, or who knows what else!) within Theropoda.

Birds as dinosaurs Ostrom, as we have seen, constructed a compelling anatomical case for birds as dinosaurs in his 1974 paper on *Archaeopteryx*. In 1986, Gauthier published his now-classic paper on saurischian monophyly, in which he addressed Ostrom's observations from a cladistic viewpoint. Gauthier's analysis forms the backbone of our treatment of bird origins in Chapter 10 and, as we've seen, overwhelmingly confirms the fundamental dinosaurian ancestry of birds. With so many important evolutionary insights afforded by the use of cladograms, there should be no wonder why we have emphasized a cladistic approach throughout this book.

The rise and fall of Dinosauria

Ideas about the rise of dinosaurs underwent considerable rethinking during the second half of the twentieth century. But before turning to these, let's first set the stage for the emergence of dinosaurs in the Triassic. From its outset some 251 million years ago, the Triassic was dominated on land by therapsids. Among these, the sleek, dog-like cynodonts were the chief predators, while the rotund, beaked, and tusked dicynodonts were the most abundant and diverse of herbivores (see Figure 13.3a). From the middle and toward the end of the Triassic, these therapsids shared the scene with squat, plant-eating, and swine-like archosauromorphs called aetosaurs and a few carnivorous crocodile-like archosaurs (see Figure 13.4).

Archosaurs began as sprawlers and ended up with either semi-erect (crocodilians and pterosaurs) or fully erect stance (dinosaurs and near dinosaurs (including birds)). Thus changes at the hip, knee, and ankle enabled a fully erect, parasagittal posture in which the legs acted not only as support pillars when standing but also provided for longer strides and more effective walking and running ability (see Chapter 12).

Toward the end of the Triassic, approximately 225 Ma, there was a great change of fortunes for amniotes. The majority of therapsids went extinct (one highly evolved group of therapsids, the mammals, of course survived). And it was dinosaurs that somehow rose to become the **dominant** terrestrial vertebrates, by which it is meant that they became the most abundant, diverse, and probably visible group of tetrapods. The pattern of waxing and waning in dominance (as one group supercedes another in evolutionary time) is called the **wedge** (Figure 14.13).

More puzzling, though, was *why* dinosaurs prevailed. Ideas boiled down to two basic concepts:

1   Dinosaurs out-competed their contemporaries, earning the right, as it were, to be the dominant terrestrial vertebrates.
2   Dinosaurs somehow survived, because their non-dinosaurian contemporaries went extinct, leaving the planet to dinosaurs.

Out-competition In the late 1960s and early 1970s, Alan Charig, Curator of Lower Vertebrates at what was then called the British Museum of Natural History, argued that those archosaurs that had the new, "improved" erect stance were then able to out-compete contemporary predatory,

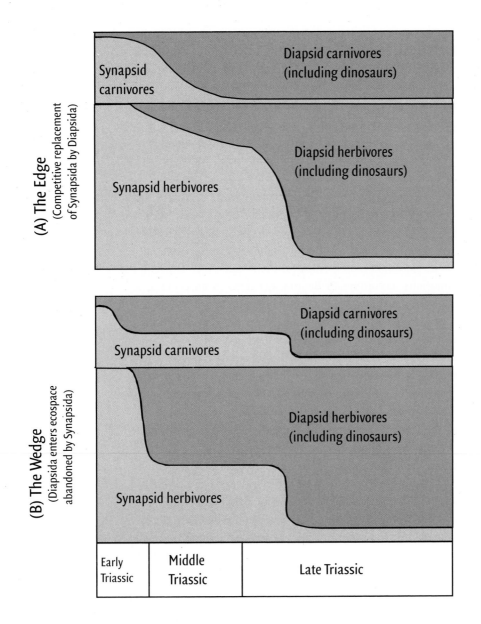

Figure 14.13. Two views of the origin of dinosaurs during the Late Triassic. (a) W*edge with the edge*: Gradual competitive replacement of synapsids, primitive archosaurs, and rhynchosaurs (both herbivores and carnivores) by herbivorous and carnivorous dinosaurs. (b) W*edge without the edge*: Rapid opportunistic replacement facilitated by extinction.

semi-sprawling therapsids for their food sources. The immediate descendants of the flashy new archosaurs were the dinosaurs. The inevitable consequence of such progressive improvements in limb posture, Charig argued, was the gradually changing pattern of faunal succession at the end of the Triassic. We can call this and any other evolutionary advantage a **competitive edge**; dinosaurs prevailed, according to Charig, by virtue of having better-designed limbs and thereby more efficient terrestrial locomotion.

At nearly the same time, Bakker was making similar arguments about the competitive superiority of endothermy in dinosaurs (see Chapter 12). He believed that, instead of limbs, it was the achievement of internally produced heat that gave dinosaurs (or their immediate ancestors) a competitive edge over contemporary and supposedly cold-blooded therapsids and rhynchosaurs. The same conclusions applied: dinosaurs won, therapsids lost, and the truth of the competitive superiority of endotherms over ectotherms could be read directly from the pattern of faunal succession at the end of the Triassic: the competitive edge produced the wedge.

Wedge without edge? Michael J. Benton of the University of Bristol is not convinced that the edge produced the wedge in the Middle to Late Triassic fossil record of the earliest dinosaurs and their predecessors. In order for edges to lead to wedges, all of the players in the game have to be present to interact with each other. And, according to Benton, they were not (note Figure 14.13). Instead, he suggests that the fossil record of the last part of the Triassic is marked by not one but two mass extinctions. The first appears to have been the more extreme and ultimately most relevant to the rise of dinosaurs. This earlier Late Triassic extinction completely decimated rhynchosaurs and nearly obliterated dicynodont and cynodont therapsids, as well as several major groups of predatory archosaurs.

Likewise, there is a major extinction in the plant realm. The important seed-fern floras (the "*Dicroidium* flora," which contained not only seed-ferns, but also horsetails, ferns, cycadophytes, ginkgoes, and conifers; see Chapter 13) all but went extinct as well, to be replaced by other conifers and bennettitaleans (see Figures 13.8 and 13.9). Dinosaurs appeared as the dominant land vertebrates only after this great disappearance of therapsids, archosaurs, and rhynchosaurs. Thus the initial radiation of dinosaurs, according to Benton, occurred in an ecological near-vacuum, with the rapid loss of the dominant land-dwelling vertebrates setting the stage for the *opportunistic* evolution of dinosaurs. No competitive edge, because there was no competition.

That there was at least one, and more than likely two, mass extinctions at the end of the Triassic Period is uncontroversial. Naturally, one of the key questions is what might have caused these extinctions. Benton has suggested that the Late Triassic extinctions may be linked with climatic changes – the regions first inhabited by dinosaurs appear to have been hotter and more arid, a change from the more moist and equable – and thence to alterations in terrestrial floras and faunas. The abrupt extinction of the *Dicroidium* flora may have caused the extinction of herbivores specialized to them and hence the predators feeding on the herbivores.

So, if Benton is right, then perhaps the archosaurian predecessors of dinosaurs may have just squeaked by, survivors not because they were somehow superior to the presumed competition, but because they happened to inherit a deserted Earth. Instead of survival having been something intrinsic to dinosaur superiority, it may have been that they simply had better luck.

Ironically (as we shall see in Chapter 16), 163 million years later the tables again turned, and mammals inherited an Earth this time deserted by the very dinosaurs who, by one means or another, had taken it from them millions of years earlier. The wedge in this case was produced without any edge.

Extraterrestrial extinction Our trip through the most recent ideas about dinosaur paleontology would not be complete without perhaps the most radical idea of all: extinction by asteroid impact (see Chapter 16). Since H. F. Osborn's time, the extinction of the dinosaurs was viewed by paleontologists as a gradual process of dwindling diversity, beginning well before the end of the Cretaceous. The prevailing view before 1980 was cleverly and succinctly summarized by University of California (Berkeley) paleontologist W. A. Clemens and colleagues who wrote, with apologies to T. S. Eliot:

> This is the way Cretaceous life ended
> This is the way Cretaceous life ended
> This is the way Cretaceous life ended
> Not abruptly but extended.[6]

Clemens and colleagues' article was aptly entitled "Out with a whimper, not a bang."

---

[6] Clemens, W. A., Jr, Archibald, J. D. and Hickey, L. J. 1981. "Out with a whimper, not a bang." *Paleobiology*, **7**, 297–298.

The revolution, of course, came in the form of the 1980 hypothesis that an asteroid came from outer space, smashed into the planet, and ultimately reset global ecosystems for all time. What a strange, wonderful, and terrifying idea – that extra-terrestrial events are important forces shaping the history of life on Earth. The conceptual revolution provoked by this vision extended far beyond the deaths of a few dinosaurs, and reverberated throughout the geosciences.[7] The asteroid and its aftermath are the subjects of our last chapter.

Today

There are more active, professional paleontologists working today than ever before in the history of the discipline, and the insights derived from the field have become particularly important, not just in the field of evolutionary biology, but also as they relate to climate science. Yet, paleontology is today at a cross-roads, with new skill sets required that would have been utterly foreign to paleontologists of a generation ago.

In evolutionary biology, it has become very clear, thanks to the work of scientists like Jacques Gauthier, that the present-day biota does not give us a complete enough picture of the past record of evolution. Deeper insights are afforded by the fossil record; indeed, how would we ever know that birds are dinosaurs if we only knew the living representatives of Dinosauria?

Yet, modern evolutionary biology has acquired powerful new tools for unraveling the course of evolution, including: (a) molecular evolution, in which the molecular difference between two organisms is measured to determine how distantly or closely they are related (see Box 11.1); (b) evolution and development (or evo-devo), in which sophisticated genetic and embryological studies are revealing the way the organisms evolve new features and produce diversity; and (c) molecular clocks, in which the rate of molecular evolution can be used to date the time that two living organisms first diverged from a single (long-extinct) ancestor (see Box 11.1). In each of these cases, deep insights into the fossil record are necessary to optimize the results. So paleontology's contributions continue to be on the forefront; yet, modern paleontologists need a far greater sophistication and specialized training – more than was ever even imagined by the old timers – in the biosciences generally and in molecular genetics and embryology in particular.

Paleontologists, too, have much to offer the world about global climate change. The Earth, of course, has gone through many episodes of global warming associated with greenhouse conditions, not the least of which was, as we've seen, during the mid Cretaceous. What was that world like? More importantly, how did the biota respond to that climate? Since the Earth has experimented with greenhouse conditions in the past, it behooves us to consider how it and its biota responded to these events – information that may give us insights about what we may expect in the future. In this case, advanced expertise in geochemistry and climate modeling could serve as an important part of the toolkit of future paleontologists.

It's true that the days of pulling dinosaurs from the ground with no thought about their implications have probably passed. But in a way, the questions are more interesting, and the stakes far greater. It's a brave, new, and exciting world out there.

---

[7] See Powell, J. L. 1998. *Night Comes to the Cretaceous*. W.H. Freeman and Company, New York, 250pp.

Box 14.9 **Young Turks and old turkeys**

"It's funny," University of Oregon paleopedologist Greg Retallack once said, "how quickly today's 'Young Turks' become tomorrow's old turkeys." The young, fire-breathing revolutionaries themselves become advocates of established dogma, as, even in their own lifetimes, their ideas are challenged and swept aside by a righteous, new, young, aggressive generation of scientists. It was that way in Richard Owen's day, and it is that way now. Here we highlight the careers of four of the currently active generation of vertebrate paleontologists.

As we have seen, the 1970s were a time of intellectual ferment in paleontology, and for the dinosaur-loving public, nobody embodied those fresh winds of change more than Robert T. Bakker (b. 1945; Figure 14.10a). In a field historically known for bookish indifference to fame, Bakker has been a ubiquitous media presence; so much so that he was recognizably caricatured as Dr Robert Burke in *Jurassic Park II*. Bearded, long-haired, and dressed for battle in his field-ready best[1] (belying degrees from Yale and Harvard), Bakker was filled with amazing ideas about birds, dinosaurs and their world. A highly competent prose stylist (described in *Harper's Magazine* as "by far the most gifted writer in his profession"[2]) and a talented illustrator, Bakker was distinctive, articulate and out to change ideas. And with his forcible advocacy of endothermy in dinosaurs, he surely has. As he has grown older, Bakker has continued to be a magnetic personality, and is the author of a thought-provoking popular treatment of dinosaurs, *Dinosaur Heresies*, as well as a novel about dinosaurs, *Raptor Red*.

Equally informal of manner and no less intellectually endowed, but somewhat lower in profile is Bakker's contemporary, John R. "Jack" Horner (b. 1946; Figure B14.9.1). Horner lacks formal advanced degrees; yet, it would be hard to find a more creative (he is a MacArthur "genius award" recipient) and accomplished paleontologist. As we have seen, it was Horner who first recognized extended parental care in *Maiasaura*; it has been Horner who has, most recently, led the field in understanding bone histology and its relationship to dinosaur growth and metabolism; it was in Horner's laboratory that proteins from *T. rex* were first extracted (see Chapters 9 and 10, and Box 11.1).

[1] Including an iconic battered hat that pre-dated that of Indiana Jones—but not that of Roy Chapman Andrews!

[2] Silverberg, R. 1981. Beastly debates. *Harper's Magazine*, October, 1981, pp. 68–78.

Figure B14.9.1. Paleontologist *extraordinaire* Jack Horner, doing what he does best.

Horner is a superb field paleontologist; rare is the summer that passes when he is not out prospecting and collecting dinosaurs. His laconic, plain-spoken manner, reputed to be the inspiration for Dr Alan Grant in the film *Jurassic Park*, belies a canny, sophisticated, approach to almost all he undertakes, whether it be building the Museum of the Rockies (Bozeman, Montana, USA) into a world-class paleontological museum and an important center for cutting-edge histological research (both living and fossil); publishing well over 100 research papers on virtually every subject pertaining to Dinosauria; or writing a best-selling account of the development of his ideas on maternal care in *Maiasaura (Digging Dinosaurs)*. Jack Horner continues to be a dominant and creative force in the field of vertebrate paleontology, using his own successes to make opportunities for other younger scientists.

Somewhat younger than Horner and Bakker is the Curator of the Division of Paleontology at the American Museum of Natural History (New York), Mark A. Norell (b. 1957; Figure B14.9.2). Norell's studies have been wide-ranging, including contributing to the development of the concept of "ghost taxa" (see Box 13.2), the discovery and description of the unusual theropod *Mononykus* (see Figure 11.9), the discovery of the first embryo of a theropod dinosaur, and the first clear "proof" that dinosaurs nested on their eggs (*Oviraptor*; see Figure 9.28). Norell has led expeditions to the Gobi

Figure B14.9.2. Mark Norell, Curator of Paleontology at the American Museum of Natural History (New York).

Figure B14.9.3. Paul Sereno in the waning sun of the Sahara Desert, digging up something not too dinosaur-like. He writes of this picture, "I have fallen prey to digging up 100 fossil humans in the Sahara – they got in the way of my dinosaurs!"

Desert for the past 18 years, during the course of which he discovered and described the dinosaurs *Shuvuuia, Apsaravis, Byronosaurus*, and *Achillonychus*, among other vertebrate fossils. He has even co-written several award-winning books, including *Discovering Dinosaurs* (1995) and *Unearthing the Dragon* (2005).

Our brief sampling of some of dinosaur paleontology's Young Turks would surely be incomplete without some recounting of the exploits of the University of Chicago's redoubtable Paul Sereno (b. 1957; see Figure B14.9.3). Sereno gives lie to the idea that they don't make paleontologists like they used to. Handsome and dashing (he was, after all, one of *People* magazine's "50 Most Beautiful People," *Newsweek*'s "100 People to Watch for the Next Millennium," and *Esquire* magazine's "100 Best People in the World"), Sereno has made it his business to travel to exotic locales (Egypt, Niger, Morocco, Argentina, China) and collect exotic fossils. Therein lies quite the list, including *Afrovenator abakensis* (a large theropod), *Carcharodontosaurus saharicus* (another large theropod, possibly larger than *Tyrannosaurus*), *Deltadromeus agilis* and *Rugops primus* (two more large theropods!), *Rajasaurus narmadensis* (a large, crested theropod), *Herrerasaurus ischigualastensis* and *Eoraptor*

lunensis (two of the most primitive dinosaurs known; see Figure 15. 5b and c), *Jobaria tiguidensis* (a 20+ m sauropod), the 13 m dinosaur-munching Saharan crocodile *Sarcosuchus imperator*, and the pisivorous large theropod *Suchomimus tenerensis*.

Despite Sereno's evident predilection for large theropods, he has made significant contributions to understanding the relationships among all dinosaurs; a look among the many cladograms in this book shows how thorough and deep his contribution to dinosaur relationships truly is. His own summary says it all:

> I see paleontology as 'adventure with a purpose.' How else to describe a science that allows you to romp in remote corners of the globe, resurrecting gargantuan creatures that have never been seen? And the trick to big fossil finds? You've got to be able to go where no one has gone before.[3]

[3] This extract is from http://www.paulsereno.org/bio.htm (last reviewed June 2012).

Summary

Paleontology is a human endeavor, and like all human endeavors, ideas have changed as the context in which those ideas developed has changed. Our earliest suggestion that humans may have seen and taken note of dinosaur fossils comes from the recognition that mythical creatures may have been inspired by observations of very large or unfamiliar-looking dinosaur fossils.

Paleontology as a science began in the Enlightenment with the recognition that observation in combination with logic and rational thinking could reveal truths about the natural world. The earliest dinosaur fossil explicitly identified as something quite unlike anything alive today was found in 1822; within 40 years, not only was a variety of extinct animals recognized, but a relative geological time scale had been constructed. It remains valid in its essentials to this very day. In 1842, English anatomist Richard Owen established the word "Dinosauria" for an extinct group of reptiles, partly to demonstrate that organisms had not evolved. His concept of dinosaurs was one of bulky, elephantine quadrupeds.

Charles Darwin's idea of evolution by natural selection was published in 1859, and, with it, the burgeoning fossil record came to be seen as making sense in an evolutionary context. The idea that other worlds had existed that were very different from our own gained broad currency. At the same time, Dinosauria became far better known, and its members seemed so disparate that they were divided into two groups (Ornithischia and Saurischia) and thought to have had separate origins within early archosaurs (which were all lumped together as a group called "Thecodontia"). The rise of dinosaurs was interpreted in Darwinian terms as the competitive success of the superior forms (dinosaurs) over inferior forms (primitive archosaurs and advanced, non-mammalian synapsids).

The first 70 or so years of the twentieth century were all about collecting, describing, and enhancing knowledge of the different forms of dinosaurs. This abruptly changed with John Ostrom's 1969–1970 interpretation of *Deinonychus* as endothermic and his re-evaluation of *Archaeopteryx* as a theropod. These revolutionary views came as the field of paleontology was revitalized as paleobiology, and as phylogenetic systematics came to be recognized as a truly scientific way of inferring relationships among even extinct forms. Phylogenetic systematics demonstrated the fundamental monophyly of Dinosauria (also affirming, parenthetically, the monophyly of Saurischia and Ornithischia), destroyed "Thecodontia," and clearly showed that birds are dinosaurs.

When, in 1980, it was postulated that an asteroid caused the end-Cretaceous extinction (which obliterated the non-avian dinosaurs), paleontologists came to recognize that extraterrestrial events can have a profound effect on earthly events; that extinctions can occur regardless of how well adapted a particular group is. It was at this time that a re-evaluation of the rise of Dinosauria was carried out, in which the success of the group was no longer ascribed to competitive superiority, but rather to extinctions that had liberated ecospace for dinosaurs to colonize.

Today, dinosaurs continue to be studied in a variety of ways (see Box 14.9): through the discovery and description of new forms; as biological entities functioning in ancient ecosystems; via analyses of the large-scale evolution of the group; through histological and even molecular analyses; and from the standpoint of evolution and development.

Selected readings

Andrews, R. C. 1929. *Ends of the Earth.* G. P. Putnam's Sons, New York, 355pp.

Bakker, R. T. 1986. *Dinosaur Heresies.* William Morrow, New York, 481pp.

Bakker, R. T. and **Galton, P. M.** 1974. Dinosaur monophyly and a new class of vertebrates. *Nature*, **248**, 168–172.

Benton, M. J. 1984. Dinosaurs' lucky break. *Natural History*, **93**(6), 54–59.

Bryson, B. 2003. *A Short History of Nearly Everything*. Broadway Books, New York, 544pp.

Cadbury, D. 2001. *Terrible Lizard*. Henry Holt and Company, New York, 384pp.

Charig, A. J. 1972. The evolution of the archosaur pelvis and hindlimb: an explanation in functional terms. In Joysey, K. A. and Kemp, T. S. (eds.), *Studies in Vertebrate Evolution*. Winchester, New York, pp. 121–155.

Desmond, A. 1975. *The Hot-Blooded Dinosaurs*. The Dial Press, New York, 238pp.

Desmond, A. 1982. *Archetypes and Ancestors*. University of Chicago Press, Chicago, IL, 287pp.

Gauthier, J. A. 1986. Saurischian monophyly and the origin of birds. In Padian, K. (ed.), *The Origin of Birds and the Evolution of Flight*. Memoirs of the California Academy of Sciences no. 8, San Francisco, pp. 1–56.

Gauthier, J. A., **Kluge, A. G.** and **Rowe, T.** 1988. Amniote phylogeny and the importance of fossils. *Cladistics*, **4**, 105–209.

Hennig, W. 1979. *Phylogenetic Systematics*, translation by D. D. Davis and R. Zangerl. University of Illinois Press, Urbana, IL, 263pp.

Lessem, D. 1992. *The Kings of Creation*. Simon and Schuster, New York, 367pp.

Mayor, A. 2000. *The First Fossil Hunters*. Princeton University Press, Princeton, NJ, 361pp.

Ottaviani, J., Cannon, Z., **Petosky, S., Cannon,C.** and **Schultz, M. 2005.** *Bone Sharps, Cowboys, and Thunder Lizards: A Tale of Edward Drinker Cope, Othniel Charles Marsh, and the Gilded Age of Paleontology*. GT Labs, New York, 168pp.

Preston, D. J. 1986. *Dinosaurs in the Attic: An Excursion into the American Museum of Natural History*. St. Martin's Press, New York, 244pp.

Shubin, N. 2008. *Your Inner Fish: A Journey Into the 3.5-Billion-Year History of the Human Body*. Pantheon Books, New York, 230pp.

Sternberg, C. H. 1985. *Hunting Dinosaurs in the Bad Lands of the Red Deer River, Alberta, Canada*. NeWest Press, Edmonton, Alberta, 235pp.

Topic questions

1 Who invented the term "Dinosauria?" What was his idea about dinosaur metabolism?

2 In relation to paleontology, what is meant by "the edge?" and "the wedge?"

3 Decribe a dinosaur as imagined by Victorian paleontologists. What changed our ideas from the Victorian conception?

4 Describe the kinds of vertebrate faunas that existed just before dinosaurs became the most abundant and diverse terrestrial vertebrates.

5 What were the dramatic "revolutions" that changed the face of paleontology after the 1960s?

6 Cite an example of how early twentieth-century imperialism moved the science of paleontology forward.

7 Give two examples of how cladograms changed paleontological thinking.

8 In what fundamental ways does the legend of the griffin differ from our understanding of *Protoceratops*?

9 Contrast what you know of the Triassic–Jurassic extinction with the Cretaceous–Tertiary extinction.

10 Explain the difference between a cladistic view of dinosaur origins and H. G. Seeley's understanding of the origins of dinosaurs.

CHAPTER OBJECTIVES

- Understand something about the origin and early evolution of dinosaurs

- Learn how we study the origin of a major group of animals

CHAPTER 15

DINOSAURS: IN THE BEGINNING

In the beginning…

How did it all begin? Was there something special, or different, or *better* about that first dinosaur, that led to all the rest? Was the appearance of the earliest dinosaur a celebrated, Baroque event, with the Glorious First Dinosaur surrounded by trumpet-bearing cherubs and allegorical mythological figures heralding its birth? Do we even know who the first dinosaur was? Shouldn't the ancestor of all dinosaurs have been something *different* and *unique*?[1] Ultimately, it is possible that the importance of that first dinosaur was just the proliferation of all its descendants. Maybe the first true dinosaur was just another Middle Triassic archosaur doing its Middle Triassic archosaur thing.

"Middle Triassic" is a clue; as we'll see below, many undoubted dinosaurs – animals that clearly belong to familiar groups such as Theropoda and Sauropodomorpha – are known from the Late Triassic from nearly everywhere around the world. So a good guess for the age of the very first dinosaur would be, perhaps, towards the end of the Middle Triassic. Still, even if such rocks and fossils were extremely abundant, the likelihood of finding the actual progenitor – the very beast that was the great granddaddy (or mommy) of a lineage – is vanishingly small; even if you did find the remains of that very animal, how would you know it was *the one* (and not, for example, its first cousin or next-door neighbor)?

So we'll resort yet again to the cladogram: the hierarchy of characters that specifies what features ought to be present in an ancestor. It is then a question of finding a creature that most closely matches the expected combinations of characters and character states. We'll search among ancient Archosauria (sometimes termed "basal archosaur") for fossils that embody the unique features, specified by the cladogram, of the hypothetical dinosaur ancestor. In the case of dinosaurs, we have several closely related small early archosaurs that are very good candidates.

Archosauria

Pseudosuchia vs. Ornithodira

Recall (Chapter 4) archosaurs: that group of reptiles bearing an antorbital fenestra (see Figure 4.12), whose living representatives are crocodiles and birds. There was a fundamental split early in the evolution of archosaurs, with two groups emerging, Pseudosuchia and Ornithodira (Chapter 4; Figure 15.1).

The former consists of a variety of ground-hugging, quadrupedal, armored behemoths, several of which are showcased in Figure 13.4. Most pseudosuchians didn't make it into the Jurassic. The exception, however, are crocodilian relatives (Crocodylomorpha), which, among the living, are all that is left to represent this group.

Ornithodira consists of two groups: Dinosauromorpha (*morph* – shape), which includes true dinosaurs and their near-relatives, and pterosaurs, the magnificent flying reptiles that aren't birds (Figures 4.13 and 13.7). Birds of course are theropod dinosaurs, and are thus the only living representatives of this great branch of archosaurs (Figure 15.1; see Chapters 10 and 11).

[1] Answers: Not sure; Not sure; No; No; Not necessarily!

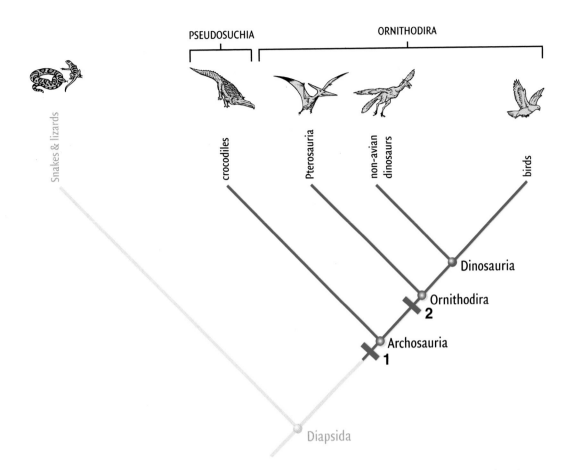

Figure 15.1. Cladogram of Archosauria, showing the fundamental split between Pseudosuchia and Ornithodira. Derived characters include : at **1**, antorbital opening; at **2**, distal ends of neural spines not expanded; second phalanx of hand digit II longer than first phalanx; tibia longer than femur; compact metatarsus.

Dinosauromorpha

Paleontology is notoriously terminology-rich, and as we get successively closer to Dinosauria, the names come thick and fast. Some paleontologists (including us) use Dinosauromorpha as a formal term encompassing dinosaurs *and* their close relatives; some use it as a paraphyletic group referring to just those close relatives.[2] Here we form an alliance and stick with it to the end ("Survivor"-style) with just the clades Ornithodira and Dinosauromorpha (Figure 15.2).

So who are dinosauromorphs? The earliest, most primitive forms, non-dinosaur (basal) dinosauromorphs, were small (about 1m or less, including tail), lightly built, likely insectivorous or carnivorous creatures, many of whom walked on all fours but ran bipedally, and could have starred in the opening scene of *Jurassic Park III*. From Argentina come *Marasuchus* (once known as "*Lagosuchus*") and *Lagerpeton*. First from Poland, and then Africa, Argentina, and North America, are the diminutive silesaurids, including *Silesaurus*, *Asilisaurus*, and

---

[2] Dinosauromorphs exclusive of dinosaurs (thus, a paraphyletic assemblage) were once called "pseudosuchians." Some paleontologists recognize a group – not quite true dinosaurs – snuggled even closer to Dinosauria than dinosauromorphs, called Dinosauriformes (*formes* – in form of). Dinosauromorphs exclusive of pterosaurs are grouped within the clade Avemetatarsalia. We'll stick to the KISS Principle, here.

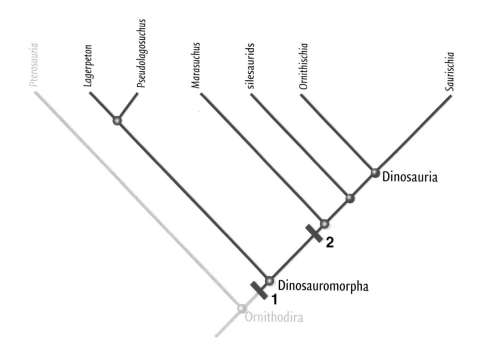

Figure 15.2. Cladogram of Dinosauromorpha. Derived characters include, at **1**, offset head of femur, straight cnemial crest, longest metatarsal > 50% of length of tibia, metatarsal V without phalanges and pointed; at **2**, elongate deltapectoral crest on humerus; extensively perforated acetabulum, tibia with transversely expanded subrectangular distal end (see also Figure 4.15).

Pseudolagosuchus, respectively. Also among the silesaurids is the poorly understood *Lewisuchus*[3] from Argentina. Only recently recognized, silesaurids apparently had a world-wide distribution. Silesaurids were evidently herbivorous, with teeth that superficially resemble those of ornithischians. Several non-dinosaur basal dinosauromorphs are shown in Figure 15.3.

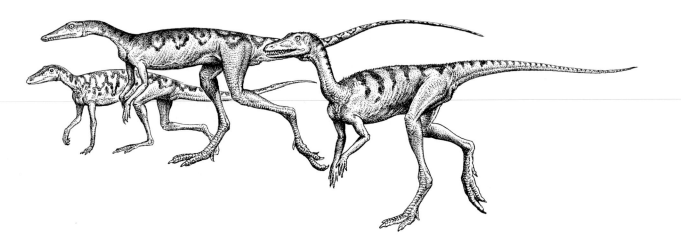

Figure 15.3. Basal dinosauromorphs – *Lagosuchus; Marasuchus; Pseudolagosuchus.*

---

[3] There is considerable evidence that *Lewisuchus* is the same animal as *Pseudolagosuchus*. If so, because *Pseudolagosuchus* is the younger (more recently named) name, all specimens of *Pseudolagosuchus* will come to be called *Lewisuchus*.

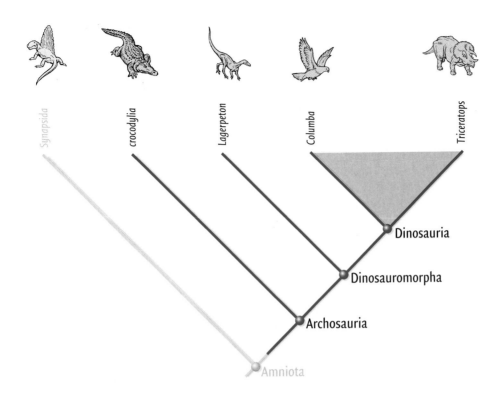

Figure 15.4. The definition of Dinosauria: cladogram showing Ornithischia, represented by *Triceratops*, and Saurischia by a pigeon (*Columba*), in Amniota with monophyletic Dinosauria – the group defined by these two organisms and their most recent common ancestor – grayed in.

Dinosaurs defined

The line between dinosaur and non-dinosaur turns out to be a finely graded one, indeed; so much so that the branching point termed "Dinosauria" could be mistaken as arbitrary. Why this point and not one node forward or back?

The decision as to what is and what is not a dinosaur is best understood by thinking in terms of *definition*: the idea of constructing the limits or extent of a particular group (Box 15.1) based upon its evolution. In this way, dinosaurs are defined as *the monophyletic group (e.g., the clade) whose members contain both pigeons (representing Saurischia) and* Triceratops *(representing Ornithischia)*. What this says is, if Dinosauria is monophyletic, it must include all the descendants of the most recent common ancestor of saurischians and ornithischians (Figure 15.4). This, of course, effectively limits who can and who can't be considered a dinosaur, whether or not a particular animal or even a particular set of characters is actually known (see Box 15.1). And this also removes some animals – the non-dinosaur basal dinosauromorphs – from Dinosauria.

Dinosauria

The oldest known dinosaur material, *Eoraptor* and *Herrerasaurus*, comes from Argentina, in sediments somewhat younger than those that produced the basal dinosauromorphs *Marasuchus*[4] and *Lagerpeton* (Figure 15.5).

---

[4] At the time, it was described as a species of "*Lagosuchus*," a name that is no longer used.

Box 15.1 A phylogenetic classification

It is ironic that although evolution is the unifying concept by which biologists understand the biota, our taxonomy, the means by which we summarize the underlying relationships of organisms, is non-evolutionary. Although this fact is perhaps absurd, it is no surprise; as we have seen, Linnaeus devised his taxonomy from a pre-evolutionary, creationist perspective. The notion of *phylogenetic relationship* was later grafted onto pre-evolution Linnaean taxa. This was bound to induce a few minor hiccups along the way – like the inconvenient facts that birds and mammals (amniotes, both!) are, evolutionarily speaking, reptiles (see Chapter 10 and Box 4.2, respectively). Oh well!

Phylogenetic systematics (cladistic methods; see Chapter 3) was invented as a means by which evolutionary events could be reconstructed, and it too involved the grafting of Linnaean taxonomic categories onto cladograms to represent biological relationships. Linnaeus identified groups using *overall similarity*; many Linnaean groups are thus paraphyletic. By contrast, cladists use *special similarities* – the shared derived (or diagnostic) characters we've emphasized throughout this book – to identify relationships and to diagnose groups.

In the early 1990s, cladists Jacques Gauthier (see Figure 14.12) and Kevin de Queiroz came to recognize that identifying groups by diagnosis really involved no property of evolution. They noted that taxon names are not necessarily based upon properties, but rather upon membership. So, for example, the group Canidae would not necessarily be based upon some diagnostic character – although such a character

might exist – but rather upon membership: it is the group that contains *Canis*.

In place of either diagnostic characters or membership properties, neither of which is directly connected with the process of evolution, Gauthier and de Queiroz established a subtle but important distinction between *diagnosis* (the quality that we have used throughout this book) and phylogenetic *definition*, which explicitly involved the process of organic evolution in the membership of a group. Returning to Canidae, a phylogenetic definition for the group might be "the group stemming from the most recent common ancestor of *Canis* (dog) and *Vulpes* (fox)." This evolutionary definition, while not telling us what characters to look for in canids, effectively delimits on the cladogram what can and what cannot evolutionarily be considered a canid.

With this definition, common ancestry – the fingerprint of organic evolution – becomes fundamental to the meaning of the name Canidae.

De Queiroz and Gauthier recognized three different types of phylogenetic definitions (Figure B15.1.1).

Node-based definitions are those that define the name of a clade based upon the most recent common ancestor of two specified organisms (Figure B15.1.1a). The canid definition above is node-based. *Stem-based* definitions identify the clade of organisms sharing a more recent common ancestor with a particular taxon, relative to another (Figure B15.1.1b). For example, you could evolutionarily define a taxon as all

---

These two animals are also by their characters the most primitive, and their very primitiveness has made them difficult to classify definitively. *Eoraptor* was once believed to be a dinosaur that was neither saurischian nor ornithischian; then it and *Herrerasaurus* came to be recognized as primitive theropods; now, *Eoraptor* is thought to be a very primitive sauropodomorph! From the same rocks also came an ornithischian: *Pisanosaurus*, a sauropodomorph: *Panphagia*, and two more theropods: *Eodromaeus* and *Sanjuansaurus*. The sedimentary rocks of the Ischigualasto Formation, from which all of these animals came, have been unreliably dated at 229–218 Ma.

From nearby rocks (in Brazil) of presumed equivalent age, come the basal saurischian *Staurikosaurus* and sauropodomorph *Saturnalia*. In rocks younger than those that produced *Staurikosaurus*, also in Brazil, the primitive dinosaur *Guaibasaurus*, originally described as a basal saurischian but reinterpreted as a theropod, was discovered (Figure 15.6).

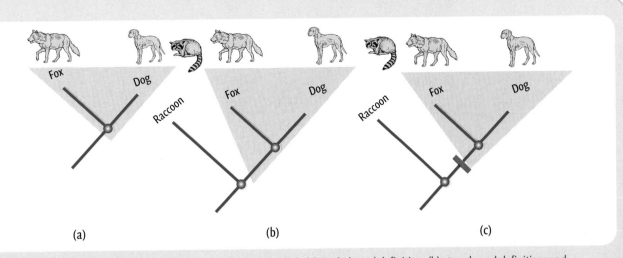

Figure B15.1.1. Phylogenetic definitions of the canids (dog family). (a) Node-based definition; (b) stem-based definition; and (c) apomorphy-based definition.

those organisms sharing a more recent common ancestor with *Canis* rather than a raccoon (*Procyon*). Finally, de Queiroz and Gauthier recognized *apomorphy-based* definitions, in which the group stems from the first ancestor to evolve a particular character (Figure B15.1.1c). This type of definition is a familiar one here; it refers to groups as we have diagnosed throughout this book.

In the same way, the phylogenetic definition of Dinosauria in Chapter 4 was apomorphy-based. The definition as given in this chapter, however, is node-based: the descendants of the most recent common ancestor of a saurischian

(*columba*, the pigeon) and an ornithischian (*Triceratops*). A stem-based definition of Dinosauria might be, all those organisms sharing a more recent common ancestor with *Columba* (a derived saurischian) than with *Lagerpeton*. This might work, unless it turned out that *Marasuchus* was closer to Dinosauria than to *Lagerpeton*. If so, these two archosaurs would have to be considered members of Dinosauria. Here, because we are not so sure about the exact relationships of *Marasuchus* and *Lagerpeton* relative to Dinosauria, the node-based definition is the more scientifically conservative approach.

The evolution of Dinosauria

The relationships among dinosauromorphs, dinosaurs, and their near relatives, pterosaurs (see Chapter 4; Figures 4.11, 4.13, and 13.7) have been somewhat confusing – mainly because the characters of basal dinosauromorphs and the early dinosaurs are too primitive, and those of pterosaurs are too advanced! This leaves us, on the one hand, with a cloud of primitive organisms lacking distinctive diagnostic characters upon which to build the phylogeny, but on the other hand with a group (pterosaurs) with so many characters that are so derived that it is hard to figure out the details of how these features evolved. Nonetheless, within the last few years, the outline of a consensus has been reached regarding the relationships of these two groups.

It had been thought, as recently as the early 1980s, that pterosaurs – otherwise highly modified for flight – might be the closest archosaurian relatives to dinosaurs, together sharing derived features as ornithodirans (see Figure 4.11). More recently, however, with the discovery

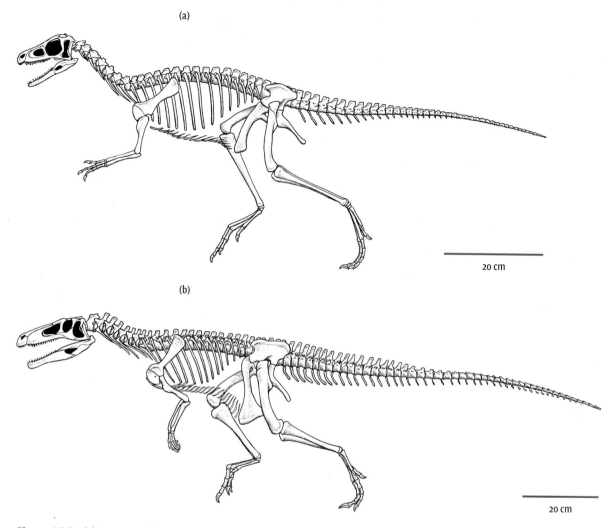

Figure 15.5. (a) *Eoraptor,* (b) *Herrerasaurus.*

of silesaurids, as well as a deepened understanding of creatures like *Lagosuchus* and *Lagerpeton,* the non-dinosaur basal dinosauromorphs are likely closest to dinosaur ancestry, with pterosaurs distant within the clade Ornithodira. Among non-dinosaur basal dinosauromorphs, silesaurids are believed to be closest to dinosaur ancestry, with *Marasuchus* and *Lagerpeton* successively more distant (Figure 15.2).

The general outlines of this consensus didn't come easily; paleontologist M. Langer and colleagues have shown that between the years 1986 and 2007, new hypotheses of the relationships of non-dinosaur basal dinosauromorphs to dinosaurs were published at the rate of about one a year! Part of the reason for this is that, as we have seen, all of these early dinosauromorphs have many primitive characters. It is also due to the abundance of new material that was discovered during those years. For example, once silesaurids were identified, paleontologists began to correctly recognize silesaurid fossils all over the world. Then, too, many of these fossils were, and are still, lacking skull material (note that the diagnostic characters listed in Figure 15.2 are not cranial). Skull material has been tremendously useful in sorting out the relationships of fossil taxa, and we can hope that when complete skulls for some of these taxa become known, their phylogenetic relationships will be clearer.

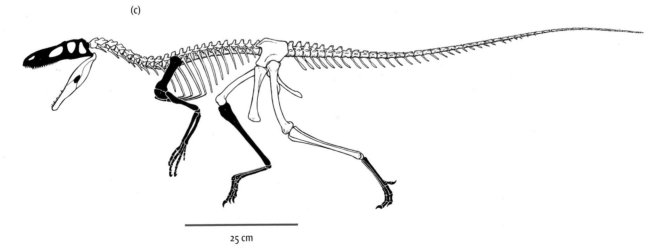

(a)

20 cm

(b)

30 cm

(c)

25 cm

Figure 15.6. (a) *Staurikosaurus*; (b) *Saturnalia*; (c) *Guaibasaurus*.

There is an interesting and perhaps surprising consequence of this phylogeny. With archosaurs like *Marasuchus* and the silesaurids close to dinosaurian ancestry, apparently dinosaurs were primitively obligate bipeds. This means that the earliest dinosaurs were creatures that were completely and irrevocably bipedal. Because the primitive stance for archosaurs is quadrupedal, and because Dinosauria is monophyletic, it follows that creatures like *Triceratops*, *Ankylosaurus*, and *Stegosaurus*, in fact, *all* quadrupedal dinosaurs, must have secondarily evolved (or re-evolved) their quadrupedal stance. They must have (phylogenetically) got back down on four legs, as it were, after having been up on two. In fact, you can see the remnant of bipedal ancestry when you look at a stegosaur or a ceratopsian, in which the back legs are quite a bit longer than those at the front. This happened early; it appears that even silesaurids returned to at least partial quadrupedality.

The very earliest dinosaurs were likely carnivorous, bipedal, and small- to medium-sized. They would have looked superficially like a theropod, and as we have seen from recent work described in Chapter 10, could have been covered with some kind of simple, down feather-like plumage. Unfortunately, the fossil record is moot (and mute) on this point. They might even have taken a tentative step or two down the road towards endothermy (Chapter 12).

Let the games begin!

Our information is limited about both the place and pace of the spread of dinosaurs. The oldest dinosaurs known (above) are all from a chunk of time of the Late Triassic called the Carnian Stage (Box 15.2). Because the Carnian-aged dinosaurs described above are all known from Brazil and Argentina, paleontologists have suggested that the cradle of Dinosauria was South America. Yet, terrestrial Carnian rocks that preserve dinosaur remains are not so common globally. So we end up asking, are the Carnian dinosaurs from South America truly the oldest dinosaurs, and thus South America was the place where dinosaurs originated? Or alternatively, are the terrestrial Carnian rocks of South America simply the only game in town, and thus they are most of what we know of the terrestrial Carnian Stage globally? We have no satisfactory answer for those questions. At this point all we can really say is that the evidence is strong that dinosaurs first appeared during the Carnian Stage, and that they might have first evolved in South America.

What a difference a Stage makes The Carnian Stage is thought to have ended at 228 Ma, which marks the beginning of the Norian Stage, the block of time that followed it (Box 15.2). By Norian time, dinosaurs clearly had acquired a startlingly pervasive global presence, and apparently populated all of the supercontinent of Pangaea (Figure 15.7).

What is striking is that even among these very early dinosaurs, ornithischians, sauropodomorphs, and theropods are recognizable, suggesting that, if we understand the geochronology correctly (Box 15.2), the evolution of these groups took place relatively quickly after the first appearance of dinosaurs.

This is not to say that the near-dinosaurs politely stepped aside the first time that a true dinosaur showed up at the party. The earliest dinosaurs were rather primitive representatives of their group, and many of them look superficially (and in some cases not-so superficially) similar to, and were living among, the basal dinosauromorphs from which they are thought to have derived. They likely did similar kinds of things ecologically. We have seen how the rich Carnian deposits of Argentina contain both early dinosaurs and non-dinosaur basal dinosauromorphs. In the Norian of the North American southwest, also, dinosaurs like the theropods *Coelophysis*

Box 15.2 **No dates; no rates**

Geochronologists, the scientists who determine the numerical ages of rocks, tell us, "No dates; no rates!" By this they mean that if we don't know when, numerically, events occurred, we can't really determine how quickly or slowly they took place. In this chapter, we tell a story of rapid early dinosaur evolution; yet, without dates, our story could be a house of cards. Numerical dates from unstable isotopes, such as those discussed in Chapter 2, are rare for this time interval (one geochronologist called the Late Triassic a "black hole"), and those few that are known are not particularly precise.

Historically, in the absence of numerical ages, relative dating was carried out using a combination of lithostratigraphy and biostratigraphy (see Chapter 2; Figure 2.3). These techniques were used to divide the Triassic into three formal Epochs: Early, Middle, and Late. Each of these, in turn, was subdivided using biostratigraphy into smaller blocks of time called Stages. The key Stages for our story are the Carnian, the Norian, and the Rhaetian, the last three Stages of the Late Triassic (Figure B15.2.1).

Numerical dates for these three Stages are very rare indeed; so while we know the sequence of time, exactly when that sequence occurred can be uncertain. For example, stratigraphers can still get a rude shock: in 2007, following a careful reanalysis of marine dates and faunas, the Norian gained almost 7 million years (at the expense of the Carnian!).

The situation is more of a problem in the terrestrial realm. Dinosaurs, as we have seen, are exclusively terrestrial beasts, and geographically dispersed dinosaur faunas around the globe have been correlated with each other. Yet, it has not been easy to securely tie these dinosaur faunas to the Norian and Carnian Stages, because these Stages are based upon marine organisms.

Still, there is progress: dates (229–218 Ma) exist for the Ischigualasto Formation of Argentina (Chapter 13), the unit that produced *Eoraptor* and *Herrerasaurus*. But we have no dates for the presumably equivalent rocks in nearby Brazil, which produced *Guaibasaurus* and *Staurikosaurus*. Recently a suite of extremely precise dates has been obtained for the Chinle Formation of the North American southwest, a rock unit that produced Late Triassic dinosaurs such as the theropod *Coelophysis* and the primitive saurischian *Chindesaurus*. For the time being, though, those dates stand alone, because their exact relationship to the Carnian and Norian Stages is poorly understood – since the numerical ages of those Stages is still not well understood. And this means that exactly how these North American faunas tie into other North

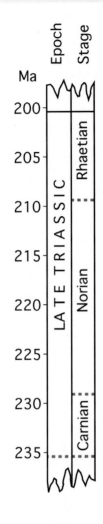

Figure B15.2.1. The Late Triassic time scale. Numerical dates from below the base of the Carnian approximate the base of the Carnian to slightly younger than 236 Ma. With the exception of a date from the upper Carnian of Italy, there are generally no reliable isotopic ages for the Late Triassic at this time. Therefore, ages for the Carnian–Norian boundary (227–228 Ma) and the Norian–Rhaetian boundary (208–209 Ma) have been interpolated using magnetostratigraphy, a technique based upon global magnetic polarity reversals detected in key stratigraphic sections. The Triassic–Jurassic boundary, on the other hand, has been precisely dated by isotopic methods at 201.3 Ma – but in the marine realm, ecologically and geographically far from dinosaurs.

American Late Triassic faunas, let alone global Late Triassic faunas, is not so clear, either. Stay tuned!

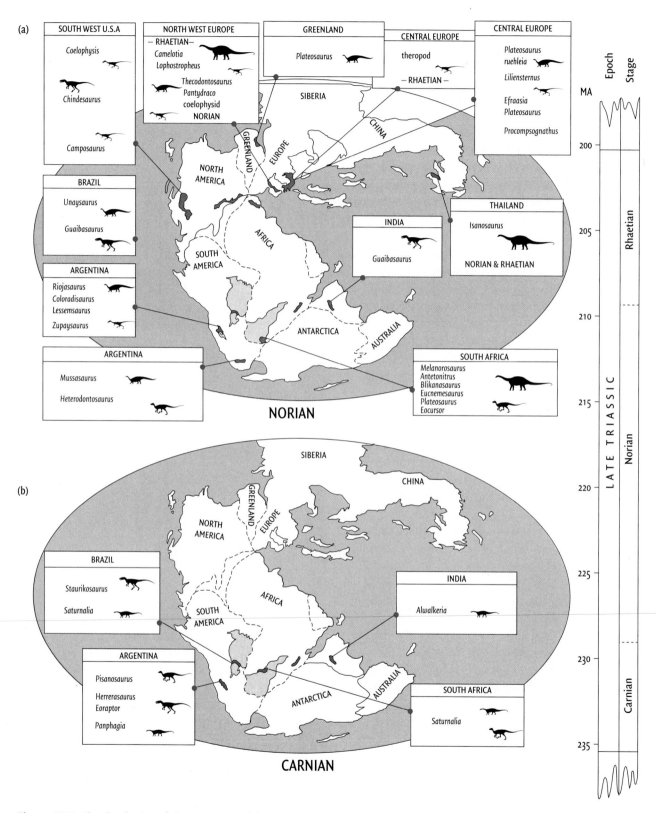

Figure 15.7. The distribution of dinosaurs around the Late Triassic supercontinent of Pangaea (a) Norian; (b) Carnian. Redrawn from Langer *et al.*, 2010.

and *Chindesaurus* lived with non-dinosaur basal dinosauromorphs such as *Dromomeron*, an animal closely related to *Lagerpeton*. Indeed, the Late Triassic was a time generally characterized by a relatively stable, long-lived global fauna consisting of dinosaurs and non-dinosaur dinosauromorphs co-existing. By the end of the Late Triassic, however, non-dinosaur dinosauromorphs phased out of the picture, and true dinosaurs were off and cursorial!

Summary

The earliest and morphologically most primitive dinosaurs were small, bipedal, carnivorous, and/or insectivorous creatures, known from deposits in Argentina and Brazil, dated at about 229–218 Ma. These animals lived with, as well as were preceded by, several groups of archosaurs – non-dinosaur basal dinosauromorphs – of similar morphology and inferred behavior. The similarities of morphology suggest that dinosaur ancestry is to be found among the basal dinosauromorphs, likely, among the recently recognized silesaurid clade. These earliest known dinosaurs, although primitive, have been affiliated with somewhat derived dinosaurian groups, such as Ornithischia, Saurischia, Theropoda, and Sauropodomorpha. This suggests that some amount of evolution must have occurred preceding 229 Ma and the appearance of early dinosaurs such as *Herrerasaurus* and *Eoraptor*.

Once they made their Carnian (early Late Triassic) or even pre-Carnian appearance, dinosaurs radiated and diversified. By Norian (middle Late Triassic) time, they had achieved a global distribution, and were starting to dominate terrestrial faunas.

Selected readings

Brusatte, S. L., Nesbitt, S. J., Irmis, R. B., Butler, R. J., Benton, M. J. and Norell, M. A. 2010. The origin and early radiation of dinosaurs. *Earth-Science Reviews*, **101**, 68–100.

Langer, M. C., Ezcurra, M. D., Bittencourt, J. S. and Novas, F. E. 2010. The origin and early evolution of dinosaurs. *Biological Reviews*, **85**, 55–110.

Nesbitt, S. J. 2011. The early evolution of archosaurs: relationships and the origin of major clades. *Bulletin of the American Museum of Natural History*, **352**, 292pp.

Topic questions

1 What is Pseudosuchia? Ornithodira? How would one tell the difference?

2 Give an example of a pseudosuchian archosaur. When did the last pseudosuchian live?

3 What do those names (Pseudosuchia; Ornithodira) have to do with Dinosauria?

4 What is a dinosauromorph?

5 Please name two non-dinosaurian dinosauromorphs.

6 What do those non-dinosauromorphs tell us about the morphology of the first dinosaurs?

7 Describe the earliest known dinosaurs. Are these also the most primitive dinosaurs? What is the difference between "earliest known" and "most primitive?"

8 People have discussed whether or not the evolution of Dinosauria was "fast" or "slow." Which do you think it was? Please explain your answer: (a) what do you mean by "fast" (or "slow"), and (b) how do you measure fast or slow evolution?

9 In a phylogenetic classification, what is meant by the words "definition" and "diagnosis?" What is the difference between the two? Why would this matter?

CHAPTER 16

THE CRETACEOUS–TERTIARY EXTINCTION:
THE FRILL IS GONE

CHAPTER OBJECTIVES

- Learn about the K/T boundary

- Learn about the K/T extinction

- Evaluate scientific hypotheses in the light of incomplete data

How important were the deaths of a few dinosaurs?

The extinction to which the non-avian dinosaurs finally succumbed after about 163 million years on Earth is called the Cretaceous–Tertiary extinction, commonly abbreviated K/T.[1] The K/T extinction involved much more than just dinosaurs. Among the "highlights" were:

- a large asteroid collided with Earth;
- the great cycles of nutrients that formed the complex food webs in the world's oceans temporarily shut down;
- many mammals went extinct;
- landscapes were deforested; and
- wildfires likely raged.

By comparison with that, how important were the deaths of a few dinosaurs?

Geological record of the latest Cretaceous

Earth gets a makeover

Mountains and volcanoes The Late Cretaceous was a time of active plate movement, mountain-building, and volcanism. The Rocky Mountains, Andes, and Alps were all entering important growth periods.

The Indian continent experienced a unique episode of volcanism that occurred from 67 to 64 Ma (from the very end of the Cretaceous into the Early Tertiary), consisting of episodes of lava flows, called the Deccan Traps, which spewed molten rock over an area of 500,000 km². Volatile gasses – carbon dioxide, sulfur oxides, methane, and possibly nitric oxides among the most prevalent – were emitted into the atmosphere, possibly affecting global temperatures and damaging the ozone layer. Yet between the lava episodes over that 3 million year period, life apparently returned to normal.

Sea level The Late Cretaceous was marked by lowered global sea levels (a regression), from highs enjoyed during mid-Cretaceous time (about 100 Ma). The regression peaked just before the K/T boundary and global sea levels began to rise as 65.5 Ma – our best date for the K/T boundary – came and went. Still, by the end of the Cretaceous, more land was exposed than had been in the previous 60 or so million years.

Seasons We have seen that the latter half of the Cretaceous seems to have been a time of gentle cooling from the highs reached in the mid Cretaceous (see Chapters 2 and 13). In North America at least, climates through the Late Cretaceous were relatively equable, based upon plant fossils.

Asteroid impact

In the late 1970s, geologist Walter Alvarez and a team of co-workers (Figure 16.1) were studying K/T marine outcrops now exposed on land near a town called Gubbio, in Italy. They

---

[1] The extinction is thus said to have occurred at the Cretaceous–Tertiary (or K/T) boundary. The "T" in K/T obviously stands for the Tertiary Period. The "K" the German word *Kreide*, or chalk, because the Cretaceous was first identified at the chalk cliffs of Dover (England).

Figure 16.1. The team of University of California (Berkeley) scientists who first successfully proposed the theory of an asteroid impact at the K/T boundary. Left to right: geochemists Helen V. Michel and Frank Asaro, geologist Walter Alvarez, and physicist Luis Alvarez.

were struck by the fact that the lower half of the Gubbio rock exposure is composed of a rock made up entirely of thin beds of the microscopically sized shells of *Cretaceous* marine organisms. The upper half of the exposure was almost exclusively of thin beds of the microscopic shells of *Tertiary* marine organisms. Between the two was a thin (2–3 cm) layer of clay, the K/T boundary.

Analyses showed that the clay layer contained unusually high concentrations of iridium, a rare, platinum group metal.[2] Instead of the expected amount at the Earth's surface, about 0.3 parts per billion (ppb), the iridium content was a whopping 10 ppb at Gubbio. So the iridium anomaly, as it came to be called, contained about 30 times as much iridium as Alvarez and his co-workers had expected to find (Figure 16.2).

Iridium is normally found at the Earth's surface in very low concentrations, but it is found in higher concentrations in the core of the Earth and from extraterrestrial sources; that is, from outer space. Given that, the Alvarez group determined that the source of the iridium had to be extraterrestrial. The deal was sealed when they found iridium anomalies at two other K/T sites, one in Denmark and the other in New Zealand. With stunning intuition, they concluded that at 65.5 Ma, an asteroid had to have smacked into Earth, delivering the iridium and, coincidentally, causing the K/T mass extinction. Luis Alvarez, Nobel Prize-winning physicist, and a member of the team, described the relationship between an asteroid impact and the iridium layer in this way:

> When the asteroid hit, it threw up a great cloud of dust that quickly encircled the globe. It is now seen worldwide, typically as a clay layer a few centimeters thick in which we see a relatively high concentration of the element iridium – this element is very abundant in meteorites, and presumably in asteroids, but is very

---

[2] It is a common misconception that iridium metal is toxic and deadly. In fact, like its chemical relatives gold and platinum, it is quite unreactive. For those with significant disposable incomes, boutique fountain pens and watches made with iridium are available.

rare on earth. The evidence that we have is largely from chemical analyses of the material in this clay layer. In fact, meteoric iridium content is more than that of crustal material by nearly a factor of 10^4. So, if something does hit the earth from the outside, you can detect it because of this great enhancement. Iridium is depleted in the earth's crust relative to normal solar system material because when the earth heated up [during its formation] and the molten iron sank to form the core, it "scrubbed out" [i.e., removed] the platinum group elements in an alloying process and took them "downstairs" [to the core]. (Alvarez, 1983, p. 627)

Because the three sites are distributed around the globe, Alvarez and co-workers calculated that the asteroid had to have been about 10 km (about 6 miles) in diameter to spread an iridium dust layer globally.

In the intervening years, a tremendous amount of work has been done to explore the possibility of an asteroid impact at 65.5 Ma. Most importantly, the number of K/T sites with anomalous concentrations of iridium at the boundary has reached well over 100 (Figure 16.3). Moreover, the iridium anomaly was discovered on land (Figure 16.4) as well as in ocean sediments, affirming that it is a global phenomenon.

Shocked quartz and microtektites also came to be recognized as part of the fingerprint left by the asteroid. "Shocked quartz" is the name given to quartz that has been placed under such pressure that its molecular structure becomes deformed (Figure 16.5). It is now recognized that the kind of pressure that can cause such deformation could only be generated by impacts; indeed, shocked quartz is now known from many different impact sites, and has come to be recognized as a diagnostic criterion for meteor impacts.

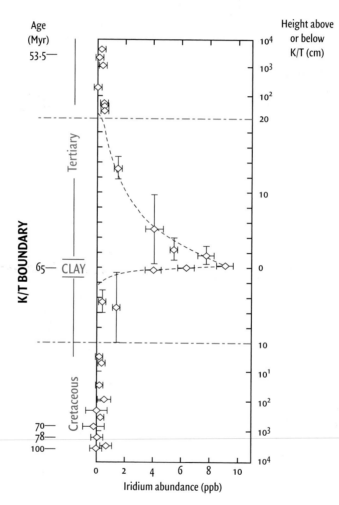

Figure 16.2. The iridium (Ir) anomaly at Gubbio, Italy. The amount of Ir increases dramatically at the clay layer to 9 parts per billion (ppb), and then decreases gradually above it, returning to a background count of about 1 ppb. On the right are numbers representing the thickness of the rock outcrop; on the left the time intervals (in millions of years (Myr)) and rock types are identified. Note that the vertical scale is *linear* close to the K/T boundary, but *logarithmic* away from the boundary, to show results well above and well below the boundary.

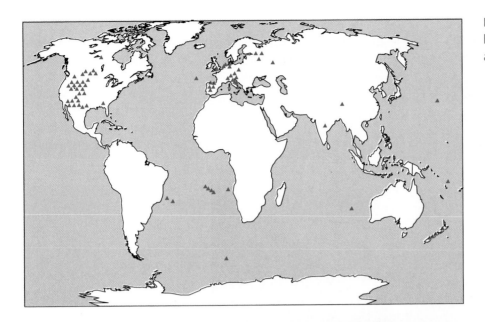

Figure 16.3. More than one hundred and three known iridium anomalies around the world.

Figure 16.4. The iridium-bearing clay layer in Montana; one of the first localities on land where anomalous concentrations of iridium were discovered. Lines bracket thin Cretaecous-Tertiary boundary claystone.

Microtektites are small, droplet-shaped blobs of silica-rich glass. They represent material thrown up into the atmosphere in a molten state due to the tremendous energy released when a meteor strikes the Earth. Quick cooling occurs while they're still airborne and then they plummet down on Earth as a rain of solid, glassy blobs.

The "smoking gun" As early as 1981, a bowl-shaped structure 180 km in diameter, buried under many meters of more recent sediments, was reported from the Yucatan Peninsula of Mexico, in the region near the town of Chicxulub (translated approximately as "devil's tail"; Figure 16.6). Ten years later, drill cores taken through the structure revealed shocked quartz: Chicxulub was a buried impact structure.

At about the same time, an approximately 1 m thick sequence of glass was discovered in Haiti, suggesting that the source of the glass had to be somewhere, relatively nearby. Its

Figure 16.5. Shocked quartz from the terrestrial K/T boundary in eastern Montana. The etched angled lines across the face of a grain of quartz represent a failure of the crystal lattice along known crystallographic directions within the mineral. Grain is 70 μm across (1 μm = 10⁻⁶ m).

chemical composition was shown to be the same as the composition of the rocks that make up the Chicxulub structure.

The pieces really started falling into place. Several years earlier (1988) evidence of a tsunami in K/T deposits in the Gulf Coast region of Texas had been reported. The Chicxulub site was well situated to produce the tidal wave deposits recognized in the sedimentary record. Finally, the Chicxulub structure was dated at 65.5 Ma, the time of the K/T boundary.

Further study of Chicxulub below the surface of the Earth, using sophisticated geophysical techniques, showed a bullseye pattern with a circular peak and large concentric rings around it, representing topography preserved in buried rocks below the surface (Figure 16.7). Interestingly enough, the northwest part of the outermost ring is broken through. The distinctive ring pattern suggests that a large asteroid, 10–15 km in diameter,[3] approached Earth from the southeast at a low angle of about 30°. The distribution of iridium, shocked quartz, and microtektites across the Western Interior of North America (north and west of the crater) reinforce the idea of a low-angle, directional impact (Figure 16.8).

What did the asteroid do to Earth when it struck? Numerous scenarios were proposed, most of them inspired by post-nuclear apocalyptic visions. Of these, only a few remain current:

- *Blockage of sunlight.* It was initially hypothesized that sunlight would have been blocked from the Earth for about three to four months. This would have caused a cessation of photosynthesis and a short-term temperature decrease (now called an impact winter).
- *Infra-red radiation pulse.* It has been theorized that tremendous amounts of energy in the form of infra-red radiation and heat must have been released immediately upon impact. The initial global heat release at ground zero might have been 50 to 150 times as much as the energy of the sun as it normally strikes Earth. One group of scientists likened this radiation at the Earth's surface to an oven left on the broil.

[3] Recall that previous estimates of size were based upon the global distribution of the ejecta; this estimate was based upon the morphology of the impact site.

- *Global wildfires.* With so much in- stantaneous heat production, fires might have broken out spontaneously around the globe. Soot-rich horizons from five K/T sites in Europe and New Zealand have been identified, in which the amount of the element carbon (the soot) was enriched between 100 and 10,000 times over background. The soot has been attributed to wildfires, perhaps the result of the infra-red heat pulse.

All of these catastrophic effects are short term, which means that they affected the globe for days, months, or at most a few years. In a longer-term sense, that is, on eological time scales (10^4–10^6 years), climates were little affected by the asteroid impact. What we know of climates in the latest Cretaceous suggests that they did not differ significantly from those in the early Tertiary.

Biological record of the latest Cretaceous

The other side of the K/T coin are the biological extinctions. No amount of comets, volcanoes, asteroids, meteors, cooling, warming, ice ages, or natural (or unnatural) catastrophes can be used to explain any extinction until we understand the anatomy of the K/T mass extinction itself (Box 16.1). The pattern of survivorship, that is, who goes extinct and who survives, becomes an important issue in understanding an extinction and determining potential causes. So what were the biological damages?

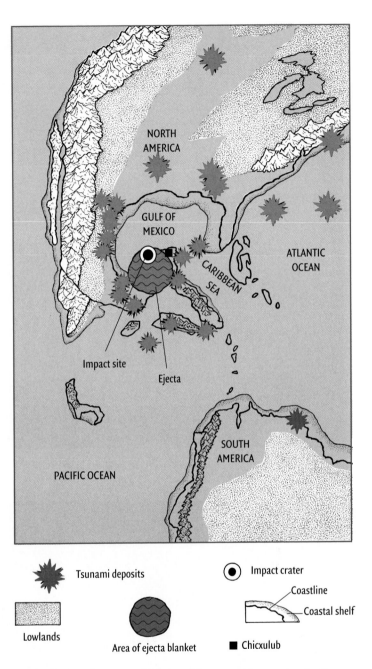

Figure 16.6. Paleogeographic map of the ground-zero region for the K/T asteroid, Yucatan Peninsula, Mexico. The geography of the region as we know it today is superimposed over the geography of the region at 65.5 Ma.

Oceans

Continental seas and shelves Because the shallow seas that covered large expanses of the continents receded before the K/T boundary, very few shallow marine deposits are preserved that record the last 2–3 million years of the Cretaceous. And because many groups of organisms lived and died in shallow continental seas and shelves, we lack data for such groups.

How well or badly fish and sharks fared remains largely conjectural, although it is appar- ent that, as a group, they did not suffer the kind of wholesale decimation seen in other groups.

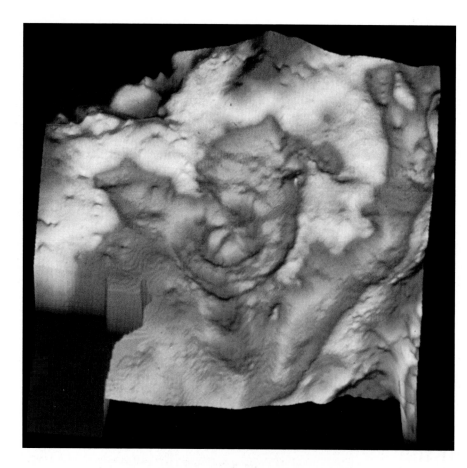

Figure 16.7. Three-dimensional geophysical reconstruction of the remnants of the Chicxulub crater. A gravimeter measures subsurface changes in gravitational attraction of rocks under the town of Chicxulub. These variations in gravitational attraction show a large-scale bullseye pattern of concentric rings, diagnostic of a meteor impact. North is toward the top of the page.

Crater showing oblique strike

Figure 16.8. Reconstruction of an asteroid impact with Earth. Planetary geologist P. H. Schultz and geobiologist S. L. D'Hondt suggest that the asteroid struck Earth at an angle of about 30°, coming from the southeast.

Box 16.1 **Extinction**

Paleontologists generally divide extinctions into two categories. The first are the so-called **background extinctions**, isolated extinctions of species that occur in an ongoing fashion. The second type are called **mass extinctions**. The latter certainly have caught media and the public's attention, and they appear to be something *qualitatively* as well as quantitatively different than background extinctions.

Background extinctions

Although background extinctions are less glamorous than mass extinctions, they are essential to biotic turnover: University of Tennessee paleobiologist M. L. McKinney has estimated that as much as 95% of all extinctions can be accounted for by background extinctions. Isolated species disappear from a variety of causes, including out-competition (the edge), depletion of resources in a habitat, changes in climate, the growth or weathering of a mountain range, river channel migration, the eruption of a volcano, the drying of a lake, the spraying of a pesticide, or the destruction of a forest, grassland, or wetland habitat.

Dinosaur populations had a species–turnover rate of around 2 million years per species. This means that each species lasted about 2 million years, before a new one appeared and the old one disappeared.[1] Although some dinosaur extinctions coincided with earlier mass extinction events (such as those at the Triassic–Jurassic and Cretaceous–Tertiary boundaries), most dinosaurs fell prey to background extinctions. By far the majority of favorite and famous dinosaurs – *Maiasaura, Dilophosaurus, Protoceratops, Deinocheirus, Styracosaurus, Velociraptor, Iguanodon, Ouranosaurus, Allosaurus* (to name a tiny fraction)

– were the victims of background extinctions. The ultimate dinosaur extinction didn't wipe out the total number of species accumulated over 160 million years, it killed only the latest-evolved representatives of the group (see Figure 13.1).

Mass extinctions

Mass extinctions involve *large numbers* of species and *many types* of species undergoing *global extinction* in a *geologically short* period of time. None of these has a truly precise definition, because there are no fixed rules for mass extinctions. Indeed, how do we know that there even were mass extinction "events" and how can we recognize them? A compilation of invertebrate extinctions through time (Figure B16.1.1) shows that, although extinctions characterize all periods (it is these that are termed background extinctions), there are intervals of time in which extinction levels are significantly elevated above background levels. Such intervals are said to contain the mass extinctions. Fifteen such intervals are recognized, of which five clearly towered above the others (Figure B16.1.1).

The 15 mass extinctions are classified into "minor," "intermediate," and "major" mass extinctions, on the basis of the amount of extinction that took place above background. In the entire history of life, only one extinction qualifies as "major"; that is, the **Permian–Triassic** (commonly called **Permo-Triassic**) extinction. The remaining four of the Big Five – including dinosaur extinction – are considered to have been "intermediate." The rest are considered "minor," although undoubtedly not to the organisms that succumbed during them.

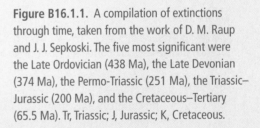

Figure B16.1.1. A compilation of extinctions through time, taken from the work of D. M. Raup and J. J. Sepkoski. The five most significant were the Late Ordovician (438 Ma), the Late Devonian (374 Ma), the Permo-Triassic (251 Ma), the Triassic–Jurassic (200 Ma), and the Cretaceous–Tertiary (65.5 Ma). Tr, Triassic; J, Jurassic; K, Cretaceous.

[1] This is a simple statistical average for all of Dinosauria. It was not necessary for an older species to disappear before its descendant appeared.

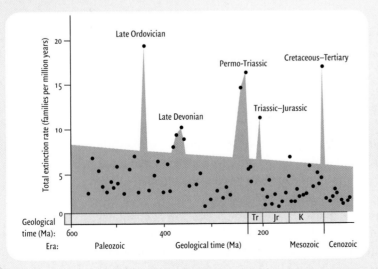

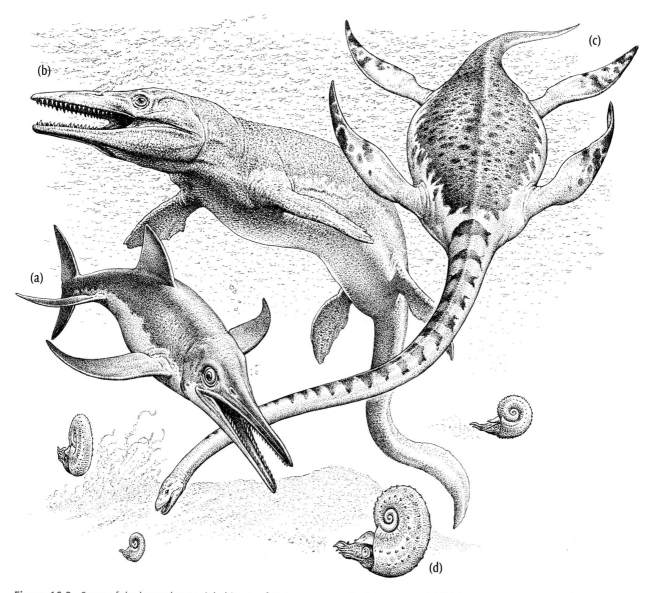

Figure 16.9. Some of the better-known inhabitants of Cretaceous seas. Vertebrates are: (a) ichthyosaur (*Platypterygius*); (b) mosasaur (*Tylosaurus*), and (c) plesiosaur (*Elasmosaurus*). (d) The shelled, tentacled invertebrates toward the bottom of the drawing are cephalopod mollusks called ammonites.

The whale- and dolphin-like marine diapsids called ichthyosaurs (Figure 16.9a) are known to have disappeared well before the K/T boundary. Not so in the case of marine-adapted lizards called mosasaurs (Figure 16.9b). Recent work suggests that these went extinct geologically abruptly, at the end of the Cretaceous. More equivocal is the record of plesiosaurs, the long-necked, Loch Ness-type, fish-eating diapsids of the Jurassic and Cretaceous (Figure 16.9c), for whom there are, Loch Ness notwithstanding, no credible post K/T records.

Among fossil invertebrates, perhaps the most famous group are the ammonites (Figure 16.9d). Ammonites lived right up to the K/T boundary, before finally going extinct. Another important group of invertebrates are the bivalves. Careful study has shown that, with one exception (which went extinct much earlier), 63% of all bivalves went extinct sometime within the last 10 million years of the Cretaceous. The record is, unfortunately, not more precise than this, but it does show that the extinction took place without regard for latitude: bivalves in temperate regions were just as likely to go extinct as those in the tropics.

Marine microorganisms Because of the richness of their fossil record, foraminifera, marine microscopic, shelled, single-celled organisms that are either planktonic (living in the water column) or benthic (living within sediments) have dominated discussions of K/T boundary events (Figure 16.10). Micropaleontologists studying foraminifera have shown persuasively, since as early as the late 1970s (and in many studies thereafter, including the observations of the Alvarez team at Gubbio) that the planktonic foraminiferan extinction was abrupt, with only a few species crossing the boundary into the Paleocene.

An entirely different group of marine microorganisms, calcareous nanofossils, also shows an abrupt extinction. It is safe to say that most (but not all!) paleontologists working with marine microorganisms are inclined toward a catastrophic view of the extinction.

The "Strangelove" oceans Some of the most unexpected results came from a series of studies of K/T oceanic primary production; that is, the amount of organic matter synthesized by

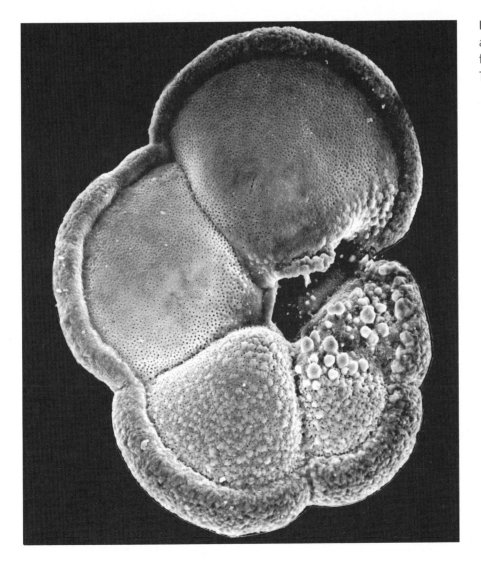

Figure 16.10. The carbonate shell of a modern planktonic (free-swimming) foraminifer, *Globorotalia menardii*. The long dimension is 0.75 mm.

organisms from inorganic materials and sunlight.

At the K/T boundary there was observed a rapid and complete breakdown in nutrient cycling between surface and deep waters, to less than 10% of what it had been. For some oceanographers, this signaled that the world's oceans were effectively all but dead. For the succeeding 1.5 million years, nutrient cycling remained at levels well below those preceding the original drop.[4]

Obviously, nutrient cycling is fundamental to oceanic health. With oceans covering 75% of the Earth's surface (or even more during the many high sea levels experienced during Earth history), it would not be an exaggeration to state that Earth's marine and terrestrial ecosystems are dependent upon these great cycles.

Figure 16.11. The K/T boundary in eastern Montana, USA. The boundary is midway up the butte, right at the dotted line. Below is the dinosaur-bearing latest Cretaceous Hell Creek Formation; above is the Tertiary Fort Union Formation. No dinosaurs have ever been found in the Fort Union Formation.

Terrestrial record

For better or worse, virtually all of what we know of the K/T boundary on land also comes from the Western Interior of North America (Figure 16.11). There, several well-studied, complete sections have provided the best insights available into the dynamics of the extinction.

Plants The plant fossil record in the Western Interior has two major components, a palynoflora (spores and pollen) and a megaflora (the visible remains of plants, especially leaves; Figure 16.12). After 15 years of intensive scrutiny, both records agree nicely with each other and both records indicate that a major extinction occurred geologically instantaneously at the K/T boundary.

Interestingly, pollen that is typical of early Paleocene time does not immediately follow the extinction of the Cretaceous pollen. Instead there is a high concentration of fern spores just after the iridium anomaly, suggesting that, immediately after the extinction of the Cretaceous plants, there was a "bloom" of fern growth, interpreted to be a pioneer community growing

---

[4] The moribund K/T oceans were called "Strangelove" oceans after Dr Strangelove, Peter George's brittle, grotesque character, played by Peter Sellers in the eponymous film, who was unconcerned about a scorched, post-nuclear world.

Figure 16.12. Plant fossils. (a) Late Cretaceous leaf. The leaf is from an angiosperm that became extinct at the K/T boundary. The specimen is from just outside Marmarth, North Dakota, USA. (b) Pollen grains belonging to the genera *Proteacidites* (1) and *Aquilapollenites* (2), both important genera in measuring the moment of the terrestrial K/T extinction. *Proteacidites* is about 30 μm across; *Aquilapollenites* is about 50 μm.

on a devastated post-impact landscape. In time, the fern flora gave way to a more diverse angiosperm flora characteristic of the early Paleocene.

Outside North America, an interesting southern high-latitude flora is known from New Zealand. There, an abrupt pollen and spore extinction as well as the fern spike are also known. In short, the pollen record suggests that the terrestrial K/T boundary was characterized by global deforestation.

The megafloral record based upon 25,000 plant specimens for the Western Interior of North America shows that while some environmental changes caused extinctions earlier than the K/T boundary, a major extinction took place precisely at the boundary, exactly correlated with the pollen extinction and iridium anomaly. The extinction of 79% of the known angiosperms suggested that, as suspected, the fern "bloom" may have been a response to the absence of flowering plants that would normally have occupied the ecosystem.

Vertebrates Some clear patterns of survivorship, that is, who survived and who did not, can be extracted from the K/T vertebrate fossil record of the US, Western Interior. Organisms that lived in aquatic environments (that is, rivers and lakes) showed up to 90% survival, whereas organisms living on land showed as little as a 10% survivorship. Thus the extinction seems not to have drastically affected aquatic organisms such as fish, turtles, crocodiles, and amphibians, but apparently wreaked havoc among terrestrial organisms such as mammals and, of course, dinosaurs. Several other survivorship patterns, not as statistically robust, also appear in the data: small vertebrates are favored over large vertebrates, ectotherms over endotherms, and non-amniotes over amniotes (Figure 16.13).

Dinosaurs Dinosaurs are difficult animals to study (Box 16.2) and for many years, no scientific study of dinosaurs at the K/T boundary was ever carried out. Inexplicably, although no data were ever published to show this, it was long accepted that dinosaurs gradually died off about 10 million years before the boundary.

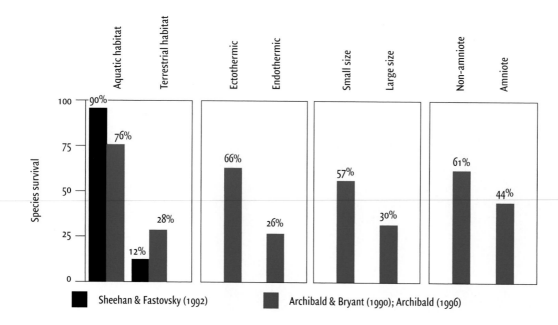

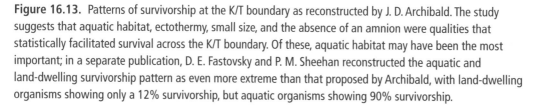

Figure 16.13. Patterns of survivorship at the K/T boundary as reconstructed by J. D. Archibald. The study suggests that aquatic habitat, ectothermy, small size, and the absence of an amnion were qualities that statistically facilitated survival across the K/T boundary. Of these, aquatic habitat may have been the most important; in a separate publication, D. E. Fastovsky and P. M. Sheehan reconstructed the aquatic and land-dwelling survivorship pattern as even more extreme than that proposed by Archibald, with land-dwelling organisms showing only a 12% survivorship, but aquatic organisms showing 90% survivorship.

While that is clearly not correct, the exact pattern of how they went out remains controversial. While the overall trajectory of dinosaur diversity through time is ever-increasing (see Figure 13.2), data obtained in 2009 suggest that globally, dinosaurs may have experienced a decline in diversity several million years before the K/T boundary. Indeed, it may have depended upon who you were and what you did: theropods may not have declined, while ceratopsians and hadrosaurs may have. These fluctuations in their biodiversity, however, were no greater than many that had occurred in the preceding 163 million years of dinosaurs on Earth. Nobody has identified anything that makes this particular fluctuation in dinosaur populations unique.

Decline in biodiversity or no, we can be sure that there were not just a few dinosaurs hanging on right before the end of the Cretaceous. Indeed, non-avian dinosaurs were still a very diverse, abundant, and important terrestrial group of vertebrates within 1 million – 500,000 years of the K/T boundary. How quickly, then, were they eliminated from Earth?

Box 16.2 Dinosaurs: all wrong for mass extinctions

What are some of the problems with reconstructing changes in dinosaur populations over time? For one thing, dinosaurs are, by comparison with foraminifera for example, large beasts and, more importantly, not particularly common.[1] For this reason, the possibility of developing a statistically meaningful database is impractical, and rigorous studies of dinosaur populations are very hard to carry out. Just counting dinosaurs can be difficult. Mostly, one doesn't find complete specimens, and adjustments have to be made. For example, if you happen to find three vertebrae at a particular site, they might be from one, or two, or three individuals. The only way to be sure that they belong to a single individual is to find them articulated. Suppose they are not; then one must speak of **minimum numbers of individuals**, in which case the three vertebrae would be said to represent one individual: that would be the minimum number of individual dinosaurs that could have produced the three vertebrae. On the other hand, if one found two left femora, then the minimum number of individuals represented would be two.

It would be nice to use all the specimens that have been collected in the last 170 years of dinosaur studies in a study of changes in dinosaur diversity. Unfortunately, dinosaur specimens are commonly collected because they are either beautiful specimens or rare; hardly criteria for ensuring that an accurate census of dinosaur populations has been performed. So any study that really is designed to get an accurate census of dinosaur abundance or diversity at the end of the Cretaceous must begin by counting specimens in the field, which is a labor-, time-, and cost-intensive proposition.

Then, of course, the taxonomic level at which to count dinosaurs can create problems. Suppose that two specimens are found; one is clearly a hadrosaurid and the other is an indeterminate ornithischian. The indeterminate ornithischian might be a hadrosaurid, in which case we should count two hadrosaurids. But then again it might not (because its identity is indeterminate), in which case calling it a hadrosaurid would give us more hadrosaurids in our survey than actually existed. On the other hand, calling both specimens "ornithischians" is quite correct, but not very informative, if we hope to track the survivorship patterns of *different types* of dinosaurs.

Finally, within the sediments themselves, problems of correlation exist. Suppose that, in Montana, we record the last (highest level) dinosaur in the Jordan area and then record the last (highest) dinosaur in the Glendive area, about 150 km away from Jordan. Can these two dinosaurs be said to have died at the same time? How could one possibly know? Suppose that in fact these dinosaurs died 200 years apart. An interval of 200 years, viewed from a vantage point of 65 million years, is literally a snap of the fingers. Yet 200 years is a long time when one is considering an instantaneous global catastrophe that ideally is measured in milliseconds.

---

[1] How rare are dinosaurs in this part of the world? Of course, we cannot know the density of dinosaurs within the rocks, but their surface density was calculated by sedimentologist P. White and colleagues, using the Sheehan et al. database (Fastovsky and Sheehan, 1997, p. 527), reported, "White and Fastovsky calculated that 0.000056 dinosaurs are preserved per m2 of outcrop. Considered more realistically, in a statistical sense one must search a 5 m wide path of exposed rock that is 4 km long to find a single dinosaur fragment identifiable to family level (or lower)." (Fastovsky, D. E. and Sheehan, P. M. 1997. Demythicizing dinosaur extinctions at the Cretaceous–Tertiary boundary. In Wolderg, D. L., Stump, E. and Rosenberg, G. D. (eds.), Dinofest International. Academy of Natural Sciences, Philadelphia, p. 527.)

In the late 1980s and through the 1990s, field-based studies were finally designed and carried out to determine the rate of dinosaur extinction. All of them took place in the American West: two on what had been a low-lying coastal plain in what is now eastern Montana and western north Dakota, and one in an intermontane basin in what is now Wyoming, which formed in the ancestors of the present-day Rocky Mountains, rising to the west (Figure 16.14) *The three studies all concluded the same thing: the dinosaur extinction was geologically abrupt.*

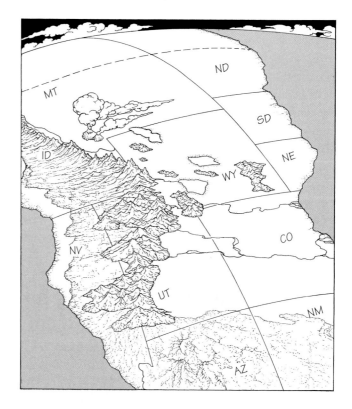

Figure 16.14. Paleogeography of the Western Interior of the USA, as it would have looked during Late Cretaceous time.

The two studies in the coastal plains of what is now eastern Montana and western North Dakota were quantitative censuses of dinosaur diversity during the last 1.5 million years of the Cretaceous. One looked at the ecological diversity; that is, the proportion of the total dinosaur population taken by each of eight families of dinosaurs. The other counted genera of all vertebrates through the last 1.5 million years of the Cretaceous in that region, looking for changes in either abundance or diversity. Both demonstrated that, within about 160,000 years of the K/T boundary, neither ecological diversity nor abundance and generic diversity changed (Figure 16.15).

The last of the three studies, carried out in what was an ancestral Rocky Mountain intermontane basin (Figure 16.16), utilized an approach very similar to the coastal plain study of vertebrate genera described above. And the results from that study were much the same as those from the other studies: the extinction of the dinosaurs was geologically abrupt. Major extinctions occurred in most groups, but particularly in dinosaurs and mammals (Figure 16.16). A key point, however, is that none of these studies can distinguish whether the extinction took every day, or whether it took only the last minute, of that last 150,000 years of the Cretaceous.

In summary, the very limited data from the Western Interior of the USA strongly indicate an abrupt end for the non-avian dinosaurs. Only time and much further study will enable us to integrate other dinosaur-bearing localities from around the world into what is already known of North America.

Extinction hypotheses

Much – indeed, most – of what has been proposed to explain the extinction of dinosaurs does not even possess the basic prerequisites for a viable, scientific theory. These minimal criteria are:

1   *The hypothesis must be testable.* As we have seen (see Chapter 3), for a hypothesis to be considered in a scientific context, it must be *testable*; that is, it must have predictable, observable consequences. Without testability, there is no way to falsify a hypothesis and, in the absence of falsifiability, we are then considering belief systems rather than scientific hypotheses. If an event occurred and left no traces that could be observed (by whatever means available), science is simply not an appropriate tool to investigate the event.

2   *The hypothesis must be parsimonious; that is, explain all the events in question.* This criterion is rooted in the principle of parsimony (again, see Chapter 3). We would like to explain an event or series of events. If each step of the event (or events) requires an additional *ad hoc* explanation, our hypotheses lose strength. They are strongest when the most parsimonious explanation is used: that explanation which explains the most observations. For this reason, if we can explain all that we observe at the K/T boundary with a single hypothesis, we have produced the most parsimonious hypothesis and it has a good chance of being correct.

Extinction hypotheses

In Table 16.1 we present about 80 years of serious, published proposals designed to explain the extinction of the dinosaurs (although see Box 16.3). The majority were published within the past 30 years. Consider each; you don't need to be a professional paleontologist to reject most of them, for most fail to meet the twin criteria for science enumerated above.

Any theory that purports to explain K/T events in a meaningful way must also explain the other events associated with the boundary. With that in mind, the hypothesis that an asteroid impact caused the events at the K/T boundary becomes the most interesting and plausible hypothesis.

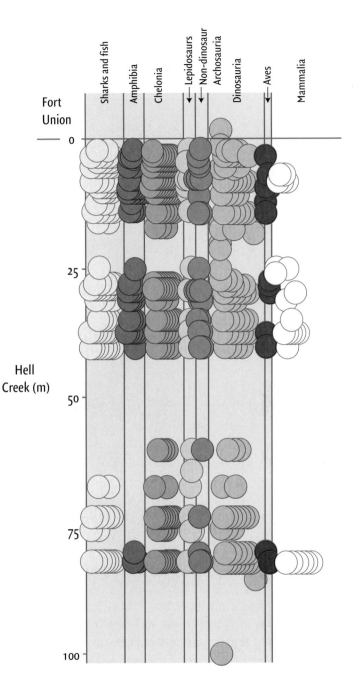

Figure 16.15. Sudden extinction of the dinosaurs. The vertical axis shows meters through the Hell Creek Formation, the uppermost unit in the Western Interior of the USA. "0" is the K/T boundary. The horizontal axis shows various vertebrate groups (including dinosaurs) that are found within the Hell Creek. Virtually all vertebrate groups are present throughout the thickness of the Hell Creek; there is no gradual decrease in the groups as the boundary is approached. The data indicate that the extinction of the dinosaurs and other vertebrates at 65.5 Ma was geologically instantaneous.

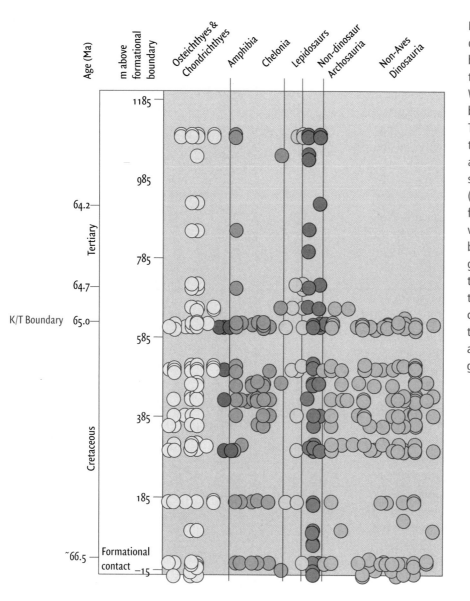

Figure 16.16. Sudden extinction of the dinosaurs in intermontane basin sedimentary deposits of the Late Cretaceous of the US Western Interior, as demonstrated by Lillegraven and Eberle (1999). The vertical axis shows meters through the K/T Ferris Formation and time (Ma). The horizontal axis shows various vertebrate groups (including dinosaurs) that were found in the study. Virtually all vertebrate groups are present below the boundary; there is no gradual decrease in the groups as the boundary is approached. After the boundary, diversity is drastically reduced. The data indicate that the extinction of the dinosaurs and other vertebrates 65.5 Ma was geologically instantaneous.

Does the idea that an asteroid impact caused the K/T extinctions have predictable consequences? Clearly, the aswer to the above question is "Yes." Firstly, if the asteroid produced global consequences, evidence for it should be visible globally. After 23 years of research, the evidence for global influence of the K/T boundary asteroid impact is overwhelming (see Figure 16.3). That being the case, what kind of predictable consequences are there in terms of the extinction?

In the case of the bivalves and plants, the fact that the extinctions took place regardless of latitude is strong evidence that those extinctions were due to a global effect, which was apparently unrelated to climate. Had climate been involved as a causal agent, one might expect to see latitudinal changes in the patterns of extinction, but as we have seen such is not the case.

Other evidence comes from the rate at which the extinction took place. If the asteroid really caused the extinction, the event should have been what W. A. Clemens (after W. S. Gilbert) dubbed a "short, sharp shock." Patterns of *gradual* extinction would falsify the asteroid impact as a causal agent, whereas patterns of *abrupt* or *catastrophic* extinction would corroborate the hypothesis. And, as we have seen, the evidence is mounting that the extinction was abrupt.

Table 16.1. Proposed causes for the extinction of the dinosaurs (based in part upon Benton, 1990)

I.Proposed biotic causes

A. *Medical problems*

(a) Slipped disks in the vertebral column causing dinosaur debilitation

(b) Hormone problems

(1) Overactive pituitary glands leading to bizarre and non-adaptive growths

(2) Hormonal problems leading to eggshells that were too thin, causing them to collapse in on themselves in a gooey mess

(c) Decrease in sexual activity

(d) Blindness due to cataracts

(e) A variety of diseases, including arthritis, infections, and bone fractures

(f) Biting insects carrying diseases that did dinosaurs in over hundreds of thousands to millions of years

(g) Epidemics leaving no trace but wholesale destruction

(h) Parasites leaving no trace but wholesale destruction

(i) Change in ratio of DNA to cell nucleus causing scrambled genetics

(j) General stupidity

B. *Racial senescence*

This is the idea, no longer given much credence, that entire lineages grow old and become "senile," much as individuals do. Thus, in this way of thinking, late-appearing species would not be as robust and viable as species that appeared during the early and middle stages of a lineage. The idea behind this was that the dinosaurs as a lineage simply got old and the last-living members of the group were not competitive for this reason

C. *Biotic interactions*

(a) Competition with other animals, especially mammals, which may have out-competed dinosaurs for niches, or perhaps ate their eggs

(b) Overpredation by carnosaurs (who presumably ate themselves out of existence)

(c) Floral changes

(1) Loss of marsh vegetation (presumably the single most important source of food)

(2) Increase in deforestation (leading to loss of dinosaur habitats)

(3) General decrease in the availability of plants for food with subsequent dinosaur starvation

(4) The evolution in plants of substances poisonous to dinosaurs

(5) The loss from plants of minerals essential to dinosaur growth

II. Proposed physical causes

A. *Atmospheric causes*

(a) Climate became too hot so they fried

(b) Climate became too cold so they froze

(c) Climate became too wet so they got waterlogged

(d) Climate became too dry so they desiccated

(e) Excessive amounts of oxygen in the atmosphere caused:

(1) changes in atmospheric pressure and/or atmospheric composition that proved fatal; or

(2) global wildfires that burned up the dinosaurs

(f) Low levels of CO_2 removed the "breathing stimulus" of endothermic dinosaurs

(g) High levels of CO_2 asphyxiated dinosaur embryos

(h) Volcanic emissions (dust, CO_2, rare earth elements) poisoned dinosaurs one way or another

Table 16.1. (cont.)

B. *Oceanic and geomorphic causes*

    (a) Marine regression produced loss of habitats

    (b) Swamp and lake habitats were drained

    (c) Stagnant oceans produced untenable conditions on land

    (d) Spillover into the world's oceans of Arctic waters that had formerly been restricted to polar regions, and subsequent climatic cooling

    (e) The separation of Antarctica and South America, causing cool waters to enter the world's oceans from the south, modifying world climates

    (f) Reduced topographical relief and loss of habitats

C. *Other*

    (a) Fluctuations in gravitational constants leading to indeterminate ills for the dinosaurs

    (b) Shift in the Earth's rotational poles leading to indeterminate ills for the dinosaurs

    (c) Extraction of the Moon from the Pacific Basin perturbing dinosaur life as it had been known for 140 million years (!)

    (d) Poisoning by uranium from Earth's soils

D. *Extraterrestrial causes*

    (a) Increasing entropy leading to loss of large life forms

    (b) Sunspots modifying climates in some destructive way

    (c) Cosmic radiation and high levels of ultraviolet radiation causing mutations

    (d) Destruction of the ozone layer, causing (c)

    (e) Ionizing radiation as in (c)

    (f) Electromagnetic radiation and cosmic rays from the explosion of a supernova

    (g) Interstellar dust cloud

    (h) Oscillations about the galactic plane leading to indeterminate ills for the dinosaurs

    (i) Impact of an asteroid (for mechanisms, see the text)

Recovery

Catastrophic events tend to leave a distinctive mark: organisms that first colonize deserted ecospace tend to speciate rapidly, to be rather small, and to adopt generalist lifestyles (rather than developing a highly specialized behavior such as exclusively meat-eating or herbivorous behaviors). Such organisms are termed disaster biotas and are known in vertebrate, invertebrate, and plant communities.

At the K/T boundary, recall that, in the plant realm, the initial colonizing flora was a short-lived growth of ferns. These have been interpreted as a disaster flora developing in a disrupted and unstable landscape.

The mammals that evolved throughout the recovery at the K/T boundary were extremely small generalists. They speciated rapidly, taking about 5 million years (Figure 16.17) to evolve a range of sizes and develop specializations (such as herbivores and carnivores). Theirs is the pattern of a disaster fauna that came through a catastrophic event and inherited deserted ecospace.

Recent phylogenies based upon the rates of molecular evolution have suggested that modern mammals' roots are to be found within the Cretaceous, implying that the mammalian radiation that characterized the Tertiary was actually well underway during the latest Cretaceous. In fact, the far-distant ancestors of modern mammals were likely around during the Late Cretaceous, but the rapid species' turnovers of the earliest Tertiary disaster faunas shows the clear mark of a catastrophic event.

Does the asteroid impact hypothesis explain all the data? In fact, there does appear to be a correlation between extinction selectivity and the asteroid as a causal agent in the extinctions. Those marine creatures that suffered the most extinctions were those that depended directly upon primary productivity for their food source. Such creatures included not only the planktonic foraminifera and other planktonic marine microorganisms, but also ammonites, other cephalopods, and a variety of mollusks. On the other hand, organisms that not only depended on primary productivity but could also survive on detritus, that is, the scavenged remains of other organisms, fared consistently better. In marine deposits, detritus-feeders were apparently less affected by the extinction.

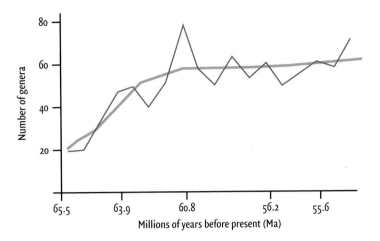

Figure 16.17. Radiation of mammals after the K/T boundary. The vertical axis shows species of mammals; the horizontal axis shows time. The dark blue line is the exact counts of genera at any particular time; the pale blue line is in the inferred, general shape of mammalian diversity. The interpretation (pale blue line) shows a rapid increase in the number of genera of mammals during the first 3–5 million years or so of the Tertiary, followed by a kind of leveling off of diversity.

In the terrestrial realm, the strong selectivity between land-dwelling and aquatic tetrapod survival (see Figure 16.13) correlates with feeding strategy: aquatic vertebrates tend to utilize detritus as a major source of nutrients, while land-dwelling veterbrates are far more dependent upon primary productivity. *The tetrapods that survived the K/T boundary were primarily aquatic detritus-feeders.* This is because river and lake systems can serve as a repository for detrital material, and organisms that live in such environments and can utilize this resource were protected against short-term drops in primary productivity. They may also have been protected from the strong infra-red radiation pulse, as well as the wildfires.

One group of terrestrial tetrapods for which there are virtually no data are *avian* dinosaurs (birds). We know that this group, like mammals, suffered very significant extinctions; yet obviously some birds survived. Their survival may have been tied into an ability to fly and seek refuge, it may have been dumb luck, or it may have been something that we have not yet recognized. The sparse fossil record of birds makes understanding the dynamics of the K/T boundary as it relates to birds very difficult.

Other hypotheses

Continental basalts Earlier in this chapter, we mentioned the Deccan Traps, an outpouring of continental flood basalts in India. These bracketed the K/T boundary, beginning at about 67.5 Ma, and occurring in three major eruptive phases until about 64.7 Ma. Could these have affected the dinosaurs in some way?

For many years, French flood basalt specialist Vincent Courtillot has documented the relationship between episodes of continental flood basalt (CFB) volcanism and mass

Box 16.3 **The real reason the dinosaurs became extinct**

Not every published hypothesis has been serious. In 1964, for example, E. Baldwin suggested that the dinosaurs died of constipation. His reasoning went as follows. Toward the end of the Cretaceous, there was a restriction in the distribution of certain plants containing natural laxative oils necessary for dinosaur regularity. As the plants became geographically restricted, those unfortunate dinosaurs living in places where the necessary plants no longer existed acquired stopped plumbing and died hard deaths. The same year, humorist W. Cuppy noted that "the Age of Reptiles ended because it had gone on long enough and it was all a mistake in the first place," a view with which many characters in the *Jurassic Park* series would have probably agreed.

The November, 1981 issue of the *National Lampoon* offered its explanation, entitled "Sin in the Sediment." The Christian right was the target:

It's pretty obvious if you just examine the remains of the dinosaurs … Dig down into older sediments and you'll see that the dinosaurs were pretty well off until the end of the Mesozoic. They were decent, moral creatures, just going about their daily business. But look at the end of the Mesozoic and you begin to see evidence of stunning moral decline. Bones of wives and children all alone, with the philandering husband's bones nowhere in sight. Heaps of fossilized, unhatched, aborted dinosaur eggs. Males and females of different species living together in unnatural defiance of biblical law. Researchers have even excavated entire orgies – hundreds of animals with their bones intertwined in lewd positions. Immorality was rampant.

In 1983, sedimentary geologist R. H. Dott Jr published a short note in which he vented his frustrations with the pollen season, suggesting that it was pollen in the atmosphere that killed the dinosaurs. He called his contribution "Itching Eyes and Dinosaur Demise."

The issues raised by the *National Lampoon* were compelling enough to again be raised in 1988 by the *Journal of Irreproducible Results*. There, L. J. Blincoe developed a new hypothesis about the "fighting dinosaurs" specimen (see Figure 9.23!):

A thorough but cursory review of fossil specimens… has revealed a unique fossil found in the Cretaceous "beds" of Mongolia in 1971. The fossil featured two different species of dinosaur, one a saurischian carnivore (*Velociraptor*), the other an ornithischian herbivore [sic] (*Protoceratops*), in close association at the moment of their deaths. Prejudiced by their preconceived notions of dinosaur behavior, paleontologists have almost unanimously interpreted this find as evidence of a life and death struggle [see Chapter 11] … However, an alternative theory has now been developed which not only explains this unusual fossil, but also answers the riddle of the dinosaurs disappearance. Quite simply, when their lives were ended by sudden catastrophe, these two creatures were locked together … in a passionate embrace. They were, in fact, prehistoric lovers.

---

extinctions through time. The relationship that he observed is really quite striking; each continental flood basalt event since the Permian seems to have been associated with a mass extinction (Figure 16.18). Could the K/T extinction have been caused through a chain of events that began with the continental flood basalts of the Deccan Traps?

It is hard to say, "Absolutely not!" particularly as recent research suggests that the mass extinction at the Permo-Triassic boundary may have been associated with continental flood basalts. Still, some problems plague the idea that continental flood basalts caused the K/T extinctions. First, the relationship between extinctions and continental flood basalts bears scrutiny: while for every flood basalt there appears to be an extinction, there is not a flood basalt for *every* mass extinction. Some extinctions do not appear on the chart. This tells us that not every mass extinction is caused by a flood basalt; why then should we assume that the K/T extinction is one of those that was?

Figure B16.3.1. O'Donnell's take on the cause of extinction of dinosaurs.

The implications of this startling interpretation are clear: dinosaurs engaged in trans-species sexual activity. In doing so they wasted their procreative energy on evolutionarily pointless copulation that resulted in either no offspring or, perhaps on rare occasions, in bizarre, sterile mutations (the fossil record is replete with candidates for this later category.[5]

For the ultimate causes of the extinction, however, we think O'Donnell's perspective published in the *New Yorker* says it all (Figure B16.3.1).

Second, you will note that the basalts began before the extinctions and continued after. In between each of the pulses of volcanism, life restored itself, as shown by horizons containing abundant, well-preserved plant and animal fossils preserved between the flows. What, then, was different about the second pulse?

Finally, other aspects of the K/T extinction do not clearly implicate flood basalts: the global fires, the selectivity of the extinction, the evidence for physical disruption at the boundary globally, and chemical signals around the Gulf of Mexico region that match those of the impact crater, and of course the presence at the K/T boundary globally of an extraterrestrial signature, namely iridium and shocked quartz. For all these reasons, then, attractive as the continental flood basalt idea is, it does not fit best the known data that characterize the K/T extinctions.

[5] Blincoe, L. J. 1988. *Journal of Irreproducible Results,* **33**, 24.

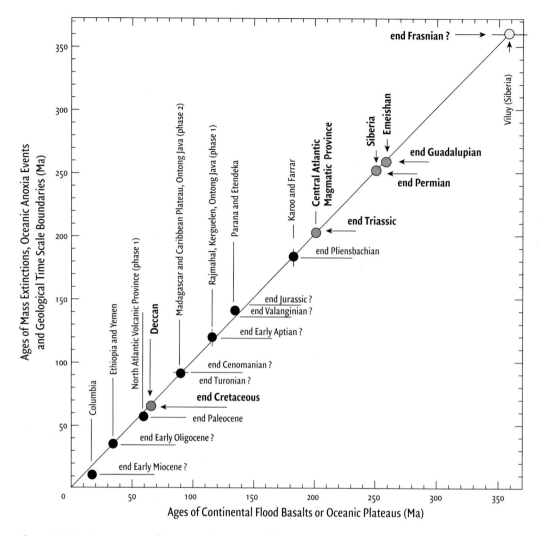

Figure 16.18. Comparison of the ages of continental flood basalt events and mass extinction events. The straight line indicates excellent correlation between the ages of the basalts and the timing of the mass extinction events. Data compiled by Vincent Courtillot.

Multiple causes It's been said that "complex extinctions require complex causes," and a number of workers have suggested that a unique suite of events fortuitously coalesced at the K/T boundary, to produce the pattern of selectivity that we have observed in the extinctions. These events include the asteroid impact, the Deccan volcanism, seal-level fall, global climatic deterioration through cooling just before the K/T boundary ... in fact, something of a laundry list of what is known about boundary events. Does this better explain the data?

Well, not exactly; however, the reason why it does not provide a satisfactory explanation for what caused the extinction is at least in part a philosophical one. Ultimately in this way of thinking, each K/T event is ascribed a role in the extinctions; ultimately, therefore, any and all K/T events must have been involved in the extinctions. Because of this, the hypothesis is unfalsifiable: if all K/T events are involved in the extinction, they were in essence involved because they were K/T events; and there is no way to eliminate any particular event. The inability to reject any particular event means that the hypothesis is in effect unfalsifiable.

Still, there was an extinction at the K/T boundary that was most likely due to events that occurred at that boundary. But if we can't know which event – because none can be absolutely eliminated –it really says nothing new about the K/T boundary: the hypothesis so general that it ultimately explains nothing. Ironically, the multi-cause hypothesis could very well be correct: a unique suite of events uniquely occurring at the K/T boundary could have uniquely caused the mass extinction. But as we have seen, the very multi-causal nature of the hypothesis precludes its testability, and therefore its significance as a scientific hypothesis.

So what happened at the K/T boundary?

Scientists who view the asteroid impact as the cause of K/T events usually envision some kind of dramatic and short-term disturbance to the ecosystem. Such a disturbance was likely a dust cloud blocking sunlight for a few months, a pulse of infra-red radiation, and global wildfires. It may never be possible to know exactly which factor(s) did which deed(s), and trying to reconstruct it in so exact a manner may be stretching the resolution of the fossil record well past its capability. Whatever the disturbance to the terrestrial ecosystem, it seems to have had, at the very least, a deadly effect on global primary productivity, which in turn seems to have decimated organisms solely dependent upon upon primary productivity. Detritus-feeders survived. Whatever else is true, the absence of dinosaurs after 163 million years of terrestrial importance was the event that allowed the mammals to evolve and occupy the place in the global ecosystem that they presently hold.

Summary

Although there has been considerable speculation about what happened to the non-avian dinosaurs at 65.5 Ma, any meaningful hypothesis must operate within the bounds of science: it must be testable and it must explain as much of the data as possible. Given those constraints, only one hypothesis matches what is known about the K/T extinction: the hypothesis that an asteroid hit the Earth 65.5 million years ago and killed many organisms, including the non-avian dinosaurs.

The evidence for a large (10 km in diameter) asteroid striking the Earth at 65.5 Ma in the Yucatan region of Mexico is now extensive, and virtually incontrovertible. Its immediate affects are likely to have been, globally, blockage of sunlight for 3–4 months and the propagation of an energy pulse and associated wildfires in the terrestrial realm as well as, locally, tidal waves and severe environmental disruption.

Key among the various biotic conseqences of these events is that the great nutrient cycles that characterize healthy oceans were severely disrupted.

The asteroid impact is consistent with the limited amount that is known of the dinosaur extinction, which is that in the Western Interior of North America, at least (the only place in the world in which rates of dinosaur extinction have been studied), the dinosaur extinction was as instantaneous as we are able to resolve at a distance of 65.5 Ma.

The patterns of terrestrial vertebrate survivorship are relatively well understood. Analyses show that vertebrates not in aquatic ecosystems were most susceptible to extinction; other inferences, less robust, suggest that larger animals, endotherms, and amniotes also were less likely to survive than smaller animals, ectotherms, and anamniotes. Non-avian dinosaurs were large organisms, and clearly dependent upon primary productivity.

One explanation for the survivorship of organisms living in aquatic food webs is that these are generally not as dependent upon primary productivity as those in land-based food chains. Because primary productivity was clearly perturbed at the K/T boundary, organisms in aquatic food webs were likely protected both by dietary preference and by the fact that the aquatic realm provided a refuge from the physical catastrophes caused by the asteroid impact.

The recovery took 2–3 million years in the oceans, and about 5 million years on land.

Selected readings

Alvarez, L. W. 1983. Experimental evidence that an asteroid impact led to the extinction of many species 65 Myr ago. *Proceedings of the National Academy of Sciences*, **80**, 627–642.

Alvarez, W. 1997. *T. rex and the Crater of Doom*. Princeton University Press, Princeton, NJ, 185pp.

Alvarez, L. W., Alvarez, W., Asaro, F. and **Michel, H. V.** 1980. Extraterrestrial cause for the Cretaceous–Tertiary extinction. *Science*, **208**, 1095–1108.

Archibald, J. D. 1996. *Dinosaur Extinction and the End of an Era*. Columbia University Press, New York, 237pp.

Archibald, J. D. 2011. *Extinction and Radiation: How the Fall of the Dinosaurs Led to the Rise of Mammals*. Johns Hopkins University Press, Baltimore, MD, 108pp.

Benton, M. J. 1990. Scientific methodologies in collision the history of the study of the extinction of the dinosaurs. *Evolutionary Biology*, **24**, 371–400.

Fastovsky, D. E. and **Sheehan, P. M.** 2005. The extinction of the dinosaurs in North America. *GSA Today*, **15**, 4–10.

Frankel, C. 1999. *The End of the Dinosaurs – Chicxulub Crater and Mass Extinctions*. Cambridge University Press, New York, 223pp.

Hartman, J., Johnson, K. R. and **Nichols, D. J.** (eds.). 2002. *The Hell Creek Formation and the Cretaceous–Tertiary Boundary in the Northern Great Plains*. Geological Society of America Special Paper no. 361, 520pp.

Johnson, K. R. and **Hickey, L. J.** 1990. Megafloral change across the Cretaceous/Tertiary boundary in the northern Great Plains and Rocky Mountains, U.S.A. In Sharpton, V. L. and Ward, P. D. (eds.), *Global Catastrophes in Earth History: An Interdisciplinary Conference on Impacts, Volcanism, and Mass Mortality*. Geological Society of America Special Paper no. 247, pp. 433–444.

Koeberl, C. and **MacLeod, K. G.** (eds.). 2002. *Catastrophic Events and Mass Extinctions: Impacts and Beyond*. Geological Society of America Special Paper no. 356, 746pp.

Lillegraven, J. A. and **Eberle, J. J.** 1999. Vertebrate faunal changes through Lanican and Puercan time in southern Wyoming. *Journal of Paleontology*, **73**, 691–710.

Maas, M. C. and **Krause, D. W.** 1994. Mammalian turnover and community structure in the Paleocene of North America. *Historical Biology*, **8**, 91–128.

Powell, J. L. 1998. *Night Comes to the Cretaceous*. W. H. Freeman and Company, New York, 250pp.

Ryder, G., Fastovsky, D. E. and **Gartner, S.** (eds.). 1996. *The Cretaceous–Tertiary Event and Other Catastrophes in Earth History*. Geological Society of America Special Paper no. 307, 569pp.

Sharpton, V. L. and **Ward, P. D.** (eds.). 1990. *Global Catastrophes in Earth History: An Interdisciplinary Conference on Impacts, Volcanism, and Mass Mortality*. Geological Society of America Special Paper no. 247, 631pp.

Silver, L. T. and **Schultz, P. H.** (eds.). 1982. *Geological Implications of Large Asteroid and Comets on Earth*. Geological Society of America Special Paper no. 190, 528pp.

Topic questions

1 What do the words regression, detritus, iridium anomaly, shocked quartz, microtektites, Chicxulub, ichthyosaurs, plesiosaurs, ammonites, mosasaurs, foraminifera, planktonic, and benthic refer to?

2 What are primary production and disaster faunas?

3 What is meant by nutrient cycling? Why is that important?

4 What is the K/T boundary? When was it?

5 What kinds of physical events took place at the K/T boundary?

6 Describe the biotic extinctions that took place at the K/T boundary.

7 Describe the results of the studies that concluded that the dinosaurs died abruptly?

8 What is meant by "geologically abrupt" when speaking of the extinction of the dinosaurs?

9 Describe the kinds of physical events that might have occurred when the asteroid hit the Earth?

10 Choose four extinction hypotheses from Table 16.1 and evaluate them in terms of the criteria for a viable scientific theory.

Glossary

Using this Glossary

The goal of this Glossary is to help to clarify language and images that may be unfamiliar to students who happen to live in the Recent (us!). Below, therefore, follows a complete listing of all the words highlighted in this textbook, as well as other words of relevance to the subject of dinosaur paleontology. Readers are provided with the chapter in which the word is highlighted; in some cases, however, readers are also referred to places where the concept(s) embodied by the word is treated. Finally, in a few cases, readers are referred to a figure as well. A relatively few words have no cross-reference, implying that the ideas they represent recur throughout the book.

A

ADP (adenosine diphosphate). A molecule involved in the energy production of a cell. It is produced when ATP breaks down to release energy.

ATP (adenosine triphosphate). A molecule synthesized by the body as a means to store energy. The energy is stored in the three phosphate bonds, one of which is then broken to release energy (ATP → ADP + inorganic phosphate + energy). *See* Chapter 12.

Acetabulum. Hip socket. *See* Chapter 4.

Acromion (acromial) process. A broad and plate-like flange on the forward surface of the shoulder blade. *See* Chapter 5.

Adenosine diphosphate. *See* ADP. *See also* Chapter 12.

Adenosine triphosphate. *See* ATP. *See also* Chapter 12.

Advanced. In an evolutionary context, shared or derived (or specific), with reference to characters. *See* Chapter 3.

Aerobic. A type of metabolism involving a complex series of oxidation steps through the citric acid cycle. *See* Box 12.2.

Akinetic. *See* Kinetic. *See also* Chapter 7.

Allometry. The condition in which, as the size of organisms changes, their proportions change as well. For example, if an ant were scaled up to the size of a 747 airplane, its features – body, head, legs, etc. – would no longer have the same proportions relative to each other. *See* Chapter 12.

Altricial. Pertaining to organisms that are born relatively underdeveloped, requiring significant parental attention for survival. *See* Chapter 7.

Alvarezsauridae. An unusual group of theropods equipped with stout carpometacarpus-like features. Their phylogenetic status is currently poorly understood. *See* Chapter 11.

Alveolus (pl. alveoli). A sac-like anatomical structure. *See* Chapter 8.

Amnion. A membrane in some vertebrate eggs that contributes to the retention of fluids within the egg. *See* Chapter 4.

Amniote. An organism bearing amniotic eggs. *See* Chapter 4.

Anaerobic. Without oxygen. *See* Chapter 12.

Analog. In anatomy, structures that perform in a similar fashion but have evolved independently. *See* Chapter 3.

Analogous. Adjective form of analog. *See* Chapter 3.

Anamniote(s). An organism whose egg has no amnion. *See* Chapter 4.

Anapsid(s). The group that contains all amniotes with a completely covered skull roof. *See* Chapter 4.

Ancestral. In an evolutionary sense, relating to forebears. *See* Chapter 3.

Angiosperms. Flowering plants. *See* Chapters 4, 5, 6, and 13.

Ankylosauria. Armored quadrupedal thyreophorans (Ornithischia) whose bodies were encased in a bony pavement of osteoderms. *See* Chapter 5.

Ankylosauridae. Along with Nodosauridae, one of the two clades that make up Ankylosauria. *See* Chapter 5.

Antediluvian. Occurring before the Biblical flood. *See* Chapter 14.

Anterior. Pertaining to the head-bearing end of an organism.

Antorbital fenestra. An opening on the side of the skull, just ahead of the eye. This is a character that unites the clade Archosauria. *See* Chapter 4.

Arboreal. Pertaining to trees; as the arboreal hypothesis, referring to the idea that bird flight evolved by birds jumping out of trees. *See* Chapter 10.

Archosauria. A clade within Archosauromorpha. The living archosaurs include birds and crocodiles. *See* Chapters 4 and 10.

Archosauromorpha. The large clade of diapsids that includes the common ancestor of rhynchosaurs and archosaurs, and all its descendants. *See* Chapter 4.

Ascending process of the astragalus. A wedge-shaped splint of bone on the astragalus that lies flat against the shin (between the tibia and fibula) and points upward. Diagnostic character of Theropoda. *See* Chapter 4.

Assemblage. In paleontology, a group of organisms. The term is used to refer to a collection of fossils in which it is not clear how accurately the collection reflects the complete, ancient formerly living community. *See* Chapters 2 and 12.

Asteroid. A large extraterrestrial body. *See* Chapter 15.

Astragalus. Along with the calcaneum, one of two upper bones in the vertebrate ankle. *See* Chapter 10.

Atom. The smallest particle of any element that still retains the properties of that element. *See* Chapter 2.

Atomic number. The number of protons (which equals the number of electrons) in an element. *See* Chapter 2.

B

Background extinctions. Continually occurring, isolated extinctions of individual species. As distinct from mass extinctions. *See* Chapter 15.

Badlands. Extremely rough country, generally carved by rivers and floods. *See* Chapter 1.

Barb. Feather material radiating from the shaft of the feather. *See* Chapter 10.

Barbule. A small hook that links barbs together along the shaft of the feather. *See* Chapter 10.

Barometric. Adjectival form of barometer. Barometers measure atmospheric pressure; the atmospheric pressure at the sea level is 1 atmosphere (1 atmosphere = 14.7 lb/in^2). Atmospheric pressure increases dramatically with depth of submersion. *See* Chapter 8.

Beak. Sheaths of keratinized material covering the ends of the jaws (synonym: rhamphotheca). *See* Chapters 4–7.

Belief system. A conceptualization that is faith-based, knowledge that does not require the application of logic, reason, and/or observation. *See* Chapter 15.

Bennettitaleae. A group of extinct cone-bearing plants. *See* Chapter 13.

Benthic. With reference to the marine realm (oceans), living within sediments. *See* Chapter 16.

Biogeography. Pertaining to the distribution of organisms in space. *See* Chapter 6.

Biomass. The sum total of the weights of organisms in the assemblage or community being studied. *See* Chapter 12.

Biostratigraphy. The study of the relationships in time among groups of organisms. *See* Chapter 2.

Biota. The sum total of all organisms that have populated the Earth.

Body fossil. The type of fossil in which a part of an organism becomes buried and fossilized as opposed to trace fossil. *See* Chapter 1.

Bone histology. The study of bone tissue. *See* Chapter 12.

Bonebeds. Relatively dense accumulation of bones of many individuals, generally composed of a very few kinds of organisms. *See* Chapter 6.

Brain endocasts. Internal casts of braincases. *See* Chapter 12.

Braincase. Hollow bony box that houses the brain; located toward the upper, back part of the skull. *See* Chapter 4.

C

Cadence. In locomotion, the rate at which the feet hit the ground. *See* Chapter 5.

Calcareous nanofossils. Fossils of extremely small planktonic microorganisms. *See* Chapter 16.

Carbohydrates. A family of 5- and 6-carbon organic molecules whose chemical bonds, when broken, release energy. *See* Chapters 5 and 12.

Carpal. Wrist bone. *See* Chapter 4.

Carpometacarpus (pl. carpometacarpi). Unique structure in all living, and in most ancient, birds, in which bones in the wrist and hand are fused. *See* Chapter 10.

Cast. Material filling up a mold. *See* Chapter 1.

Cellular respiration. The breakdown of carbohydrates through a regulated series of oxidizing reactions. *See* Chapter 12.

Cenozoic. An Era, lasting from 65.5 Ma to present. *See* Chapters 2, 13, and 15.

Centrum. The spool-shaped, lower portion of a vertebra, upon which the spinal cord and neural arch rest. *See* Chapter 4.

Cerapoda. The ornithischian clade of Ceratopsia + Pachycephalosauria + Ornithopoda. *See* Part II.

Ceratopsia. Beaked dinosaurs of Asia and North America who, together with Pachycephalosauria, make up Marginocephalia. *See* Chapter 6.

Character. An isolated or abstracted feature or characteristic of an organism. *See* Chapter 3.

Cheek. The muscular, fleshy organs along the side of the jaw that retain food in the mouth. *See* Part II.

Cheek teeth. Teeth that lie against the cheeks; in mammals, the cheek teeth consist of the premolars and molars. *See* Part II.

Chronostratigraphy. The study of geological time. *See* Chapter 2.

Clade. Group of organisms in which all members are more closely related to each other than they are to anything else. All members of a clade share a most recent common ancestor that is itself the most basal member of that clade. Synonymous with "monophyletic group" and "natural group." *See* Chapter 3.

Cladogram. A hierarchical, branching diagram that shows the distribution of shared, derived characters among selected organisms. *See* Chapter 3.

Clavicle. Collarbone. *See* Chapter 4.

Coelophysoidea. The theropod clade containing *Coelophysis* and its close relatives. *See* Chapter 9.

Co-evolution. The idea that two organisms or groups of organisms may have evolved in response to one another. *See* Chapter 13.

Collect. To obtain fossils from the Earth. *See* Chapter 1.

Computed tomography (CT). A medical scanning technique using computers to construct a two-dimensional image using X-ray imaging around an axis of rotation. It has proven to be an extremely successful means of resolving incompletely prepared fossils as well. *See* Chapter 12.

Continental effects. The effect on climate exerted by continental masses. *See* Chapter 2.

Convergent. In anatomy, pertaining to the independent invention (and thus, duplication) of a structure or feature in two lineages. The streamlined shape of whales, fish, and ichthyosaurs is a famous example of convergent evolution. *See* Chapter 9.

Coprolite. Fossilized feces. *See* Chapters 1 and 9.

Coracoid. The lower (and more central) of two elements of the shoulder girdle (the upper being the scapula). *See* Chapter 4.

Coronoid process. A bony enlargement or process at the back of the lower jaw for the attachment of jaw-closing musculature. *See* Part II.

Cretaceous. A Period lasting from 146 Ma to 65.5 Ma.

Cretaceous/Tertiary boundary. That moment in time, 65.5 million years ago, between the Cretaceous Period and the Tertiary Period. *See* Figure 2.4 and Chapter 15.

Cretaceous/Tertiary extinction. The event, 65.5 million years ago at the Cretaceous/Tertiary boundary, in which all the non-avian dinosaurs, as well as many other terrestrial vertebrates, became extinct. *See* Chapter 16.

Crop. To cut short; in anatomy, to bite off. *See* Part II.

Crurotarsi. A clade of archosaurs including crocodilians and their close relatives. *See* Chapter 4.

Curate. In paleontology, to incorporate, preserve, and catalog specimens into museum collections. *See* Chapter 1.

Cursorial. Pertaining to running; as the cursorial hypothesis, referring to the idea that bird flight evolved by birds running along the ground. *See* Chapters 4 and 10.

Cycadophyte. A bulbous, fleshy type of gymnosperm. *See* Chapter 13.

D

Deccan Traps. Interbedded volcanic and sedimentary rocks in western and central India of Cretaceous–Tertiary age. *See* Chapter 16.

Deltoid crest. A large process at the head of the humerus. *See* Chapter 10.

Dense Haversian bone. A type of Haversian bone in which the canals and their rims are very closely packed. *See* Chapter 12.

Dental battery. A cluster of closely packed cheek teeth in the upper and lower jaws, whose shearing or grinding motion is used to masticate plant matter. *See* Chapter 6.

Denticles. Small bumps or protuberances generally associated with teeth. *See* Chapters 6 and 7.

Derived. In an evolutionary context, pertaining to characters that uniquely apply to a particular group and thus are regarded as having been "invented" by that group during the course of its evolutionary history. *See* Chapter 3.

Detritus. Loose particulate rock, mineral, or organic matter; debris. *See* Chapter 16.

Developmental biology. The study of how organisms develop from a fertilized egg to an adult. *See* Chapter 14.

Diagnostic. In phylogeny, a feature that uniquely pertains to a group of organisms. Diagnostic features permit the identification of groups of organisms because, uniquely, all members of that group possess the feature (and ideally all other groups do not). *See* Chapter 3.

Diapsida. The large clade of amniotes that includes the common ancestor of lepidosauromorphs and archosaurs, and all its descendants. *See* Chapter 4.

Diastem(a). A gap. *See* Part II.

Digitigrade. In anatomy, a position assumed by the foot when the animal is standing, in which the ball of the foot is held high off the ground and the weight rests on the ends of the toes. Opposite of plantigrade. *See* Chapter 8.

Dinosauria. A clade of ornithodiran archosaurs. *See* Chapter 4.

Disarticulated. Dismembered. *See* Chapter 1.

Disaster biota. Organisms that colonize a landscape immediately after an ecological disaster. Disaster biotas tend to have three characteristics: small size, high rates of speciation, and generalist life strategies. *See* Chapter 16.

Distal. In anatomy, in the direction away from the central part (or core) of the animal. *See* Chapters 3 and 12.

Diversity. The variety of organisms; the number of *kinds* of organisms. *See* Chapter 13.

DNA hybridization. A molecular biological technique that measures the difference between two comparable strands of DNA. *See* Chapter 11.

Down. A bushy, fluffy, type of feather in which barbules and vanes are not well developed, used for insulation. *See* Chapter 10.

Dynamic similarity. A conversion factor that "equalizes" the stride rates of vertebrates of different sizes and proportions, so that speed of locomotion can be calculated. *See* Chapter 12.

E

Ecological diversity. The proportion of an ecosystem that is occupied by a particular lifestyle, such as feeding type or mode of locomotion. For a simple example, one might study an ecosystem by dividing it into herbivores, carnivores, and omnivores. *See* Chapter 15.

Ecospace. Niches that are available in an ecosystem. Simple categories of ecospace, for example, include "carnivore," "herbivore," and "scavenger." More refined categories could include "grazing" versus "browsing" herbivores, and large and small carnivores. *See* Chapter 16.

Ectothermic. Regulating temperature (and thus metabolic rate) using an external source of energy (heat). The opposite of endothermic. *See* Chapter 12.

Electron. A negatively charged subatomic particle. Electrons reside in clouds around the nucleus of an atom. *See* Chapter 2.

Element. In chemistry, an atom distinguished by the number of protons in its nucleus; in anatomy, discrete part of the skeleton, that is, an individual bone. *See* Chapters 2 and 4.

Enantiornithes. A group of sparrow-like, relatively common Mesozoic avialians. *See* Chapter 11.

Encephalization. That condition in which an organism bears a head structure that is distinct from the rest of the body and that contains a brain. *See* Chapter 12.

Encephalization Quotient (EQ). An estimate based on brain size and body weight, designed to determine the intelligence of an extinct organism, relative to a living organism whose intelligence can be ascertained. *See* Chapter 12.

Endemic. An organism or fauna is said to be endemic to a region when it is restricted to that region. *See* Chapter 13.

Endemism. The property of being endemic. *See* Chapter 13.

Endosymbionts. Organisms that live within another organism in a mutually beneficial relationship. *See* Chapter 8.

Endothermic. Regulating temperature (and thus, metabolic rate) using an internal source of energy. The opposite of ectothermic. *See* Chapter 12.

Enlightenment. The Enlightenment was a seventeenth- and eighteenth-century political, social, and philosophical movement that, among other qualities, emphasized the primacy of logic, reason, and observation for understanding the natural world. *See* Chapter 14.

Epicontinental sea. Relatively shallow (at most, a few hundred meters) marine water covering a continent (synonym: epeiric sea). *See* Chapter 2.

Era. A very large block of geological time (hundreds of millions of years long), composed of Periods. *See* Chapter 2.

Erect stance. In anatomy, the condition in which the legs lie parasagittal to (alongside) the body and do not extend laterally from it. *See* Chapter 4.

Esophagus. A tube located between the neck vertebrae and the trachea, leading from just behind the mouth down to the stomach. In humans it is about 15 cm long; in sauropods it could have reached 10–15 m in length. *See* Chapter 8.

Estivate. In zoology, to spend summers in a state of torpor. *See* Chapter 7.

Euornithopoda. A monophyletic group containing the more derived members of Ornithopoda. *See* Chapter 7.

Eurypoda. The ornithischian clade Stegosauria + Ankylosauria. *See* Chapter 5.

Eustatic. Global. *See* Chapter 2.

Evo-Devo (evolution and development). The coupling of the sciences of evolutionary biology and developmental biology to attempt to understand how new features in organisms have evolved. *See* Chapter 14.

Evolution. In biology, descent with modification. *See* Appendix 3.1.

Extraterrestrial. From outer space. *See* Chapter 16.

F

Fauna. A group of animals presumed to live together within a region.

Femur. The upper bone in the hindlimb (thigh bone). *See* Chapter 4.

Fibula. The smaller of the two lower leg bones in the hindlimb; the bone that lies alongside the shin bone (tibia). *See* Chapters 4 and 10.

Fit. Adjectival form of fitness. Fitness is the degree to which natural selection acts upon an organism to assure the presence of its genetic makeup in the succeeding (descendent) generation. *See* Chapter 3.

Flight feather. Elongate feather with well-developed, asymmetrical vanes; usually associated with flight. *See* Chapter 10.

Flood basalt. Episodic lava flow from fissures in the Earth's crust.

Flux. A measure of change; rate of discharge times volume.

Folsom projectile point. A style of tool-making (likely a spearhead) known from the Great Plains of the USA, first found near Folsom, New Mexico, and dating from 9500–8000 BCE (Before the Common Era). *See* Chapter 14.

Footplate. On the pubis, an anterior–posterior enlargement of the distal end. Generally associated with theropods. *See* Chapter 10.

Footprint. Trace fossil left by the feet of vertebrates. *See* Chapter 1.

Foramen magnum. The opening at the base of the braincase through which the spinal cord travels to connect to the brain. *See* Chapters 4 and 5.

Foraminifera (sing. foraminifer). Single-celled, shell-bearing organisms that live in the oceans. *See* Chapter 16.

Fossil. Technically, anything buried; generally refers to the buried remains of organisms. *See* Chapter 1.

Fossilization. The process of becoming a fossil. *See* Chapter 1.

Fossorial. Burrowing. *See* Chapter 7.

Frill. In ceratopsians, a sheet of bone extending dorsally and rearward from the back of the skull, made up of the parietal and squamosal bones. *See* Chapter 6.

Furcula. Fused clavicles (collarbones); the "wishbone" in birds and certain non-avian theropods. *See* Chapter 10.

G

Gamete(s). A mature sex cell; either male or female. *See* Chapter 13.

Gastralia. Belly ribs. *See* Chapter 10.

Gastrolith. Smoothly polished stone in the stomach, used for grinding plant matter. *See* Chapter 5.

Genasauria. The ornithischian clade of Thyreophora + Cerapoda. *See* Part II.

General. In phylogenetic reconstruction, referring to a character that is non-diagnostic of a group; in this context, synonymous with primitive. *See* Chapter 3.

Generalist. In ecology, unspecialized. Basically willing and able to eat anything to stay alive. In terms of diet, at least, humans are generalists; a meat-eater, like a great white shark, would be considered a specialist. *See* Chapter 16.

Generic. Adjective from the word genus, the second-smallest grouping in the Linnaean classification (a genus (pl. genera) is composed of species, the smallest formal category in the Linnaean classification). *See* Chapter 4.

Genotype. The total genetic makeup of an organism. *See* Chapter 3.

Geochemistry. Chemistry particularly as related to geological problems. *See* Chapter 14.

Geochronology. The science of determining the age (in years) of the Earth. Adjectival form of the word geochronological. Practitioners of geochronology are, not surprisingly, geochronologists. *See* Chapter 2.

Ghost lineage. Lineage of organisms for which there is no physical record (but whose existence can be inferred). *See* Chapter 13.

Gigantothermy. Modified mass homeothermy, which mixes large size with low metabolic rates and control of circulation to peripheral tissues. *See* Chapter 12.

Girdle(s). Something that encircles; in this case attachments for the limbs that partially encircle the trunk. *See* Chapter 4.

Gizzard. A muscular chamber just in front of the glandular part of the stomach. *See* Chapter 8.

Glycogen. A complex, carbohydrate-based molecule used by the body for energy storage. *See* Chapter 5.

Gnathostome. A vertebrate with a jaw (formal term: Gnathostomata). *See* Chapter 4.

Gondwana. A southern supercontinent comprising present-day Australia, Africa, South America, and Antarctica. *See* Chapter 2.

Gregarious. Highly socialized; regularly moving in flocks or herds. *See* Chapters 5 to 8.

Gut. Stomach, intestine, and bowels. *See* Part II.

Gymnosperms. A group of seed-bearing, non-flowering plants, including pines and cypress. Gymnosperms are not a monophyletic grouping, unless angiosperms are also included within the group. *See* Chapters 5 and 13.

H

Half-life. The amount of time that it takes for 50% of a volume of unstable isotope to decay. *See* Chapter 2.

Hard part. In paleontology, all hard tissues, including bones, teeth, beaks, and claws. Hard parts tend to be preserved more readily than soft tissues. *See* Chapter 1.

Haversian canal. In bone histology, a canal composed of secondary bone. *See* Chapter 12.

Hemal. Referring to blood. In bone anatomy, hemal arch. A vertebral process straddling the ventral side of the vertebral column and pointing ventrally (oriented opposite to the neural arch). *See* Chapter 10.

Hierarchy. As applied here, the ordering of objects, organisms, and categories by rank. The military and the clergy are both excellent examples of hierarchies; in these, rank is a reflection of power and, one hopes, accomplishment. Another hierarchical system is money, which is ordered by value. *See* Chapter 3.

Histology. The study of tissues. *See* Chapters 12 and 14.

Homeotherm. Organism whose core temperature remains constant. *See* Chapter 12.

Homologous. Two features are homologous when they can be traced back to a single structure in a common ancestor. *See* Chapter 3.

Horn core. The horn in many animals, cows and sheep being two familiar examples, are constructed of a bony central part, the core, covered by a layer, or sheath, made of keratin. Ceratopsian dinosaurs had this type of horn. *See* Figure 6.21.

Hornlets. Small horns. *See* Chapter 9.

Humerus. The upper arm bone. *See* Chapter 4 (especially Figure 4.5).

Hyoid bone(s). Paired elongate bones in the throat that form a support for the tongue. *See* Chapter 5.

Hypothesis of relationship. A hypothesis about how closely or distantly organisms are related. *See* Chapter 3.

I

Ichnofossil. Impression, burrow, track, or other modification of the substrate by organisms. *See* Chapter 1.

Ichthyosaurs. Dolphin-like marine reptiles of the Mesozoic. *See* Figure 16.9.

Ilium. The uppermost of three bones that make up the pelvis. *See* Chapter 4.

Impact ejecta. The material thrown up when an asteroid strikes the Earth. *See* Chapter 15.

Impact winter. The idea that a temporary period of climatic cooling caused by sunlight blockage by ejecta would follow an impact. Based on climatic data from large volcanic eruptions such as Krakatoa. *See* Chapter 16.

Interspecific. Among different species. *See* Chapter 6.

Intraspecific. Within the same species. *See* Chapters 5 and 6.

Iridium (Ir). A non-toxic, platinum-group metal, rare at the Earth's surface. *See* Chapter 16.

Iridium anomaly. High concentrations of iridium in a single place; the name comes from the idea that since Ir levels at the Earth's surface are generally low, high concentrations would be considered anomalous. *See* Chapter 16.

Ischium. The most posterior of three bones that make up the pelvis. *See* Chapter 4.

Isotopes. In chemistry, elements that have the same atomic number but different mass numbers. *See* Chapter 2.

J

Jacket. In paleontology, a rigid, protective covering placed around a fossil, so that it can be moved safely out of the field. Commonly made up of strips of burlap soaked in plaster. *See* Figure 1.10.

Jurassic. A Period lasting from 200 Ma to 146 Ma.

K

K-strategy. The evolutionary strategy of having few offspring, which are cared for by the parents. The symbol *K* stands for the carrying capacity of the environment. *See* Chapter 7.

K/T (boundary). Common abbreviation for that moment in time, 65.5 Ma, which marks the boundary between the Cretaceous and the Tertiary. *See* Chapter 16.

Keel. A flange or sheet of bone, as in the keeled sternum of birds; named for its resemblance to the keel on a sailboat. *See* Chapter 10.

Keratin. A protein that forms the basis of nails, horns, hooves, feathers, and hair. *See* Part II.

Kinetic. With reference to skull anatomy, movement between bones of the skull. *See* Chapter 7.

L

Lactic acid. An organic acid produced as a byproduct of muscle exertion; $CH_3CH(OH)COOH$. *See* Chapter 12.

LAG. *See* Lines of arrested growth. *See also* Chapter 12.

Lambeosaurines. The hollow-crested hadrosaurid dinosaurs.

Land bridge. A corridor of land between two continents that allows the passage of organisms from one continent to the other. Such corridors have occasionally existed, for example, between South America and North America, as well as between North America and Asia. *See* Chapter 13.

Laurasia. A northern supercontinent. *See* Chapter 2.

Lepidosauromorpha. One of the two major clades of diapsid reptiles; the other clade is Archosauromorpha. *See* Chapter 4.

Lines of arrested growth (LAGs). Lines that are inferred to represent times of non-growth, visible in the cross-section of bones. *See* Chapter 12.

Lithostratigraphy. The general study of all rock relationships. *See* Chapter 2.

Lower temporal fenestra. The lower opening of the skull just behind the eye; *see* Temporal fenestra. *See also* Chapter 4.

M

Mandible. The lower jaw. *See* Chapter 4.

Mandibular foramen. A fenestra in the posterior portion of the mandible. *See* Part II.

Marginocephalia. The clade of dinosaurs that includes the most recent common ancestor of pachycephalosaurs and ceratopsians and all of its descendants. *See* Chapter 5.

Mass extinctions. Global and geologically rapid extinctions of many kinds, and large numbers, of species. *See* Chapter 15.

Matrix. In paleontology, the rock that surrounds fossil bone. *See* Chapter 1.

Megaflora. The visible remains of plants, especially leaves. *See* Chapter 16.

Mesotarsal. A linear type of ankle in which hinge motion in a fore–aft direction occurs between the upper ankle bones (the astragalus and calcaneum) and the rest of the foot. *See* Chapter 4.

MDT. *See* Minimal divergence time. *See also* Chapter 13.

Mesozoic. An Era lasting from 251 Ma to 65.5 Ma. *See* Chapter 2.

Metabolism. The sum of the physical and chemical processes in an organism. *See* Chapter 12.

Metacarpal. Bone in the palm of the hand. *See* Chapter 4.

Metapodial. A general name for metacarpals and metatarsals. *See* Chapter 4.

Metatarsal. Bone in the sole of the foot. *See* Chapter 4.

Micropaleontology. The study of microscopic organisms, such as marine plankton. A person who specializes in micropaleontology is a micropaleontologist. *See* Chapter 16.

Microtektite. A small, droplet-shaped blob of silica-rich glass thought to have crystallized from impact ejecta. *See* Chapter 16.

Minimal divergence time (MDT). The minimal amount of time missing between the two descendent species and their common ancestor; calculated by comparing phylogeny and age of fossils. *See* Chapter 13.

Minimum number of individuals (MNA). A technique for estimating how many individual organisms are represented in a locality. If we find two left thigh bones (among other elements), then we know that at least two different individuals are represented. If we found six theropod teeth, however, we would not know whether they came from one or more than one (up to six) different individuals; the *least* number of individuals that could have produced those six teeth is one. *See* Chapter 16.

Mold. Ichnofossil that consists of the impression of an original fossil. *See* Chapter 1.

Molecular clock. The use of the rates of mutation in certain molecules to determine how long ago two organisms diverged from a common ancestor. *See* Chapters 11 and 14.

Molecular evolution. The idea that molecules can evolve at particular rates. *See* Chapter 11.

Monophyletic group. A group of organisms that has a single ancestor and contains all of the descendants of this unique ancestor (synonymous with clade and "natural group"). *See* Chapter 3.

Monotypic. Of one type; generally, composed of one species. *See* Chapter 7.

Morphology. The study of shape. *See* Chapter 3.

Mosasaurs. Late Cretaceous marine-adapted lizards. *See* Chapter 15.

N

Naris (pl. nares). Opening in the skull for the nostrils. *See* Chapter 4.

Natural selection. The process by which certain members of a population are more effectively able to assure the representation of their genes in the succeeding generation (e.g., the descendent generation). *See* Chapter 3.

Neoceratosauria. The theropod clade *Coelophysis*, *Ceratosaurus*, and near relatives. *See* Chapter 9.

Neural arch. A piece of bone that straddles the spinal cord; generally with a central process that rises dorsally. *See* Chapters 4 (especially Figure 4.5) and 10.

Neutron. Electrically neutral subatomic particle that resides in the nucleus of the atom. *See* Chapter 2.

Node. A bifurcation or two-way split point in a phylogenetic diagram (cladogram). *See* Chapter 3.

Nodosauridae. Along with Ankylosauridae, one of the clades making up Ankylosauria. *See* Chapter 5.

Non-avian dinosaurs. All dinosaurs *except* birds. *See* Chapters 1, 9, and 10.

Non-diagnostic. In phylogeny, a feature that does not uniquely pertain to a group of organisms. Non-diagnostic features do not permit the identification of a particular group of organisms because members outside of that group also possess the feature. *See* Chapter 3.

Notochord. An internal rod of cellular material that, primitively at least, ran longitudinally down the backs of all chordates. May be thought of as a precursor to the vertebral column. *See* Chapter 4.

Nuchal ligament. An elastic ligament running dorsally in the neck from the back of the head to a posterior cervical vertebra. In sauropods, it likely helped to support the head. *See* Chapter 8.

Nucleus. Central core of the atom. *See* Chapter 2.

Nutrient cycling. As used here, the movement of nutrients from the shallow surface waters to the bottom of the ocean. *See* Chapter 16.

Nutrient turnover. The amount of nutrients that pass through the system; in the case of bone development, the amount of metabolic activity associated with bone growth. *See* Chapter 12.

O

Obligate biped. Tetrapod that must walk or run on its hind legs. *See* Chapters 4, 9, and 15.

Occipital condyle. A knob of bone at the back of the skull with which the vertebral column articulates. *See* Chapter 4.

Occiput. The back of the skull. *See* Chapter 6.

Occlusion. Contact between upper and lower teeth; necessary for chewing. Teeth that contact each other between the upper and lower jaws are said to occlude. *See* Part II.

Olfactory bulbs. An enlarged part of the brain that deals with the sense of smell. *See* Chapter 5.

Ontogeny. Biological development of the individual; the growth trajectory from embryo to adult. *See* Chapters 6 and 11.

Opisthopubic. The condition in which at least part of the pubis has rotated backward to lie close to, and parallel with, the ischium. *See* Part II.

Orbit. Eye socket. *See* Chapter 4.

Ornithischia. One of the two monophyletic groups comprising Dinosauria. *See* Chapter 4.

Ornithodira. The common ancestor of pterosaurs and dinosaurs, and all its descendants. *See* Chapter 4.

Ornithopoda. Biped and quadrupedal herbivorous ornithischian dinosaurs. *See* Chapter 7.

Ornithothoraces. A group of Mesozoic birds including Aves. *See* Chapter 11.

Ornithurae. Hesperornithiformes, Ichthyornithiformes, and Aves. *See* Chapter 11.

Ornithuromorpha. The lineage of ornithothoracean birds, including Aves. *See* Chapter 11.

Osteichthyes. Bony fishes that include ray-finned and lobe-finned gnathostomes. *See* Chapter 4.

Osteoderm. Bone within the skin; may be small nodule, plate, or a pavement of bony dermal armor. *See* Chapter 5.

Oxidation. Bonding of oxygen. *See* Chapter 12.

P

Pachycephalosauria. Dome-headed ornithischians of North America and Asia who, together with Ceratopsia, make up Marginocephalia. *See* Chapter 6.

Palate. The part of the skull that separates the nasal cavity (for breathing) from the oral cavity (for eating); usually strengthened by a paired series of bones. *See* Chapter 4.

Paleobiology. The discipline of paleontology, arising in the 1970s, that actively sought to understand the biology of fossil organisms. *See* Chapter 14.

Paleobotany. The study of ancient plants. A person who studies ancient plants is called a paleobotanist. *See* Chapter 13.

Paleoclimate. Ancient climate. *See* Chapter 2.

Paleoenvironment. Ancient environment.

Paleontology. The study of ancient life; distinguished from anthropology, which is the study of humans, and archaeology, which is the study of past civilizations. A paleontologist is someone who studies paleontology.

Paleozoic. An Era lasting from 543 Ma to 251 Ma. *See* Chapter 2.

Palpebral. A rod-like bone that crosses the upper part of the eye socket. *See* Part II.

Palynoflora. Spores and pollen. *See* Chapter 16.

Pangaea. The mother of all supercontinents, formed from the union of all present-day continents. *See* Chapter 2.

Parasagittal stance. Stance in which the legs are held under the body. *See* Chapter 4.

Parascapular spine. An enlarged spine over the shoulder. *See* Chapter 5.

Parsimony. A principle that states that the simplest explanation that explains the greatest number of observations is preferred to more complex explanations. *See* Chapter 3.

Patellar groove. A groove at the distal end of the femur to accommodate the patella (knee cap). *See* Chapter 11.

Pectoral girdle. The bones of the shoulder; the attachment site of the forelimbs. *See* Chapter 4.

Pectoralis (muscle). The muscle that drives the wing's powerstroke in bird flight. *See* Chapter 10.

Pedestal. In paleontology, a pillar of matrix underneath the fossil. *See* Figure 1.10.

Pelvic girdle. The bones of the hips; the attachment site of the hindlimbs. *See* Chapter 4.

Perforate acetabulum. A hole in the hip socket; a diagnostic character of Dinosauria. *See* Figure 4.5.

Period. Subdivision of an Era, consisting of tens of millions of years. *See* Chapter 2 (especially Figure 2.4).

Permineralization. The geological process in which the spaces in fossil bones become filled with a mineral. *See* Chapter 1.

Permo-Triassic. Relating to events at the Permian/Triassic boundary (251 Ma). *See* Chapter 16.

Phalanx (pl. phalanges). Small bone of the fingers and toes that allows flexibility. *See* Chapter 4.

Phanerozoic. That interval of time from 543 Ma to the present; it also refers to the time in Earth's history during which shelled organisms have existed. *See* Chapter 2.

Phenotype. The physical appearance and features of an organism. *See* Chapter 3.

Photosynthesis. The process by which organisms use energy from the sun to produce complex molecules for nutrition.

Phylogenetic. Pertaining to phylogeny. *See* Chapter 3.

Phylogenetic systematics. The method of determining organismic relationships that uses parsimony to select among competing hierarchical distributions of shared, derived characters (that is, cladograms). *See* Chapter 3.

Phylogeny. The study of the fundamental genealogical connections among organisms. *See* Chapter 3.

Phylum. A grouping of organisms whose makeup is supposed to connote a very significant level of organization shared by all of its members.

Phytosaur. Long-snouted, aquatic, fish-eating member of Crurotarsi. *See* Figure 13.4.

Planktonic (or planktic). Living in the water column. *See* Chapter 16.

Plantigrade. A foot position in which the bottom of the foot (the tissue below the metatarsals) lies flat on the ground. Opposite of digitigrade. *See* Chapter 8.

Plesiosaurs. Long-necked fish-eating reptiles with large flippers that inhabited Mesozoic seas. *See* Figure 16.9.

Pleurocoel. A well-marked excavation on the sides of a vertebra. *See* Chapter 8.

Pleurokinesis. Mobility of the upper jaw. *See* Chapter 7.

Pneumatic. Having air sacs or sinuses; the state of having such is called pneumaticity. *See* Chapter 8.

Pneumatic foramina. Openings for air sacs to enter the internal bone cavities. *See* Chapters 8 and 10.

Poikilotherm. Organism whose core temperature fluctuates. *See* Chapter 12.

Precocial. The condition in which the young are rather adult-like in their behavior. *See* Chapter 7.

Predator : prey biomass ratios. The ratio of the total estimated weight of predators to the total estimated weight of their prey in a particular ecosystem. *See* Chapter 12.

Predentary. The bone that caps the front of the lower jaws in all ornithischians. *See* Part II.

Preparation (prep) lab. Where preparation takes place. *See* Figure 1.11.

Prepare. To clean a fossil; to get it ready for viewing by freeing it from its surrounding matrix. *See* Chapter 1.

Prepubic process. A flange of the pubis that points toward the head of the animal. *See* Chapters 9 and 10.

Primary bone. Bone tissue that was deposited or laid down first. *See* Chapter 12.

Primary productivity. The sum total of organic matter synthesized by organisms from inorganic materials and sunlight. *See* Chapter 16.

Primitive. *See* Ancestral. *See also* Chapter 3.

Process. In relation to anatomy, part of a bone that is commonly ridge-, knob-, or blade-shaped and sticks out from the main body of the bone. *See* Chapter 4.

Productivity. The amount of biological activity in an ecosystem. *See* Chapter 1.

Prosauropoda. Monophyletic group of Late Triassic and Early Jurassic saurischian dinosaurs; the world's first high-browsing herbivores. Once thought to be ancestral to Sauropoda, now believed to share a common ancestor with sauropods. *See* Chapter 8, especially Figure 8.19, for diagnostic characters.

Prospect(ing). To hunt for fossils. *See* Chapter 1.

Proton. Electronically charged (+1) subatomic particle that resides in the nucleus of the atom. *See* Chapter 2.

Proximal. In anatomy, in the direction toward the central part (or core) of the animal.

Pterosauria. Flying ornithodiran archosaurs, closely related to (but not included within) dinosaurs. Uniquely, the wing was supported by an extraordinarily elongate digit IV. *See* Chapter 4.

Pubis. One of the three bones that make up the pelvic girdle. *See* Chapter 4.

Pull of the Recent. The inescapable fact that as we get closer and closer to the Recent, fossil biotas become better and better known. *See* Chapter 13.

Pygostyle. A small, compact, pointed structure made of fused tail bones in birds. *See* Chapter 10.

R

r-strategy. The evolutionary strategy where organisms have lots of offspring and no parental care. The letter r is obtained from the symbol r for growth rate. *See* Chapter 7.

Radiometric. The dating method to determine unstable isotopic age estimations. *See* Chapter 2.

Radius. One of the two lower arm bones; the other is the ulna. *See* Chapter 4.

Recent. The time interval that encompasses 13,000 years ago to the present. *See* Chapters 2 and 13.

Recurved. Curved backward, like a scimitar. *See* Chapter 9.

Regression. Retreating of seas due to lowering of sea level. *See* Chapter 16.

Relationship. The phylogenetic closeness of two organisms; that is, the genealogical nearness or distance of their most recent common ancestor. *See* Chapter 3.

Relative dating. The type of geological dating that, although not providing ages in years before present, provides ages relative to other strata or assemblages of organisms. *See* Chapter 2.

Remodel. In bone histology, to resorb or dissolve primary bone and deposit secondary bone. *See* Chapter 12.

Renewable resources. Resources that can be synthesized or manufactured, either naturally or artificially. *See* Chapter 1.

Replace. To exchange the original mineral with another mineral. *See* Chapter 1.

Reptilia. The old Linnaean category for turtles, lizards, snakes, and crocodiles. Reptilia as formulated by Linnaeus and as commonly used is not monophyletic; only the addition of birds to these four groups constitutes a monophyletic group. *See* Chapter 10.

Respiratory turbinate. A thin, convoluted or complexly folded sheet of bone located in the nasal cavities of living endothermic vertebrates. *See* Chapter 12.

Rhamphotheca. Cornified covering on the upper and lower jaws (for example, a beak). *See* Part II and Chapter 5.

Robust. (1) In the context of hypothesis testing, a hypothesis is said to be robust when it has survived repeated tests; that is, despite meaningful attempts, it has failed to be falsified. (2) In anatomy, strong and stout.

Rock. An aggregate of minerals.

Rostral. Referring generally to the rostrum, or snout region of the skull; in ceratopsians, a unique, diagnostic bone at the tip of the snout on the skull. *See* Chapter 6 (especially Figure 6.19).

Rostral bone. A unique bone on the front of the snout of ceratopsians, giving the upper jaws of these dinosaurs a parrot-like profile. *See* Chapter 6.

S

Sacrum. The part of the backbone where the hip bones attach. *See* Chapter 4.

Sarcopterygii. Lobe-finned fish. *See* Chapter 4.

Saurischia. One of the two monophyletic groups that Dinosauria comprises; the other is Ornithischia. *See* Chapter 4 and Part III.

Sauropoda. Monophyletic group of long-necked, long-tailed, quadrupedal herbivorous saurischians. *See* Chapter 8, especially Figure 8.20, for diagnostic characters.

Scapula (pl. scapulae). The shoulder blade. *See* Chapters 4 (especially Figure 4.5) and 10.

Sclerotic ring. A ring of bony plates that support the eyeball within the skull. *See* Figure 4.6 and Chapters 7 and 10.

Scute(s). A bony plate embedded in the skin; synonym: osteoderm. *See* Part II.

Seasonality. Highly marked seasons. *See* Chapter 2.

Secondarily evolve. To re-evolve a feature. *See* Chapters 4 and 15.

Secondary bone. Bone deposited in the form of Haversian canals. *See* Chapter 12.

Secondary palate. A shelf of bone, above the palate, over which air can be directed so that air is not mixed with the food during chewing. All mammals and crocodiles have secondary palates; some turtles do as well. *See* Chapter 5.

Sedimentary rock. A rock that generally represents the lithification, or hardening, of sediment. *See* Chapter 1.

Sedimentologist. Someone who studies sedimentary rocks and processes. *See* Chapter 1.

Seed. A capsule that contains gametes, as well as nutrients. These are generally encased in a protective pod. *See* Chapter 13.

Semi-lunate carpal. A distinctive, half-moon-shaped bone in the wrist. *See* Chapter 9.

Sexual dimorphism. Size, shape, and behavioral differences between sexes *See* Chapter 5.

Sexual selection. Selection not between all of the individuals within a species, but between members of a single sex. *See* Chapter 6.

Shaft. (1) The hollow main vane of a feather. (2) The title and eponymous lead male character in the first and most famous of the 1970s "blacksploitation" films. *See* Chapter 10.

Shocked quartz. Quartz that has been placed under such pressure that the crystal lattice becomes compressed and distorted; correctly termed "impact metamorphism." *See* Chapter 16.

Sigmoidal. Having an "S" shape; *see* Chapter 10.

Sinus. A cavity. *See* Chapter 9.

Skeleton. The supporting part of any organism. In vertebrates, the skeleton is internal and consists of tissue hardened by mineral deposits (sodium apatite). Such tissue is called "bone." *See* Chapter 4.

Skull. That part of the vertebrate skeleton that houses the brain, special sense organs, nasal cavity, and oral cavity. *See* Figure 4.6.

Skull roof. The bones that cover the top of the braincase. *See* Chapter 4 (especially Figures 4.6 and 4.9).

Soft tissue. In vertebrates, all of the body parts except bones, teeth, beaks, and claws. *See* Chapter 1.

Specialization. In biology, the idea that an organism is adapted for a particular circumstance (for example, diet, climate, ecosystem, a particular host (if it is a parasite), or any other aspect of its existence). Such an organism is termed a specialist. *See* Chapter 16.

Speciate. Evolve new species; diversify. *See* Chapter 16.

Species-specific. Appying only to a particular species (the one under discussion). *See* Chapter 5.

Specific. (1) Diagnostic of a monophyletic group; uniquely evolved. (2) Adjectival form of the word "species," the smallest formal category in the Linnaean classification. Operationally (among living organisms), if two creatures breed and produce viable offspring, they are generally said to be the same species. *See* Chapter 4.

Sphenopsid. A primitive type of vascular plant. *See* Figure 13.8.

Sprawling stance. Stance in which the upper parts of the arms and legs splay out approximately horizontally from the body. *See* Figure 4.16.

Stable isotope. An isotope that does not spontaneously decay. *See* Chapter 12.

Stapes. The middle-ear bone that transmits sound (vibrations) from the tympanic membrane to a hole in the side of the braincase (allowing auditory nerves of the brain to sense vibration). *See* Chapter 4.

Stegosauria. Thyreophorans (Ornithischia) with paired rows of bony osteoderms running along the back. *See* Chapter 5.

Sternum. The breastbone. *See* Chapters 4 and 10.

Strata. Layers of rock. *See* Figure 2.2.

Stratigraphy. The study of the relationships of strata and the fossils they contain. *See* Chapter 2.

Subatomic. Smaller than atom-sized. *See* Chapter 2.

Substitution. In molecular biology, the notion that one of the bases in the pairs that compose a DNA molecule can be substituted for another base or base-pair. *See* Chapter 11.

Superposition. The geological principle in which the oldest rocks are found at the bottom of a stack of strata and the youngest rocks are found at the top. *See* Chapter 2.

Supracoracoideus (muscle). The muscle that drives the wing's recovery stroke in bird flight. *See* Chapter 10.

Survivorship. The pattern of survival measured against extinction. *See* Chapter 16.

Synapsida. The large clade of amniotes, including mammals, diagnosed by a single temporal opening. *See* Chapter 4.

Synsacrum. A single, locked unit consisting of the sacral vertebrae. *See* Chapter 10.

T

Tarsal. Ankle bone. *See* Chapter 4.

Tarsometatarsus. The name for the three metatarsals fused together with some of the ankle bones. *See* Chapter 10.

Taxon (pl. taxa). A group of organisms, designated by a name, of any rank within the biotic hierarchy. *See* Chapter 4.

Tectonic. Referring to tectonics, that branch of geology dealing with the development and form of the surficial and near-surficial parts of the Earth. *See* Chapter 2.

Temnospondyl. Belonging to the group Temnospondyli, large, carnivorous largely Paleozoic amphibians, whose behavior must have been something like that of a crocodile. *See* Chapters 12 and 13 (especially Figure 13.6).

Temporal. (1) Referring to time. (2) In anatomy, the side region of the skull, above the jaw. *See* Chapters 2 and 4.

Temporal fenestra (pl. fenestrae). Openings in the temporal region of the skull. *See* Chapter 4.

Terrestrial. Referring to land; that is, not marine. *See* Chapter 1.

Testable hypothesis. A hypothesis that makes predictions that can be compared and assessed by observations in the natural world. *See* Chapter 3.

Tetanurae. Literally, "stiff tails"; the clade of theropods with tails stiffened as a result of overlapping caudal zygapophyses. *See* Chapter 9.

Tethyan. Referring to the ancient water mass that eventually became today's Atlantic Ocean. *See* Chapter 2.

Tetrapoda. A monophyletic group of vertebrates primitively bearing four limbs. *See* Chapter 4.

Thecodontia. A paraphyletic taxon that at one time was used to unite the separate ancestors of crocodilians, pterosaurs, dinosaurs, and birds. *See* Chapters 4 and 10.

Therapsida. The clade of synapsids that includes mammals, some of their close relatives, and all of their most recent common ancestors. *See* Chapters 12, 13 (especially Figure 13.3), and 14.

Thermoregulation. Control of body temperature. *See* Chapter 5.

Thorax. In vertebrates, the part of the body between the neck and abdomen. *See* Chapter 8.

Thyreophora. Armor-bearing ornithischians; stegosaurs, ankylosaurs, and their close relatives. *See* Part II and Chapter 5.

Tibia. One of the two lower bones in the tetrapod hindlimb; the other is the fibula. *See* Chapters 4 and 10.

Trace fossil. Impressions in sediment left by an organism. *See* Chapter 1.

Trachea. The windpipe. *See* Chapter 8.

Trackway. Group of aligned footprints left as an organism walks. *See* Chapter 1.

Transgression. Advancing of seas due to raising sea level. *See* Chapter 15.

Triassic. A Period lasting from 251 Ma to 200 Ma. *See* Chapter 2.

Triosseal foramen. The hole formed by the coracoid, furcula, and scapula, through which the tendon for the supracoracoideus connects to the humerus for the wing recovery stroke in birds. *See* Chapter 10.

Tubercles. Small bumps or protuberances. *See* Chapter 9.

Turn (a fossil). To separate the fossil from the surrounding rock at the base of the pedestal and to rotate it 180°. *See* Figure 1.10.

Tympanic membrane. Eardrum. *See* Chapter 4.

U

Ulna. One of the two lower arm bones; the other is the radius. *See* Chapter 4.

Unaltered. When original mineralogy is unchanged. *See* Chapter 1.

Ungual phalange. An outermost bone of the fingers and toes. See Chapter 4.

Unstable isotope. An isotope that spontaneously decays from an energy configuration that is not stable to one that is more stable. *See* Chapter 2.

Upper temporal fenestra. The opening in the skull roof above the lower temporal fenestra; see Temporal fenestra. *See also* Chapter 4.

V

Vane. The sheet of feather material that extends away from the shaft. *See* Chapter 10.

Vascular. Pertaining to vessels that conduct fluids. In animals, a region that is vascular has a lot of blood vessels; a vascular plant is a plant that has tubes (xylem and phloem) that conduct water and nutrients. *See* Chapter 13.

Vertebrae. The repeated structures that compose the backbone and that, along with the limbs, support the rest of the body. *See* Chapter 4.

Vertebrata. The group of all animals containing vertebrae. *See* Chapter 4.

W

Weathering. The physical or chemical breaking down of earthly materials (for example, minerals, rocks, and bones). *See* Chapter 1.

Wedge. The evolutionary pattern of waxing and waning dominance among groups of organisms. *See* Chapter 14 (especially Figure 14.13).

Z

Zygapophysis. A fore-and-aft projection from the neural arches (of vertebrae). *See* Chapter 9.

FIGURE CREDITS

B14.5.1   Sternberg, C.H., 1985, Hunting Dinosaurs in the Bad Lands of the Red Deer River, Alberta Canada, NeWest Press, Edmonton, p. 235

B14.6.1   Courtesy of the Library, American Museum of Natural History

B14.6.2   Photograph courtesy of D. E. Fastovsky

B14.8.1   Courtesy of D. B. Weishampel

B14.9.1   Photograph courtesy of Jack Horner

B14.9.2   Photograph courtesy of Mark Norrell

B14.9.3   Photograph courtesy of Paul Sereno

16.1   Photograph courtesy of W. Alvarez

16.2   Redrawn from Alvarez et al., 1980

16.4   Courtesy of Milwaukee Public Museum

16.5   Photograph courtesy of B. F. Bohor, U.S. Geological Survey

16.6   After Claeys et al., 1996

16.7   Image courtesy of M. Pilkington, Geological Survey of Canada, and A. Hildebrand, University of Calgary

16.10   Courtesy of S. D'hondt

16.11   Photograph courtesy of D. E. Fastovsky

16.12a-c   Electron micrographs courtesy of D. J. Nichols, U.S. Geological Survey

16.13   Data from: Archibald, J. D. 1996. *Dinosaur Extinction and the End of an Era.* Columbia University Press, New York, 237pp; Sheehan, P. M. and Fastovsky, D.E. 1992. Major extinctions of land-dwelling vertebrates at the Cretaceous-Tertiary boundary, eastern Montana. *Geology,* **20**, 556–560; Archibald, J. D. and Bryant, L. 1990. Differential Cretaceous-Tertiary extinctions of non-marine vertebrates: evidence from northeastern Montana. In Sharpton, V. L. and Ward, P.D. (eds.), *Global Catastrophes in Earth History: An interdisciplinary Conference on Impacts, Volcanism, and Mass Mortality.* Geological Society of America, Special Paper no. 247, pp. 549–562

16.15   Redrawn after Pearson et al., 2002

16.16   Redrawn after Lillegraven and Eberle, 1999

16.17   Redrawn after Maas, M.C. and Krause, D. W. 1994

B16.1.1   From Raup, D. M. and Sepkoski, J. J. 1984. Periodicity of extinctions in the geological past. *Proceedings of the National Academy of Science,* **81**, 801–805

B16.3.1   Drawing by O'Donnell © The New Yorker Collection 1992, Mark O'Donnell from cartoonbank.com. All Rights Reserved

16.18   Data compiled by Vincent Courtillot

INDEX OF GENERA

Note: All roots are Greek or Latin unless otherwise indicated.

INDEX OF SUBJECTS